Understanding Game-based Approaches for Improving Sustainable Water Governance

Understanding Game-based Approaches for Improving Sustainable Water Governance

The Potential of Serious Games to Solve Water Problems

Special Issue Editors

Wietske Medema
Chengzi Chew
Jan Franklin Adamowski
Igor Mayer
Arjen Wals

MDPI • Basel • Beijing • Wuhan • Barcelona • Belgrade

MDPI

Special Issue Editors

Wietske Medema
Department of Bioresource Engineering, McGill University
Canada

Chengzi Chew
Serious Games at DHI
Denmark

Jan Franklin Adamowski
Faculty of Agricultural and Environmental Sciences, Department of Bioresource Engineering, McGill University
Canada

Igor Mayer
Academy for Digital Entertainment, NHTV Breda University of Applied Sciences
The Netherlands

Arjen Wals
Department of Social Sciences, Wageningen University
The Netherlands

Editorial Office
MDPI
St. Alban-Anlage 66
4052 Basel, Switzerland

This is a reprint of articles from the Special Issue published online in the open access journal *Water* (ISSN 2073-4441) from 2018 to 2019 (available at: https://www.mdpi.com/journal/water/special_issues/Game-based-Water-Governance).

For citation purposes, cite each article independently as indicated on the article page online and as indicated below:

LastName, A.A.; LastName, B.B.; LastName, C.C. Article Title. *Journal Name* **Year**, *Article Number*, Page Range.

ISBN 978-3-03928-762-8 (Pbk)
ISBN 978-3-03928-763-5 (PDF)

Contents

About the Special Issue Editors

Wietske Medema (Ph.D.) has spent the bulk of her career working with government organizations, industry, non-government organizations, conservation groups, and civic organizations in North America, Europe, and Southeast Asia. Her work focuses primarily on building and strengthening collaborative networks and communities of practice by helping organizations identify and design social and technical solutions to build capacity, strengthen relationships, create common ground, facilitate experiments, and coordinate expertise. Her work focuses on finding ways to help translate innovative and new ideas into practical solutions, providing expertise in both research and practice.

Chengzi Chew (Commercial Lead) was the head of Serious Games and was extensively involved in the use of serious games in water management. He is currently the commercial lead for DHI's applied research and innovation.

Jan Franklin Adamowski (Professor) is a Full Professor in the Department of Bioresource Engineering in the Faculty of Agricultural and Environmental Sciences at McGill University. He is also the Director of the Integrated Water Resources Management Program, and Associate Director of the Brace Centre for Water Resources Management. He has published over 300 research papers in refereed journals, as well as three books, several chapters in books, and several monographs. Throughout his research career, he has conducted research with his students in over 30 countries. Prof. Adamowski came to McGill University in 2009 after working as a Post-Doctoral Associate at Massachusetts Institute of Technology in the USA. He completed his graduate studies in the US and Europe at several universities (Cambridge, MIT, London Business School, HEC Paris, NHH Bergen, and Warsaw Technical University) and his undergraduate degree in Canada at the Royal Military College of Canada, where he served as a Navy Officer. Prof. Adamowski's research explores engineering, as well as socio-economic and management problems, in the field of water resources to develop innovative and useful approaches to help reduce the vulnerability, as well as enhance the resilience, adaptive capacity, and sustainability of water resource systems around the world in the face of increasing challenges, uncertainty, and climate change variability.

Igor Mayer (Professor) is a lector of Serious Games, Innovation & Society at Breda University of Applied Sciences (BUas), the Netherlands. This professorship is affiliated with the Academy for Games and Media. In 2017, he was awarded an Hai Tian (Sea Sky) scholarship at Dalian University of Technology (DUT), Dalian, China. In 2019, he was awarded a scholarship by the National Natural Science Foundation of China (71774024). His main research and professional interests are the development, use, and evaluation of gaming-simulation, serious games, and, more recently, virtual, mixed, and augmented reality for policy analysis, organization, and management. His pending research has the title "Playful Organizations & Learning Systems". Over the years and in various partnerships, he has initiated, managed, and participated in a large number of serious gaming-related research and development projects. He has been a partner in many European projects, part of FP7, H2020, Eranet, Interreg, and Erasmus funding programs. One featured project is the MSP Challenge (www.mspchallenge.info) with pending EU / Interreg funded projects in the North Sea, Baltic, and Celtic regions, and invited gameplay sessions around the globe. He has published more than 140 journal and conference papers and has been the promotor of six completed Ph.D. theses, three of which are 'award winning'.

Arjen Wals (Professor) is a Professor of Transformative Learning for Socio-Ecological Sustainability at Wageningen University in The Netherlands. He also holds the UNESCO Chair of Social Learning and Sustainable Development. His teaching and research focus on designing learning processes and learning spaces that enable people to contribute meaningfully to sustainability. A large part of his work is focused on how to create conditions that support (new) forms of learning that take full advantage of the diversity, creativity, and resourcefulness that is all around us, but so far remain largely untapped in our search for a world that is more sustainable. In 2014 he was the lead author of an article published in Science on the role of citizen science in bridging science education, environmental education, and sustainability. He is the editor and co-editor of a number of popular books including Higher Education and the Challenge of Sustainability (Kluwer Academic, 2004), Creating Sustainable Environments in our Schools (Trentham, 2006), Social Learning towards a Sustainable World (Wageningen Academic, 2007), Learning for Sustainability in Times of Accelerating Change (2012), and of Routledge's International Handbook on Environmental Education Research (2013). He has (co)authored over 250 publications, of which 60 are in international peer reviewed journals. Recently he contributed as a senior policy advisor to UNESCO's Global Education Monitor 2016 Report 'Education for People and Planet' on the role of education in helping realize the UN's Sustainable development Goals.

water

MDPI

Editorial

The Potential of Serious Games to Solve Water Problems: Editorial to the Special Issue on Game-Based Approaches to Sustainable Water Governance

Wietske Medema [1,*], Igor Mayer [2], Jan Adamowski [1], Arjen E.J. Wals [3] and Chengzi Chew [4]

1 Department of Bioresource Engineering, McGill University, 21 111 Lakeshore, Ste Anne de Bellevue, QC H9X3V9, Canada; jan.adamowski@mcgill.ca
2 Academy for Digital Entertainment, Breda University of Applied Sciences, Monseigneur Hopmansstraat 1, 4817 JT Breda, The Netherlands; i.s.mayer@hotmail.com
3 Department of Social Sciences, Education and Learning Sciences, Wageningen University, 6706 KN Wageningen, The Netherlands; arjen.wals@wur.nl
4 Serious Games at DHI, Agern Allé 5, 2970 Hørsholm, Denmark; czc@dhigroup.com
* Correspondence: wietske.medema@mcgill.ca; Tel.: +1-(310)-480-5204

Received: 10 November 2019; Accepted: 25 November 2019; Published: 5 December 2019

Abstract: In this editorial, the authors (and guest editors) introduce the Special Issue titled Understanding Game-based Approaches for Improving Sustainable Water Governance: The Potential of Serious Games to Solve Water Problems. The authors take another look at the twelve contributions, starting from the subtitle question: what is the potential? The authors summarize the insights and give directions for future research.

Keywords: game-based learning; integrated water resource management (IWRM); natural resource management; simulation; serious game; social learning; stakeholder collaboration; sustainability; water governance

---

1. Introduction

In 2018 and 2019, Water (MDPI) published twelve articles in a Special Issue titled Understanding Game-Based Approaches for Improving Sustainable Water Governance: The Potential of Serious Games to Solve Water Problems. This editorial gives an introduction to the Special Issue, with insights and directions for future research.

A grant provided by the Social Sciences and Humanities Research Council of Canada (SSHRC) in 2015 (see below) brought the Guest Editors together in the UPSWING project: Understanding game-based approaches for improving sustainable water governance and stakeholder collaboration in the Great Lakes–St. Lawrence River Basin. One of the proposed actions in the project was to collect and bundle a wide range of experiences with serious games for water management.

From the literature, we already knew that the terminology around serious games is extensive: applied games, simulation-games, serious games, interactive and social simulations, gamified approaches, even Virtual and Augmented Reality (VR/AR) and more. Furthermore, game characteristics, genre and styles are often used to describe the approach: a role-playing game, a computer-supported interactive simulation, digital game, board game, etc. Having worked in water management, learning and games for quite a few years, we also knew that the topic could be approached from different angles: i.e., the subject matter, the application area, the design of the game, the game technology, the research or evaluation methodology and the learning. In short, it would not be easy to bring these diverse aspects together coherently.

In 2017, we issued a call for papers. In the following years, more than twenty manuscripts were submitted, twelve of which were published after peer review and revision. The majority of papers are empirical case studies. Two papers are an inventory and analysis of existing games. Three papers are conceptual and reflective; seven papers are case and field studies. One paper synthesizes the results [1]. The Special Issue was closed in September 2018. Table 1 gives a short overview of the papers in the Special Issue.

Table 1. Overview of papers in the Special Issue.

Game Analysis
1. Galván-Pérez et al. evaluate twenty educational videogames about water sustainability on quality of their content, game-play and educational value [2]. They conclude that water games in the simulation genre, like managing a city or region, are highly appropriate for learning. 2. Rodela, Ligtenberg and Bosma focus on an inventory of serious games in natural resource and environmental governance to conceptualize and discuss different uses: for research, learning and intervention [3]. They use examples from the literature and their own experiences with a game about shrimp farming in the Mekong Delta, Vietnam, to develop a framework for the design and evaluation of serious games as learning-based intervention.
Policy-Making and Social Learning
1. At the science-policy interface, Zhou and Mayer reconstruct the different frames that policymakers and analysts in the Netherlands and China have about the use and usefulness of models, simulations and games [4]. They demonstrate that social learning is only one of five different perspectives on the use of serious games, with others focusing on bureaucratic alignment, gaining stakeholder support and reducing uncertainty or emotions. 2. Marini et al. aim to understand the relationship between social learning and serious games in a context of IWRM [5]. They argue that Schwarz' theory on transcendental values (i.e., beyond self-interest) provides a foundation for understanding social learning in IWRM, managing competing values and working towards long-term collaboration. This gives guidance for the design and use of serious games. 3. Aubert, Medema and Wals take up the question 'why' games are beneficial (or not) for social learning in IWRM [1]. They develop a framework for design and suggest opportunities for future research.
CASE and Field Studies
4. Magnuszewski et al. present the initial results of piloting a social-simulation game, called Lord of the Valley [6]. They study the interrelations and interactions among the various stakeholders in a simulated river basin, observed during three sessions in Poland. 5. Gomes et al. report on the design, use and results of a social simulation game on the topic of drinking water, used for capacity building in a peri-urban community in Bangladesh [7]. The game combines game-theoretic models, with a board game and role play, and has been played with actual stakeholders living in a village near Kulna. They conclude that the game-based intervention successfully increased the understanding of the local stakeholders about the problems, solutions and institutions. 6. Ferrero, Bichai and Rusca report on the learning of students at IHE, Delft, the Netherlands in a role-playing game on drinking-water safety plans, in particular for public-health protection in drinking-water safety [8]. They demonstrate how and why the game helps the players to understand the governance process and the negotiation among stakeholders that it requires. 7. Susnik et al. show how serious games are designed and used for twelve case studies in the ongoing SIM4NEXUS project [9]. This project aims to integrate water-food-energy-climate for a resource-efficient Europe. The gaming approach combines system dynamics with the game AquaRepublica as a front end. The authors present the design and experiences of a pilot study in Sardinia, Italy. 8. Khoury et al. report on the design and use of a serious game for flood mitigation in a United Kingdom village [10]. They connect their approach to 'citizen science', shared vision planning and give evidence for an informative and transformative effect of the game-play. The serious game uses a 3D virtual table. 9. Keijser et al. used a board game for Maritime Spatial Planning (MSP) at the local, national and transnational level in a European context [11]. The authors organized 19 game sessions, with players of different backgrounds and familiarity to the topic. They report on the satisfaction, learning and uptake of the MSP Challenge board game, measured through questionnaires and observations. 10. Jean et al. aim to understand games as planning support systems [12]. They evaluated a computer-supported simulation-game using Maritime Spatial Planning, used with students and a few practitioners in Canada. The game was evaluated with audio–visual analysis of player interactions of individual and group learning.

2. Starting Points

The Special Issue was initiated from three perspectives:

1. A growing awareness on the urgency of water resources issues and the complexity of integrated water (resource) management;
2. An emerging paradigm that claims that the governance of natural resources, including water, benefits from a social learning approach;
3. A growing interest in the use of simulation games as a way to instrumentalize social learning.

Although each of the three perspectives are fairly well described in the literature, the questions at their intersection are not well conceptualized and studied. We had four main questions from the start:

1. How are social learning and game-based learning connected? What are the theories and concepts that tie them together?
2. What games are applied in sustainability, natural resource management and water? For what purpose are they used?
3. What is the function or role of games at the science-policy interface? What do policy makers and stakeholders themselves think about using simulations, games and play for policy making?
4. Does serious game-play have an impact on social learning in water management, and how can we observe or measure that?

3. Why Serious Games for Water Management?

This Special Issue postulates that water management and serious games are connected through complexity and social learning. The line of reasoning, nicely reflected in the twelve contributions, can be summarized as follows.

3.1. *The Urgency of the Matter; Games as Drivers for Change*

In a global context of climate change, degradation of ecosystems and depletion of natural resources, the availability and access of water for nature, consumption and production, are of imminent concern. In all parts of the world, droughts, floods and drinking-water quantity and quality are high on the societal and political agendas. To prepare the way for the change that is urgently needed, persuasive communication, education and involvement of stakeholders in society is necessary. Here, games and other media can be of use, for instance to reach a younger generation, or to bring stakeholders into a dialogue. Therefore, many organizations concerned with sustainability and water have taken the initiative to develop games with a persuasive or learning message. This is well reflected in the two journal articles about serious and persuasive games for sustainability, natural resource management and water [2,3]. The authors of the field-study articles also clearly convey this sense of urgency for sustainability.

3.2. *The Complexity of Water Management: Seeing the Bigger Picture*

Water (resource) systems are of an intricate complexity. They are a messy entanglement of hydrological aspects, geographical scales, time horizons, cause–effect relations, ecologies, human activities and much more. Problems with water are very *grass roots*, but their *root causes* are often regional to global, and solutions require long-term planning at different levels of governance. This is nicely reflected for instance in the Gomes et al. article on drinking water in peri-urban communities in Bangladesh [7]. All authors point at the inherent complexity of water management, in similar terms, to come up with a clear need: to see the bigger picture.

3.3. *Stakeholders: Managing Competing Values in the Political Arena*

The water resource and distribution questions 'who gets what, when and how, and at what costs' is rife with controversies and prone to cause conflict. The arena of water management consists of actors who play in their own interests, within or against the rules. On top of that, there are sector and stakeholder interdependencies, power balances and so on. Institutions in charge of water management are multi-level and sectoral, which may stand in the way to integral approaches. Nearly all field studies emphasize that water management relies on stakeholder collaboration. The article by Marini et al. nicely address the competing values of stakeholders and how games should take this into account—by going beyond self-interest [5]. The question of course remains whether we can expect such magic from playing a game.

3.4. Uncertainty: Operating at the Science-Policy Interface

One aspect of complexity is uncertainty about the impact of human activities on future states. Uncertainty can have many reasons, such as unavailability or incompleteness of data, imperfect calculation and simulation models, human incomprehension and limitations to science. How is evidence from science brought into a stakeholder driven and political process? This is reflected in the papers on how games work at the science-policy interface [4] and understanding games as planning support systems [5]. One of the challenges in serious games, is how to connect playful interaction with data, realistic models and simulations into the game. Several papers [9,10] demonstrate the integration of data, simulations and games.

3.5. Reality as a Game: Reshaping Rule-Based Interactions

Some theories suggest that water management can be looked at as a game. See for instance the paper by Gomes et al. that starts from game-theory [7]. At best, the zero-sum (win or lose) game is reshaped by the actors, to facilitate the sharing of the natural resource to prevent its imminent depletion. The rules of the game are formed by legislation, institutions (politics, economics) and culture. This game of course is bounded by the laws of nature (hydrology, ecology) although lack of understanding, insufficient data and uncertainty, gives room to play with that also [9]. If water management is like a game, the quality of the outcome depends upon the rule-based interactions among the players [8]. All field studies approach the serious games from the perspective of reshaping the rules for stakeholder–player interactions.

3.6. Social Learning: Instrumentation and Tooling

One strand of theory suggests that the quality of interactions among the stakeholders relies on what the stakeholders understand about the system, each other and the relationship between the rules, interaction and outcomes. In other words, the game of water management can be changed through social learning among the players. In this regard, others have claimed that communities of practice, simulations and games can be the tooling for that. Several papers, such as by Marini et al. [5] and Aubert et al. [1] stress that we first need to understand better the nature of social learning in a context of water management, and how this connects to the design and use of games for learning. One of the main challenges is how to make stakeholders learn. What are the best practices? The articles in the Special Issue give some clues and answers, but a coherent conceptualization still needs to be developed.

3.7. Learning from Practice: Evidence-Based Serious Gaming

A follow-up question is: how do we assess the effectiveness of the interventions aimed at social learning? How do we know what works, and under what conditions? A systematic scientific assessment of game-based approaches is essential to develop a deeper understanding of how and why a game-based approach fulfills the purpose and objectives it was designed for in the first place [1]. It is important to distinguish studying whether game-based approaches have worked (i.e., practice evaluation) from studying why and how game-based approaches have (or have not) worked (i.e., theory-based research assessment). In order to study and understand why and how a game-based approach has (or has not) worked, it is essential to use theoretical frameworks. Several authors make this point.

3.8. The Innovation Potential: Data, Intelligence and Immersion

We live in a data- and media-rich world, with emerging technologies giving possibilities to shape new forms of interaction. Games are among them. How can we harvest the richness of emerging digital technologies (AI, media, games, VR, AR) to create new and better forms of interaction for the good of society? A few articles take a digital, immersive turn [5,9,10], but this can be expected to become a stronger line of development in games for water management. In the contributions, the games are straightforward and practical. Board games and role-playing are predominant. One or two papers report on the use of advanced game technologies, such 3D and game engines. VR/AR in combination with big data and AI was not used. Perhaps at the time of publication, the technology was not advanced enough. It is likely that innovations in this area are reported from the engineering, design and validation perspective.

4. Future Directions

A more critical and reflective discussion of the limitations and even restrictions of games for social learning is necessary. The following points give directions for future research.

4.1. Risk of Saturation

We know remarkably little about who plays serious and persuasive games, how many people play them, why they play them, and what this delivers. When so many organizations launch persuasive and serious games in the area of sustainability, the target audience may get weary of them. The young generation is very demanding in terms of game-play. Not all of the games developed can be of equally high quality. In an educational context, students may take the wrong message.

We firmly believe that games can be powerful tools, but they should be used wisely and modestly; or they may turn against us. Game research can help to guide where and how it is wise to use them, and where not. A more critical and realistic view on serious games is needed. Perhaps we should develop and use less of them but with better quality and more impact.

4.2. Ethical Considerations and Power Imbalance

There has been serious neglect of the ethical considerations of developing and using games to induce change, especially among vulnerable target groups, such as children, excluded or disempowered groups in society. Like all forms of participation, the question can be asked: who is invited to play these games and thus, who is not? Power may come from different sources but having access to capital and control over natural resources is an important one. The games for a social learning paradigm suggest that a structural power imbalance can be overcome by interaction and that this creates mutual understanding. However, for this to happen an inherent rationality of the players, as well as equality and fairness in the game-play, is necessary. When a fight over natural resources like water is played out hard, social learning through games can turn into manipulation. It may falsely suggest that a level playing field exist and a win–win solution can be achieved. As with any form of empowerment and participation, the stakeholder–players in a serious game may feel cheated upon.

We therefore believe that social learning with or without serious games should not only be studied through social psychology, design and engineering, but also through ethics and political theory: power, participation and discursive–deliberative democracy are suggested starting points for further research.

4.3. Pitfall of Mistaken Identity

The social learning paradigm is heavily based on a constructivist world view. However, in sustainability, the realism part—evidence, data and science—is equally important. The combination of social learning and games may lead to negotiated nonsense. What happens if the 'wrong' message is passed on through serious games? What if authorities or participant–players skew the point the simulation games are trying to make? What if the players take the simulation game 'too literally' and cannot distinguish fact from fiction? What if the simulation games are not real enough and are only perceived as an enjoyable exercise after which everyone continues to think and do as before? These questions are relevant to those who want to use games for social learning in sustainability. Moreover, the intervention models on which serious games are based are seldom explicated or critically examined. The methodologies for measuring the impact of serious games are not very well established. Negative results and experiences with serious games are not reported.

We believe that a stronger instrumentation theory of social learning is necessary, with evidence-based evaluations, and frameworks and methods to do so. This is important because the use of even more immersive technologies, such as VR and AR, big data and intelligence for sustainability and natural resource management are around the corner.

Author Contributions: All authors contributed to recruiting and reviewing papers for this editorial. W.M. primarily led and coordinated this effort. I.M. and W.M. have written this editorial with input of the other authors.

Acknowledgments: Thanks to all of the contributions to the special issue, as well as to the anonymous reviewers who have contribute to the development of the articles. This special edition is the result of a Partnership Development Project funded through the Social Sciences and Humanities Research Council of Canada (SSHRC) Partnership Development grant (SSHRC 890-2014-0056) held by Jan Franklin Adamowski and contributed to by each of the guest editors of this special edition.

Conflicts of Interest: The authors declare no conflict of interest.

References

1. Aubert, A.H.; Medema, W.; Wals, A.E.J. Towards a framework for designing and assessing game-based approaches for sustainable water governance. *Water* **2019**, *11*, 869. [CrossRef]
2. Galván-Pérez, L.; Ouariachi, T.; Pozo-Llorente, M.; Gutiérrez-Pérez, J. Outstanding videogames on water: A quality assessment review based on evidence of narrative, gameplay and educational criteria. *Water* **2018**, *10*, 1404. [CrossRef]
3. Rodela, R.; Ligtenberg, A.; Bosma, R. Conceptualizing serious games as a learning-based intervention in the context of natural resources and environmental governance. *Water* **2019**, *11*, 245. [CrossRef]
4. Zhou, Q.; Mayer, I. Models, simulations and games for water management: A comparative q-method study in The Netherlands and China. *Water* **2017**, *10*, 10. [CrossRef]
5. Marini, D.; Medema, W.; Adamowski, J.; Veissière, S.; Mayer, I.; Wals, A. Socio-psychological perspectives on the potential for serious games to promote transcendental values in IWRM decision-making. *Water* **2018**, *10*, 1097. [CrossRef]
6. Magnuszewski, P.; Królikowska, K.; Koch, A.; Pająk, M.; Allen, C.; Chraibi, V.; Giri, A.; Haak, D.; Hart, N.; Hellman, M.; et al. Exploring the role of relational practices in water governance using a game-based approach. *Water* **2018**, *10*, 346. [CrossRef]
7. Gomes, S.; Hermans, L.; Islam, K.; Huda, S.; Hossain, A.; Thissen, W. Capacity building for water management in peri-urban communities, Bangladesh: A simulation-gaming approach. *Water* **2018**, *10*, 1704. [CrossRef]
8. Ferrero, G.; Bichai, F.; Rusca, M. Experiential learning through role-playing: Enhancing stakeholder collaboration in water safety plans. *Water* **2018**, *10*, 227. [CrossRef]
9. Sušnik, J.; Chew, C.; Domingo, X.; Mereu, S.; Trabucco, A.; Evans, B.; Vamvakeridou-Lyroudia, L.; Savić, D.; Laspidou, C.; Brouwer, F. Multi-stakeholder development of a serious game to explore the water-energy-food-land-climate nexus: The SIM4NEXUS approach. *Water* **2018**, *10*, 139. [CrossRef]
10. Khoury, M.; Gibson, M.J.; Savic, D.; Chen, A.S.; Vamvakeridou-Lyroudia, L.; Langford, H.; Wigley, S. A serious game designed to explore and understand the complexities of flood mitigation options in urban–rural catchments. *Water* **2018**, *10*, 1885. [CrossRef]

11. Keijser, X.; Ripken, M.; Mayer, I.; Warmelink, H.; Abspoel, L.; Fairgrieve, R.; Paris, C. Efficacy of the MSP challenge board game for stakeholder engagement in maritime spatial planning. *Water* **2018**, *10*, 724. [CrossRef]
12. Jean, S.; Gilbert, L.; Medema, W.; Keijser, X.; Mayer, I.; Inam, A.; Adamowski, J. Serious games as planning support systems: Learning from playing maritime spatial planning challenge 2050. *Water* **2018**, *10*, 1786. [CrossRef]

water

MDPI

Review

Socio-Psychological Perspectives on the Potential for Serious Games to Promote Transcendental Values in IWRM Decision-Making

Dianna Marini [1], Wietske Medema [1,*], Jan Adamowski [1], Samuel P. L. Veissière [2], Igor Mayer [3] and Arjen E. J. Wals [4]

1 Department of Bioresource Engineering, McGill University, Quebec City, QC H9X3V9, Canada; dianna.marini@mail.mcgill.ca (D.M.); jan.adamowski@mcgill.ca (J.A.)
2 Division of Social and Transcultural Psychiatry, McGill University, Quebec City, QC H9X3V, Canada; samuel.veissiere@mcgill.ca
3 Academy for Digital Entertainment, NHTV Breda University of Applied Sciences, 4817 JT Breda, The Netherlands; i.s.mayer@hotmail.com
4 Department of Social Sciences, Education and Learning Sciences, Wageningen University, 6708 PB Wageningen, The Netherlands; arjen.wals@wur.nl
* Correspondence: wietske.medema@mcgill.ca; Tel.: +1-310-480-5204

Received: 4 June 2018; Accepted: 8 August 2018; Published: 17 August 2018

Abstract: Modern day challenges of water resource management involve difficult decision-making in the face of increasing complexity and uncertainty. However, even if all decision-makers possessed perfect knowledge, water management decisions ultimately involve competing values, which will only get more prominent with increasing scarcity and competition over resources. Therefore, an important normative goal for water management is long-term cooperation between stakeholders. According to the principles of integrated water resource management (IWRM), this necessitates that managerial decisions support social equity and intergenerational equity (social equity that spans generations). The purpose of this discussion is to formulate preliminary recommendations for the design of serious games (SGs), a potential learning tool that may give rise to shared values and engage stakeholders with conflicting interests to cooperate towards a common goal. Specifically, this discussion explores whether SGs could promote values that transcend self-interest (transcendental values), based on the contributions of social psychology. The discussion is organized in the following way. First, an introduction is provided as to why understanding values from psychological perspectives is both important for water management and a potential avenue for learning in SGs. Second, a review of the description of values and mechanisms of value change from the field of social psychology is presented. This review highlights key psychological constraints to learning or applying values. Based on this review, recommendations are made for SGs designers to consider when developing games for water management, in order to promote transcendental values. Overall, the main conclusions from exploring the potential of value change for IWRM through SGs design are as follows: 1-SGs design needs to consider how all values change systematically; 2-SGs design should incorporate the many value conflicts that are faced in real life water management, 3-SGs could potentially promote learning by having players reflect on the reasoning behind value priorities across water management situations, and 4-value change ought to be tested in an iterative SGs design process using the Schwartz's Value Survey (SVS) (or something akin to it).

Keywords: serious games (SGs); water management; value change; transcendental values; social equity; sustainability; Schwartz's Value Survey (SVS); Integrated Water Resource Management (IWRM); psychosocial perspectives; decision-making processes

1. Introduction

Globalization, pollution, scarcity, social inequity, and climate change are issues that demand the attention of researchers and practitioners across various disciplines. There is a growing sense of urgency to implement innovative policies and management strategies, as our global natural resources are facing increasing pressures from population demand and the uncertain consequences of climate change. In response to complexity across natural systems and socio-political domains, there has been a shift from traditional reductionist approaches towards management strategies that integrate both socio-political and scientific dimensions. This is often referred to as "integrated natural resource management" (INRM), [1–3]. INRM strategies have evolved to incorporate adaptive management strategies. Adaptive management (AM) demands continuous and purposeful progress through monitoring and adaptation, to address the unpredictability of our natural resources [3,4].

Integrated water resource management (IWRM) offers an approach to management (under the INRM umbrella) that specifically addresses modern global freshwater challenges [3]. IWRM is defined by the Global Water Partnership as "a process that promotes the coordinated development and management of water and land resources, in order to maximize economic and social welfare in an equitable manner without compromising the sustainability of vital ecosystems" [5]. IWRM moves beyond the traditional management approach of "command and control", which assumes the predictability of stable water sources, as well as that these resources should be manipulated through technological engineering solutions in order to increase supply [3,6,7]. Its principles are based on public participation, economic efficiency, social equity, and ecological sustainability, and a key feature of IWRM is to manage freshwater resources at the scale of the watershed, involving land management, various stakeholders, and transcending administrative and political boundaries. The practice of IWRM has further evolved to incorporate adaptive management strategies that involve the continuous monitoring, learning, and improvement of methods and policies to address a continuously changing environment [3] (p. 25). Despite a growing popularity of IWRM principles and practice across the globe, key challenges remain that need to be addressed [8–10]. Barriers to IWRM implementation are often related to inefficient governance structures, lack of scientific and predictive knowledge, lack of engagement or cooperative action between stakeholders, and the limited capacity of IWRM management organizations [11,12]. Other difficulties in implementing IWRM are due to the complexity of coordinating socio-political boundaries with natural resource management that involve multiple temporal and spatial scales [13]. Scholars (across many disciplines) are exploring ways to address these challenges in order to achieve the fairness and sustainability of freshwater resources.

Collaborative forms of governance, such as IWRM and AM, are considered essential to solving complex sustainability problems and, as mentioned, require the bridging of various physical, administrative, and socio-political boundaries [14]. Human behavior changes with time, and the advancement of information and communication technologies (ICT) offers innovative means to help span these boundaries and contribute to collaborative solutions to sustainability problems, while fundamentally transforming interactions and relationships between governments, firms, non-government organizations (NGOs), and civil society [15,16]. Although the emerging field of ICT-enabled boundary spanning is still in its infancy, there is a general agreement that advanced ICT (e.g., interactive apps, virtual platforms and communities, serious games, and civic hackathons) provide novel tools to expanded collaboration across boundaries [17], as well as mechanisms to accelerate transformational change in (a) perception and meaning; (b) underlying norms and values; (c) social networks and patterns of interaction; and (d) power structures [18].

Serious games (SGs) that combine computer simulations with role-play as an integrated method for complex policy- and decision-making are particularly promising [7,19] and will be the focus of this research. Over the last decade, this type of game simulation has become more prevalent [20] in education, including teaching water management (e.g., [21,22]), water governance and policy (e.g., [23,24]), and other common-resource management fields (e.g., [25,26]). Although the professional and academic debate on the potential of serious games has quite a tradition (e.g., [19,20,27,28]), there are diverging opinions regarding the impact of such games on policy and decision-making [29], as well as on behavior change and learning outcomes [30].

Gaming technology is increasingly employed to support human learning and foster innovation [31,32]. SGs are defined as games that are designed for an educational value beyond entertainment [33], and are increasingly explored in the context of these modern socio-environmental challenges. From this perspective, SGs are discussed in terms of their potential to address the sustainable governance of natural resources by supporting individuals or groups, as well as providing spaces for collaboration and knowledge co-creation [7,11,12,34,35]. SGs have more recently been explored to support social learning and collaboration in IWRM [11,36]. The educational goals of IWRM games include the development of both soft and hard skills and can also be used to directly support decision-making. For example, one study developed a game to enhance decision-making skills for optimal water system design problems that resulted in observable improvements in the participants' ability to identify appropriate solutions [36]. Alternatively, softer skills in IWRM game design could include policy formulation and conflict resolution in transboundary management, such as in the Shariva (Shared River) game [24]. Additionally, the balancing of economic and environmental goals in sustainable watershed management is addressed in the Aqua Republica game, which can be played by individuals or groups [12].

It is interesting to note how the Aqua Republica game simulation requires its players to address conflicting goals in IWRM decision-making, such as economic prosperity versus environmental sustainability. For instance, short term economic gain, such as building a factory, results in longer term environmental degradation. On the other hand, population increase demands a certain level of economic growth to meet their energy demands. Players are thus challenged to make decisions that optimize both economic and environmental impacts. Here, participants learn policy and technological tools to sustain population growth, while preserving the environment, and are made aware of the negative environmental impacts that result from the sole pursuit of prosperity. However, a key challenge remains on how to promote the pursuit of social equity and sustainability in decision-making beyond the context of the game, in the long-term, across situations. To achieve this, it can be assumed that stakeholders must be continuously motivated to pursue goals beyond their self-interest in real life. These underlying motivations can be understood as values.

From the discipline of social psychology, values are relatively stable cognitive constructs that guide an individual's perceptions, attitudes, and behaviors, and that transcend situational boundaries [37,38]. The study of values may give insight into the potential learning outcomes of SGs that address important IWRM challenges. First, as values guide behavior and perceptions in a way that is not situation-specific, influencing a participant's values from game play could potentially result in tangible changes, beyond the context of the game. Moreover, human values have been shown to be predictive of cooperative attitudes and behavior, such as pro-environmental and charitable actions [39–42]. Also, values are thought to be invoked when reflecting on difficult decision-making that involves trade-offs or novel decisions [43]. Furthermore, values are tied to emotion, an important and often neglected variable when studying human perception, decisions, and actions, especially relating to issues of risk, uncertainty, and the management of natural resources [44–46]. For instance, it is argued that effective behavioral change (specifically in the context of adapting behaviors to mitigate the effects of climate change) should involve methods that invoke more personal and emotional responses [46]. It should come as no surprise then that values, shared values, and cooperation have already been studied in the context of game design [47,48]. Notably, a fundamental difference in values has already been recognized as a constraining factor in learning from SGs play for natural resource management [12].

Values can be evaluated through the perspective of many disciplines, such as philosophy, anthropology, economics, and management [43,49]. Therefore, it is helpful to review the theory of values from a selected discipline. There are several benefits to focus on for the social psychological study of values, which will be the focus of this discussion. The psychological perspective is particularly relevant in understanding the complex dynamics by which humans interact and shape their environments. This is best described by Bandura, as follows:

"Psychology is the one discipline that uniquely encompasses the complex interplay between intrapersonal, biological, interpersonal, and sociostructural determinants of human functioning. Psychology (is) best suited to advance understanding of the integrated biopsychosocial nature of human and how they manage and shape the everyday world around them" [50].

As values have been extensively studied in the field of social psychology, it is possible for researchers to empirically test them [43,44,51]. Therefore, game designers could potentially validate the effectiveness of their games in promoting certain values. For instance, changes in values from game play could potentially be monitored throughout the iterative design process by accessing well-tested psychological tools (i.e., Schwartz's Value Survey). Moreover, an important social psychological theory on values, Schwartz's "Theory of Basic Human Values", identifies specific values that have been empirically validated across 82 countries [37,51], and its methods of measurement have been tested in diverse cultures across the globe [49]. Therefore, these tools can potentially be used to assess the effectiveness of IWRM games that involve participants from different socio-cultural backgrounds. This is especially important when considering how cross-collaboration across socio-political boundaries are core to the practice of IWRM. All in all, the social psychological study of values may offer tools for game designers to iteratively monitor the effectiveness of their games in promoting IWRM principles, and involve a diversity of participants from different nations across the globe.

Interestingly, the development of certain values has already been presented as an important learning target for the IWRM audience. For instance, a more explicit deliberation of the underlying values (a "values approach") of governing water resources has been argued to ensure that the management outcomes are actually socially and environmentally sustainable [52–55]. However, even if one diligently pursues the IWRM principles, in practice, values often conflict [56,57]. For example, the increased pressure from population demand and urbanization can make it difficult to balance social equity and sustainability, as it can be difficult to supply the demand without investing in infrastructure that leads to long-term negative impacts on the environment [7] (p. 90).

Conflicting values that arise between stakeholders are being discussed as key water management challenges [56,58]. In practice, conflicts may arise between different applications of water usage (agriculture versus fisheries or energy) and competing groups (transboundary conflicts) [3,57,58]. Importantly, the development of shared values is an important outcome for social learning processes [12], known as "transformation", where a collection of individuals formulate a common purpose and work towards a shared goal [59]. Social learning refers to the process of learning through others in social settings (that is, of outsourcing information to others, or to a cumulative cultural repertoire of skill, knowledge, and attitudes) [60], where learning outcomes may result in changes in the understanding for a small or large group [61]. Social learning outcomes can also be categorized as cognitive enhancement (such as gaining knowledge) or moral development. The desired "normative" outcomes of social learning involve developing a sense of solidarity with the community (43), synonymous with "cooperative values" or "shared values", where the well-being of the group is being prioritized over individualistic aims. Changing values is seen as an important outcome of the social learning processes, both at the individual and societal level. Value change is an important potential outcome of social learning processes, namely "double and triple loop social learning". Double loop learning involves challenging values on an individual or societal level, while triple loop learning involves a change in governance systems as a result of changing values [12,61]. In other words, triple loop learning includes double loop learning and expands this into a transformation of the governing system.

Evidently, values are already being discussed in the context of IWRM learning and capacity building. However, values as a term or concept can be alluded to from various perspectives. Within the study of social psychology, which specific values would be desirable learning outcomes to incorporate in the design of SGs for IWRM? As seen in the IWRM definition, IWRM aims for the pursuit of social equity and environmental sustainability. Social equity aims to protect the welfare of all human beings, and environmental sustainability ensures that the welfare of future generations are also protected (thus, sustainability pursues intergenerational equity, which is social equity that spans generations). Within this field, the term "transcendental values" specifically refers to values whose underlying motivations are beyond self-interest [51]. As stakeholders are required to make decisions that pursue goals beyond their self-interest in order to align with IWRM principles (social equity and environmental sustainability), we argue that the promotion of "transcendental values" is a desirable learning outcome for SGs. Therefore, the following discussion reviews how the study of values in psychology contributes to the understanding the mechanisms and constraints involved in the promotion of transcendental values. This first requires understanding the theory of values, how values can be changed or influenced, and whether or not transcendental values can be promoted (see Figure 1 below for an overview of the structure of the discussion).

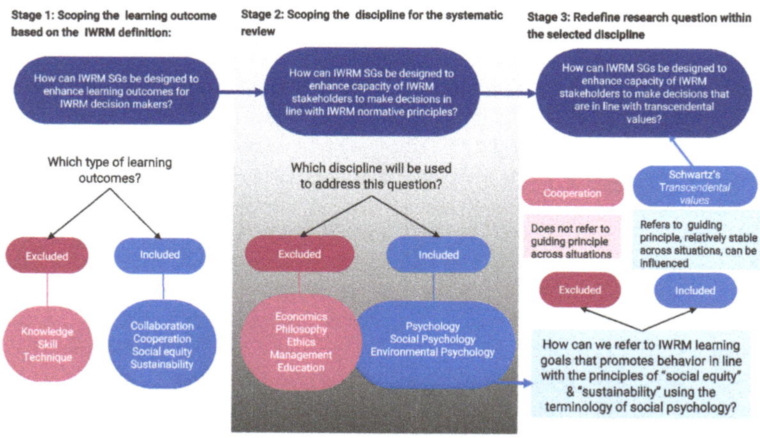

Figure 1. Conceptualization of the following research question: how can serious games (SGs) for integrated water resource management (IWRM) be designed to enhance the capacity of IWRM stakeholders to make decisions that are in line with IWRM's normative principles?

2. Method

A systematic review of values and changing values from the field of social psychology was carried out as part of this review. As values can be understood differently in many fields, it is necessary to first scope the review within a particular discipline. This review includes peer-reviewed literature within the field of social psychology. The discussion of values from other disciplines, such as economics or philosophy, were excluded. Scopus and Google Scholar were used to search the following terms: value(s), value change, change values, learn values, mechanism value change, environment values, cooperative values, values social psychology, shared values, and transcendental values. The review on the theory of values was used to recommend which specific values from the field of social psychology are desirable learning outcomes for IWRM in SG. Following this, the review of the mechanisms and constraints on how values can be changed or learned was used to develop recommendations for SGs design so as to achieve the desired learning outcomes. A brief review on the relevant topics in SGs applied to IWRM was carried out to incorporate the findings into the results and discussion. Search terms to review these topics were "serious games and water management" and, "social learning and serious games". Discussions on SGs for IWRM that focus on hard skills (technical, knowledge transfer, technical skills, and knowledge-based skills) were not included. Insights from the review are then used to make recommendations for SG design for the promotion of transcendental values in the context of IWRM (see Figure 2 below for an overview of the overall methodology for this review paper).

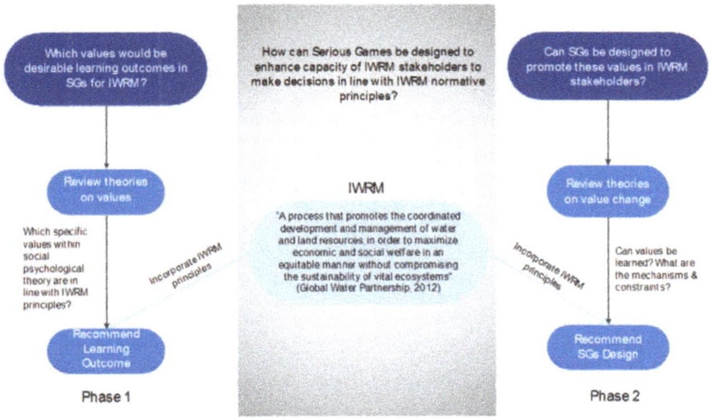

Figure 2. Overall methodology to answer the following research question: how can serious games be designed to enhance the capacity of IWRM stakeholders to make decisions in line with IWRM normative principles?

3. Understanding Values from a Socio-Psychological Perspective

3.1. Behaviorism

In order to contextualize the theory of values from social psychology, we will briefly discuss the study of learning in psychology. Behaviorism is primarily focused on human learning through the objective study of observable behavior, which is often reduced to stimulus–response relationships [62]. From this perspective, values are often tied to expected rewards. Values are implicated in classical conditioning and associative learning. Associative learning can be understood within the theory of classical conditioning (Pavlov, Watson), where a behavioral response is learned for a neutral stimulus by pairing it with a stimulus that had a pre-established behavioral response [63]. As an evolution to Pavlov and Watson's classical conditioning, Skinner introduced "operant conditioning", which adds a dimension of personal agency to these earlier theories. His principles have also been used to explain complex behaviors through "behavioral shaping", and modifying behavior through rewards such as tokens [64]. Several scientists have looked into the neural mechanisms and correlates of associative learning based on behaviorism. For instance, several models utilizing algorithms based on behavioral theories have been applied to understand the neural mechanisms behind learning reward value [65]. In addition, damage to reward centers in the brain (such as the amygdala) can devalue the conditioned stimulus (CS) and thus impede associative learning [63]. Interestingly, humans have been shown to value expected rewards subjectively. For instance, people will favor immediate rewards over long-term rewards, even if the overall monetary gain will increase in the long-term [39] (p. 271).

Some argue that learning values from a behaviorist approach (rewards and punishment) is too narrow and reductionist (oversimplified to stimulus and response mechanics). A common criticism is that behaviorism solely focuses on studying behavioral responses rather than understanding internal mental processes. At the time of its inception (and given the methods of the time), this was an extremely useful method in psychology. An important evolution from these theories was when Bandura posited that people also learn from watching others get rewarded and punished (observational learning), and demonstrated this phenomenon in the 1960s [50]. In the 'Bobo Doll experiment', Bandura observed how children interacted with a doll after having watched adults interact with them in an aggressive manner [66]. Bandura's "social learning theory" (SLT) is built on the classical/operant conditioning models of learning with the addition of the internal cognitive processes that mediate learning between stimulus and response. These cognitive processes allow for individuals to learn new behaviors from others through observation. An important contribution of Bandura's learning model is that people actively process information. An observation of another's behavior will be learned (replicated), depending on the mental processes of (1) attention; (2) retention (memory); (3) reproduction ability (self-efficacy); and (4) motivation. Moreover, when someone identifies (identification) with a model (mode of behavior, like aggression), they tend to move beyond the imitation of a single behavior, and adopt a broader range of the attitudes, beliefs, and values of that model [66].

Although Bandura's theory has evolved significantly, it is still heavily based in behaviorism. Not only does the theory of behaviorism represent a critical stepping-stone for Bandura's SLT, but it is also important in the pedagogical theories of instructional design (ID), such as giving praise to students as reinforcement [67]. Like Bandura, in addition to behaviorism, ID incorporated other theoretical constructs to account for complexity in human learning, such as active information processing, as described by Bandura's SLT [67]. An important theoretical cluster in ID is constructivism, which attempts to counter the over-emphasis of humans as passive recipients of environmental inputs in learning, and incorporate human agency, such as broad complex learning goals, as well as the ability to construct and test knowledge against reality. Notably, Bandura also recognized the over-emphasis of the environment in his SLT and subsequently developed social cognitive theory (SCT) to account for more human agency [50]. All in all, the behaviorist theories of learning have had a great influence on the subsequent learning theories that involve values. Behavior theories and understanding are still relevant, although they are not thought to sufficiently explain human learning and behavior. There has also been a movement towards the incorporation of learning in a social context, as well as the incorporation of human agency in learning models.

3.2. Social Psychological Theories of Values

Although values are discussed in the influential learning theories of behaviorism and SLT, more in-depth discussions can be found from social-psychological contributions. Values have been of interest for a long time in influencing human socio-psychological processes and behavior, but are generally seen as too abstract and vague to measure or describe in detail [68,69]. Values in social psychology can be broadly defined as concepts about desirable states (abstract ideals) that serve as guiding principles for one's life [43,44]. It is important to note that values are distinct from other psychological constructs, namely: attitudes (affective orientation to something specific), traits (not necessarily desired or reflected upon influencers of behavior), norms (ought to statements based on societal demands), needs (biological requirements), preferences (ranking outcomes of a decision), beliefs (understandings of the world), worldviews (generalized beliefs), and roles (behavioral decisions based on social situation) [43]. This distinction is especially important when, for example, discussing an ecosystem valuation that works to put a monetary value to ecosystem services. In these discussions, the valuation of a certain ecological concept, such as water quality, is referred to as a "context value", however this would be an "attitude" or "preference" according to strict social-psychological terminology. Such distinctions are also relevant in comparison to the management or economic literature. For instance, structured decision making (SDM) is a practical approach to improve decision-making for environmental management, and value-focused thinking guides decision making in strategic management. Both recognize how values are core to decision-making processes, and advise stakeholders to deliberate on how "values" are implicated when considering alternative decision outcomes [70,71]. For example, different stakeholders may ascribe different "values" or 'importance' to the potential consequences of preserving a spiritual heritage site, constructing a power plant, or preserving wildlife. Again, the term "value" in this sense is context specific, and therefore, should be referred to as an attitude or preference within the language of psychology. Additionally, the importance of considering social and environmental sustainability in business model design is a process referred to as developing "shared value" [41]. Moreover, valuation in economic theory (i.e., monetary valuation) expresses a "preference" and not a Schwartz value. This review paper focuses on the socio-psychological terminology of values, which are broader constructs that guide behaviors, attitudes, and preferences across a person's life situations. Importantly, values are different from traits because there is an element of choice in deciding what is important [38].

The content of values has also been thoroughly explored from a sociocultural framework. For instance, in cross-cultural psychology, Hofstede made a significant contribution by operationalizing values at the cultural level in order to study the relative differences. Hofstede became interested in values after recognizing stable differences in workers' values (as opposed to attitudes). In this sense, values reflected desires (such as the ideal personality of one's boss), whereas attitudes reflected someone's understanding of an actual situation (one's attitude towards their actual boss) [72]. Hofstede's values are categorized as the desirable (general ideological statements) or the desired (importance attached to a job such as cooperation). In Hofstede's "cultural dimensions theory", values are at the core of what he calls the "mental programs" of collective cultures. His value theory is based on sixty different countries, where he found four distinct dimensions, namely: power distance, uncertainty avoidance, individualism, and masculinity [73]. He posits that these dimensions represent issues faced by all societies, and their relative differences refer to different learned society responses to these issues [72,74]. Importantly, Hofstede stresses that these dimensions are "ecologically derived", based on social systems and not individuals, and thus his findings cannot be reduced to explain/predict an individual's preferences [74].

Schwartz's published work in "Advances in Experimental Social Psychology", is recognized as a significant advancement in the socio-psychological understanding of values [37,68]. Here, Schwartz describes his comprehensive "Theory of Basic Human Values", which was based on the work of several scholars who preceded him, namely: Hofstede (described above); Allport, 1960 (who first attempted to describe values); and Rokeach, 1973 (who determined that the relative priority of values were important). Schwartz's theory describes 56 distinct values (expanded from Rokeach) that were empirically evaluated in 82 countries [37,51,68]. According to this theory, all values have the following six main features: (1) they are beliefs linked to emotion; (2) refer to desirable goals that motivate action; (3) they transcend situations; (4) serve as criteria and standards for decision-making (often unconsciously); (5) are hierarchical (relative importance contrary to norms and beliefs); and (6) their relative importance guides action [51]. There are ten basic values that are found across all cultures, which are distinguished in terms of their underlying motivations (see Table 1).

Table 1. Schwartz's 10 values and underlying motivations (adapted from Schwartz [51]).

Value	Underlying Motivation	Description
Power	Self-enhancement	Social status, prestige control, or dominance over people and resources
Achievement	Self-enhancement	Personal success and demonstration of competence according to social standards
Hedonism	Self-enhancement/openness to change	Pleasure sensuous gratification for oneself
Stimulation	Openness to change	Excitement novelty challenge
Self-direction	Openness to change	Independence agency in choosing and acting, creating, and exploring
Benevolence	Self-transcendence	Preserving and enhancing the welfare of "in-group" members
Universalism	Self-transcendence	Understanding, appreciation, tolerance, and protection for the welfare of all people and for nature, contrary to the 'in-group' focus of benevolence
Conformity and tradition	Conservation	In general, to subordinate the self to socially imposed expectations Self-restraint of actions that may upset or harm others Avoid violating social expectations or norms Respect, commitment, and acceptance of the one's culture or religion
Security	Conservation	Safety, harmony, and stability of society that one identifies with, of relationships and of self Security for the self and those with whom one identifies with (family or nation)

Schwartz postulates that these 10 basic values are found cross-culturally, as they relate to human needs both as living organisms (organismic needs) and as social beings (social beings). For example, the value of hedonism expresses an organismic need to experience pleasure, and cooperation expresses a social need for cooperation [51]. Importantly, as will be discussed later in Section 4.2, on value change, the way in which situational factors influence changes in value prioritization is influenced by which kind of need (material or social) the value addresses [38]. Moreover, these ten basic values are conceptualized by Schwartz on a circular continuum, where the more similar values are in terms of their underlying motivations, the closer they are positioned to each other [51] (p. 9).

3.3. Cross-Cultural Evidence for Schwartz's Theory of Values

Schwartz's theory of basic values is a well-developed and supported theory in the psychological study of human values. Each basic human value that Schwartz has characterized in this theory has been empirically confirmed to express the basic value they were postulated to express [75], and this type of study has been replicated many times over the years since Schwartz first presented his theory in 1992. Studies have assessed the theory with data from hundreds of samples and tens of thousands of participants in 82 countries around the world, using either the Schwartz Value Survey (SVS) or Portrait Value Questionnaire (PVQ) methods of measurement [51]. The samples included highly diverse geographic, cultural, linguistic, religious, age, gender, and occupational groups, with representative national samples from 37 countries [51,76–79]. These studies provide evidence that the oppositions of self-transcendence to self-enhancement values and of openness to change to conservation values are virtually universally present. Moreover, each of the ten basic values are distinguished in at least 90% of the samples. These findings show that people in most cultures respond to ten types of values distinctly, and that the broader value orientations captured by adjacent values are discriminated nearly universally. They strongly support the idea that human values form the motivational continuum postulated by the theory.

3.4. Schwartz's Value Theory Applied to IWRM: Determining Transcendental Values as a Desired Learning Outcome

In taking a closer look at the description of Schwartz's basic values in Table 1, we can identify the values most in line with IWRM principles and thus a potential learning target for SGs design. As discussed, the ultimate aim of IWRM is to pursue social equity and intergenerational equity, thus protecting the welfare of all human beings that depend on water resources now and in the future. This means that resources are to be managed in an equitable manner, throughout space (across socio-political boundaries) and across time (protecting future generations towards sustainability). Therefore, the basic value that most closely reflects the goals of IWRM is universalism, which is described as the "understanding, appreciation, tolerance, and protection for the welfare of all people and nature" (Schwartz 2012). Importantly, this is distinguished from benevolence, which concerns the welfare of only the "in-group" members, such as one's family or nationality.

We can also explore how Schwartz's values are relevant for current IWRM challenges and conflict. First, water management involves managing limited resources across political boundaries. Transboundary conflicts may involve a value conflict of power between nations, as well as benevolence (preserving the "in-group" of a nation's welfare at the expense of an "out-group"). According to Schwartz's model on the relationships between values, universalism and benevolence both share the underlying motivation of self-transcendence. Therefore, it is clear that these values conflict with power or achievement, whose underlying motivation is self-enhancement (opposite to self-transcendence). For instance, a community that decides to build a very prosperous industrial plant that is harmful to the surrounding ecosystem can be said to value achievement over universalism. However, it may be less obvious to identify the possible tensions between benevolence or universalism. Nonetheless, the types of dilemmas faced by water managers may involve different decisions, depending on the prioritization of benevolence and universalism. A clear example of this involves deciding whether or not to increase the water supply for irrigation to provide food in one's nation (benevolence), while compromising the future of sufficient water supply for another nation downstream. Second, conflict in water management may also occur between water uses, which may involve conflicting values between power (dominance and control over water) versus security (preserving ecosystem health). On the other hand, the values of power and security in water management may be completely aligned. For instance, power over water sources, such as building a dam, may provide the necessary water for irrigation, and thus "food security" for a nation. Such conflict achievement may be relevant for industries that utilize water resources and are strongly motivated to achieve economic success and status. Third, a key IWRM challenge involves uncertainty, which is addressed in adaptive management strategies. Adaptive management strategies involve continuous evaluation and change, which may involve prioritizing the value of "openness to change" over "conservation".

Evidently, there are many potential value conflicts involved in water management challenges. Overall, it seems that promoting universalism (over conflicting values of power and achievement) is the most relevant for the IWRM goals and is transferrable across different situations. Moreover, water is a common resource that is depleted or negatively impacted if each person rationally pursues their own self-interest without regard for the collective (the so-called "tragedy of the commons" [49]. In this regard, by virtue of managing water, IWRM conflicts with the underlying motivation of self-enhancement, which further argues for IWRM to prioritize its opposing motivation—self-transcendence. However, with the recognition that there may be complex dilemmas between benevolence and universalism in the water management context, the discussion will refer to promoting "transcendental" values as a learning outcome, referring the underlying motivation of both benevolence and universalism.

3.5. Evolutionary Approaches from Behavioral Economics to Cognitive Anthropology

A general consensus from cognitive science to behavioral economics similarly casts 'values' as relatively stable pre-reflective processes that guide individual decision-making and action. In the so-called two-systems account of reasoning, spearheaded by Kahneman and Tvertsy, human thinking is characterized by evolutionarily older "fast", automatic, intuitive processes (the so-called System 1), and evolutionary younger, "slow", deliberate reasoning modulated by language (the so-called System 2) [80–82]. On this account, human motivations and behavior are primarily modulated by automatic System 1 mechanisms, such as heuristics and biases like base-rate neglect (extrapolating from immediately salient cues without considering context), the availability heuristics or frequency bias (forming a judgment based on what comes to mind easily and assumptions that are culturally widespread), the confirmation bias (attending to select cues to confirm what one already believes), or a variety of emotionally-driven processes [80,81]. Rational utility maximization, on these views, can only be derived through effortful, often counter-intuitive reasoning.

Debates are ongoing regarding the extent to which humans engage in rational decision-making, as well as the role of culture and cultural differences in facilitating this process. In psychologist Jonathan Haidt's classical experiments on moral intuitions, for example, subjects in a variety of national and socioeconomic samples were shown to reach decisions on what is right or wrong that were invariably consistent with those of their respective groups [83]. In what came to be known as the "moral dumbfounding" paradigm, subjects consistently followed their gut-feelings to reach a moral position, offered post hoc rationalizations to justify their feelings as just right or wrong, and were unable to explain their motives in more detail. Cultural psychologists and cognitive anthropologists have adopted this model to describe the universal process through which people outsource value judgments (moral and otherwise) to intuitions, primed by culturally-specific values [84]. In Haidt's experiments, college-educated students were significantly more likely than groups from other socioeconomic status to voice utilitarian values, (in Haidt's interpretation) simply because utilitarian values are commonplace and normative among western educated people. In a similar vein, experiments in developmental psychology have shown that children are intrinsically motivated to enforce the norms they can intuit from their social groups [85].

This renewed attention to culture in cognitive science has prompted a re-evaluation of rational decision-making and game-theoretic models that has shown that compared to apes, humans across cultures typically favor obedience to group norms, and group fitness to individual maximization [60,86–88]. In more recent accounts, humans have been argued not to possess any rational "reasoning" abilities at all, but rather to display the selective capacity to make intuitive inferences to question other people's motives in communicative and justificatory situations that involve arguments with others [89]. Taken together, these different "cultural learning" approaches suggest that values are best studied and experimentally manipulated by focusing on the social context in which they arise and are enacted-that is, by targeting the universal evolved biases through which humans outsource their preferences to what they expect relevant agents to also expect of the world. This might be best achieved by conceptualizing cultural groups, as people united by shared expectations that selectively pattern the salience and valence of what one attends to in the world (i.e., regimes of attention in processes of niche construction) [90,91].

The prestige bias [60,86] and in-group or out-group biases (see Section 4.3.3 for further explanation) are good candidate targets for this process. In this model, people can be motivated to shift their value systems (e.g., toward ecologically sustainable practices) by looking up to prestigious agents and cultural forms (e.g., celebrities, fashionable musical genres, and leaders) from their own group. Marking target behaviors with prestige through credibility enhancement displays (CREDs) [60,92], can help direct attention toward features of the world (e.g., locally produced foods), behaviors, and intended outcomes (e.g., recycling; attending community meetings) that will become positively valued through their association with high-status agents. Competing with a perceived out-group with an opposite set of values (often one in which the valence equation is simply reversed – e.g., recycling is bad) will also help reinforce.

4. Mechanisms and Constraints in "Learning" Transcendental Values

Figure 3 below provides an overview of the remaining discussion that focuses on determining the desired learning outcome for IWRM within the field of social psychology, as well as the structure of the subsequent sections of this review paper.

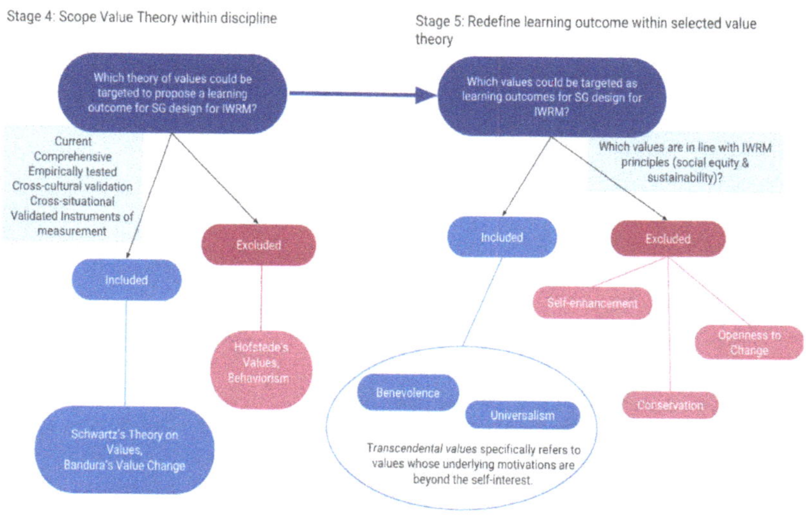

Figure 3. Determining the desired learning outcome for IWRM within the field of social psychology.

In order to explore the potential learning mechanisms behind promoting transcendental values in SGs play for water management stakeholders, it is important to distinguish between the mechanisms of value change and value-congruent behavior. Value change refers to a change in the prioritization of values, for instance, changing the prioritization of self-enhancing values (achievement) compared to self-transcendence values (benevolence) [68]. Value-congruent behavior refers to making decisions or actions that are in line with one's values. Notably, value incongruent behaviors and decisions are possible, as persons may not always perceive how a decision or behavior conflicts with their value prioritization (limit to value perception). Additionally, other psychological factors (such as norms) may dominate influence on behaviors [68]. For instance, time-constraints could cause one to override their benevolent value prioritization, leading persons to consciously avoid helping a person in need when running late for an appointment [68]. Therefore, the normative learning outcomes for water management decision-making can be categorized in the following way: (1) values are being applied (activated) in decision-making; (2) and transcendental values are prioritized.

4.1. Value-Perception and Value Congruent Behavior

To help ensure value-congruent action, people first need to notice and understand how their values are implicated in a situation. Otherwise, other psychological processes or constructs may dominate decision-making behaviors. Indeed, values' prediction of attitudes or behavior is significantly improved when people explicitly view the situation as being connected to their values [68]. Maio has explored this process, making the distinction between typical and atypical value instantiations [68] (p. 29). A value instantiation refers to a concrete application of a value. Typical value instantiations refer to situations where the relevance of a value is typically/commonly perceived, such as the value of social equality in the context of two persons of different races applying for a job. In contrast, an atypical value instantiation refers to situations where a value is less commonly perceived, such as left and right-handed people's job applicants. In this example, both typical and atypical instantiations involve the same value, however, someone who prioritizes social equity may not notice the value relevance in the atypical instantiation. According to Maio, prior instantiations of typical value applications will affect the process of value application for other typical scenarios by increasing one's ability to perceive value in similar situations [68].

The theory of atypical and typical value instantiation could be applied to SGs for water management and potentially represent long-term learning outcomes. For instance, the "rules of the game" could demand explicit identification or discussion of values involved in water conflict scenarios that represent typical or atypical value instantiations. Through this process, game players could potentially learn to more readily identify the values at play in different water management situations. Consequently, values could potentially be activated in decision-making beyond the game context, if the scenarios encountered in the game are similar to those in real life. Although this does not guarantee decision-making based on transcendental values, it potentially leads to more value-congruent decision-making. Interestingly, explicit value deliberation has recently been explored as a potential for developing shared values in ecosystem services [93]. Also, social learning processes involve critical self-reflection on one's perspectives and goals, and thus, could include a reflection of one's values in the group settings involved in water management decisions [12,61,94].

Interestingly, the theory of atypical versus typical instantiation helps clarify the difference in value-based decision making in management literature versus social psychological theory of values. For instance, an environmental scientist and spiritual leader, based on their experiences, may be said to value (management literature) certain outcomes of a water management decision over others (such as preserving heritage site versus preserving wildlife). From the domain of psychology, both the scientist and the spiritual leader may share the same values (equally prioritizing transcendence over self-interest), even if they have different preferences over the two possible outcomes. The theory of typical value instantiation may explain why the environmentalist may more easily identify how self-transcendence is implicated in preserving the wildlife, while the leader may more easily identify how self-transcendence is implicated in preserving the heritage site.

4.2. Value Change

From the perspective of social psychology, values are relatively stable psychological constructs as they are integral components of our self-concepts, maintained by emotion and past experiences [38,69,95]. Nonetheless, values have been shown to change both on an individual and societal level in certain circumstances. For instance, a change in the socio-environment that enables or suppresses one's opportunity to express a value may cause value change in order to adapt to the new circumstance. Studies on both MBA and law students, for example, found a decrease in students' mean prioritization of benevolent values with an increase in achievement values, after being exposed to the high-achieving environment of their program [38,51,96]. Also, a change from rural to urban environments was shown to foster individualistic (self-enhancing) values in Japanese immigrants [97]. Evidently, values can be suppressed if individuals cannot pursue them, based on the constraints imposed by their culture or environment.

Importantly, the direction of value change may depend on the underlying motivation being suppressed or expressed. For instance, suppressing self-enhancing values that relate to human needs lower on Maslow's hierarchy (materialism and security) may cause an increase in their prioritization, whereas transcendental values are enhanced with the opportunity to express them [38,51,98]. Second, certain values can be prioritized based on primers, such as how or what language is used in discourse. For example, using "I" versus "we" in storytelling may affect the subsequent value prioritization of collective (transcendental) values. Also, the English language acts as a primer of individualistic values compared to Chinese in SVS surveys [38]. Third, emotional triggers have been shown to lead to the prioritization of certain values. For instance, anxiety provokes self-enhancement whereas non-anxious emotions may lead to transcendental values [38,96]. Fourth, as values are generally supported by affect (unconscious truisms) rather than strong cognitive rationalization, conscious reflection and reasoning about values may cause values to change [96].

Bardi [96] was one of the first to focus on the mechanisms behind long-term value change on the individual level (referred to as intra-individual value change) using Schwartz's framework. According to her work, the understanding of value change could be categorized in the following way: individual versus group value change, systemic value change, short-term versus long-term change, and automatic versus effortful change mechanisms.

Firstly, Bardi studied the changes in terms of how individuals ranked the importance of their values (rank order change). This is distinct from the study on how values change in a group, where researchers look at the changes in the mean importance of a value of a population over a certain period of time (mean level change). Mean value changes were the primary focus of studying value change from other disciplines such as sociology and political science [38,96]. The mean level changes observed in groups and cultures across time are thought to be the result of shared experiences and changing environments, such as economic development, educational changes, and societal changes [96]. For instance, Inglehart's book, "The Silent Revolution", discusses a mean value change due to the post-modern western culture, whose values are moving away from materialism and self-enhancement, and moving towards transcendence and self-direction [98]. On the other hand, rank order value changes are observed when an individual's values change based on their distinct personal experiences.

Secondly, Bardi proposed that individual values change in a systematic way, according to Schwartz circular model, whereby, "compatible values change in the same direction and conflicting values change in opposite directions" [96]. Thus, an individual change in ranking one value will predictably affect the whole system of values in a way that is consistent with Schwartz's circular model. For example, an increase in achievement will also cause an increase in power, as both of these values share the underlying motivation of self-enhancement. Bardi supported this proposition in four longitudinal studies of rank value change in individuals [38].

Thirdly, Bardi looked into the possible mechanisms that transition from short-term to long-term changes in values. According to her model, more permanent changes in one's values result from a change in one's schema (mental model), after repeated environmental cues. For instance, one's values can change permanently following a change in one's life situation, such as becoming a parent or moving to a new environment with a different culture [38].

Fourthly, the mechanisms of value change can be automatic or effortful. The automatic processes of value change are facilitated by unconscious cues, which can be the result of primers (clues to alternate schemas), adaptation (to new environment or life situation), or identification (with role model, group, or peer). Interestingly, the value change resulting from adaptation relates back to behaviorist reinforcement theories, where a new social environment can frustrate one's values and consistently reward an alternate value. For example, moving to an individualistic culture (where positive rewards are associated with self-enhancing behaviors) can result in decreasing collectivist values. On the other hand, effortful changes in values require conscious cognitive reflection and re-evaluation, which can also occur through adaptation and identification. Additionally, effortful changes in values can result from attempts to maintain consistency in one's positive self-concept. For instance, subjects that receive negative feedback by peers after filling an SVS, (such as being characterized as selfish), have been shown to cause a subsequent re-prioritization of values. This relates back to the theory of cognitive dissonance, whereby individuals are motivated to maintain consistent self-concepts [99]. Finally, one can change values through effortful mechanism if they are convinced by direct persuasion. However, direct persuasion (such as through the media or education) is a particularly tricky route towards value change, as they are central to one's self-concept, and direct attacks to one's values often trigger defense mechanisms [38].

4.3. Some Social Psychological Limitations to "Learning" Transcendental Values

In order to manage the expectations of potential SGs to result in transformative learning outcomes, it is important to consider key challenges to influencing value change and value-congruent behaviors. The challenges described here are thought to be important in the context of water management, which involve stakeholders with individual differences that have both personal and group identities. However, it is recognized that there are many more challenges from the psychological framework that could be discussed. Here, we briefly overview the psychological mechanisms to resist change, individual differences in capacity to embrace transcendental values, and challenges of cooperation between members that hold different group identities.

4.3.1. Resistance to Changing Self-Concept

On an individual level, psychological defense mechanisms resist value change. Therefore, direct messages that try to persuade a change in values are not likely to have an effect. In fact, resistance to change attitudes from direct persuasion have been shown, where attitudes are less central to self-concepts than values [38]. Also, Maio, 2010, showed that people did not change values from cognitive elaboration if they were aware in advance of attempts of value change, and thus prepared rationalizations to their values [38,68].

4.3.2. Individual Differences in Capacity to Embrace Transcendental Values

There may be individual differences in the capacity to truly embrace transcendental values. Firstly, from the perspective of behaviorism, there may be biological differences in one's ability to experience reward, based on pro-social behaviors [39] (p. 277). For instance, neurobiological markers are associated with charitable action, and persons that reciprocated in a "trust game" were found to have significantly higher levels of oxytocin (hormone implicated in social bonds) in their blood [39] (p. 278). Secondly, from a psychopathological perspective, there may be individual differences in capacity for both cognitive perspective taking (theory of mind, ToM) and affective concern for others (affective empathy, AE). Interestingly, research carried out on autistic individuals (low ToM and normal AE) and psychopaths (high ToM and low AE), and persons with Down's Syndrome (high ToM and AE, and low IQ) have revealed that these abilities may operate independently from one another [100–102]. Therefore, training on perspective taking, (such as through role play) has shown that an enhanced cognitive ability to understand others' perspectives may not necessarily result in increased affective concern (AE) in all individuals equally (and thus not likely enhance transcendental values). Moreover, training on perspective taking for persons with a limited capacity for affective empathy has also been shown to result in a greater manipulative power [102]. Although some studies show that perspective taking can induce concern for others and the environment [41], it is important to recognize these potential limitations. Thirdly, there may be individual differences in a person's ability to disengage from their moral principles and values. Moral agency requires self-regulatory mechanisms in order to result in moral action, and persons may differ in their ability to self-regulate (self-sanction) [50].

Also, there are certain factors that can cause moral disengagement, namely: masking an immoral act as serving a moral purpose, diffusing or displacing of responsibility, weakening their perceived control upon an outcome, and dehumanizing a potential victim. Not surprisingly, persons that are prone to morally disengage are less likely to engage in prosocial behavior [50].

4.3.3. Intergroup Conflict

Intergroup conflict may be an important potential barrier to learning or fostering transcendental values during SG play. Conflict between members of a group can be understood from the study of intergroup relations, which looks at the conditions where individual behavior is made based on their membership to a group and attempts to explain their complex behavior, defined by Sherif [103] as "whenever individuals belonging to one group interact, collectively or individually, with another group, in terms of their group identification, we have an instance of intergroup behavior" [104]. The theory is characterized by three main principles, intergroup accentuation (accentuating in-group similarities and out-group differences), in-group favoritism (favoring in-group members, such as positive evaluations or allocation of resources), and social competition (competing for status/distinctiveness in social hierarchy) [105]. When the intergroup schema is activated, these three principles come into play. Notably, even without prior membership to a group, an arbitrary categorization of group members induced by an experimental setting (as sorting research subjects into teams of red and blue T-shirts) can motivate individuals to behave on the basis of their group membership rather than on individual or interpersonal relations [104,105]. As an intergroup schema can be activated arbitrarily in experimental settings, it is likely to be activated during game play if players are put into teams.

Intergroup conflict is a key player in the complexity between the pursuit of benevolence (towards 'in-group') versus universalism (for all), and is seen as one of the reasons benevolence (i.e., between in-group members) is prioritized over universalism in many cases [51]. An example of this in the context of IWRM could be someone benevolently investing their resources to build a dam to increase water supply for a community that is facing water-scarcity (in-group) at the expense of the downstream consequences to another community in the future (out-group), which also can also include species health within the ecosystem (such as fish migration). Moreover, conflict between group members is extremely relevant in the context of natural resource management, where the stakeholders involved are likely belong to well-defined social groups with histories of conflict (especially in transboundary water management between conflicting nations). For these reasons, a more in-depth discussion on intergroup conflict as a barrier is discussed below.

Empirical studies in intergroup relations have looked into factors that facilitate or constrain the reduction of conflict between group members. Firstly, making different groups cooperate rather than compete results in less discrimination and conflict between them. However, the effectiveness of a cooperative context (i.e., introducing subordinate goal) in reducing conflict depends on the salience of the existing group identities [105]. Secondly, personal contact with an out-group member may reduce conflict and prejudice. However, this requires that attention is drawn to the individual's personal characteristics rather than their group identity [105] (p. 293). Thirdly, the need for positive social identity (in social identity theory) suggests complementary roles towards a common goal for a positive interpersonal contact experience between members of an out-group. Complementary roles in achieving a goal allow for individuals to preserve their positive distinctiveness (social competition) while cooperating. This process is particularly interesting in the water management context, as it does not call for changing one's understanding of their group membership, but alters the negative understanding of interdependence between the in-group and out-group [105] (p. 29).

In a water management context, this would involve different stakeholders (i.e., industry, environmentalists, and government) to preserve their distinct group identity (which can be related to their professional background), but have a positive experience in cooperating, by engaging in complementary roles towards achieving a sustainability goal. If such an experience in a game setting could be salient/positive enough, it may lead to permanent changes in perceptions about the negative interdependence between them. In this way, game play could potentially have a positive long-term effect towards cooperation between stakeholders, without compromising or threatening their self-concepts and social identities. However, there are concerns for stable change in any of the above three processes mentioned, as they are deemed unstable in the context of optimal distinctiveness theory [105].

5. Recommendations for SGs Design

A recent study by Aubert et al. [20] provides an in-depth review of 43 SGs on water-related issues, and categorizes these games according to the level of technology used (i.e., from no low-tech to fully immersive high-tech) and the degree of verisimilitude (i.e., from modeling complex reality, using scientific models and real-world data to not using any scientific models or real-world data). These 43 serious games span purposes ranging from the following: (a) SGs that broadcast a message to teach and raise awareness on water related issues; (b) SGs that present direct or indirect exchanges of information (e.g., data, knowledge and worldviews); and (c) to SGs that reproduce a real-world situation with accurate reality, to provide a training experience for professionals. Direct (i.e., two-sided information) exchange games involve simulation games, often played in a workshop format with a scientific game facilitator, that aim to structure problems or develop scenarios while informing a scientific model, as well as allowing participants to develop an understanding of other players' perspectives [25,27]. Indirect exchange games allow for learning or awareness raising as well as data collection. The focus of this study is on what Aubert et al. [20] refer to as "hybrid games", which combine role playing games with computer simulations, and allow players to experience the impact of their decisions or actions over time, while developing an understanding of the complex interactions of social, environmental, and economic factors. These types of games enhance discussion and learning among players, thus enhancing social learning, and generally require a facilitator or game leader to introduce the game context and rules, encourage collaboration, and facilitate a debriefing phase [19,106].

Based on the psychological mechanisms and constraints of value change discussed in this paper, there are several factors that could be incorporated into the design of these "hybrid" SGs for water management stakeholders. Such design elements could potentially lead to the improved application of social equity and sustainability in IWRM decision-making through the enhancement of transcendental values. Of course, there is considerable research that remains to be conducted in terms of its potential effectiveness. The recommendations made here are preliminary, yet are thought to warrant further exploration and research. The concluding remarks will be organized by relating the key features of values from social psychology, and then relating them to SGs design. These relationships are discussed in Table 2 (SGs design opportunities) and Table 3 (SGs game design constraints).

Table 2. Serious games (SGs) design opportunities for "learning" transcendental values.

Key Features of Social Psychology	SGs Design Opportunities
Values are hierarchical in nature It is the relative prioritization of values to one another that is important, not values in isolation.	SGs design could incorporate scenarios of conflicting values where players are forced to prioritize their values in decision-making. This is possible in the water management context that often involves conflict of values (such as economic efficiency versus intergenerational equity).
Values change as system, not in isolation Schwartz's circular model is an accurate model of the mental representations of relationships between compatibilities and conflicts between different values.	If the game fosters self-enhancing values, such as achievement, then self-transcendental values, such as cooperation, will be reduced. SGs design can influence values not targeted in the design. SGs design needs to consider all values at play in game design if it aims to target one value. This is especially interesting in the context of game play that often involves achievement objectives for players, whereby the 'rules of the game' could demand cooperation or competition in order to succeed. Also, game design could involve threats to one's security, which could conflict with cooperation.
Values are primarily based on emotion (a) Emotions are the primary source of information for consensually important values and play a stronger role than past-behavior and beliefs. (b) There are different emotional consequences when violating one's peripheral versus central values. Violating peripheral values results in anxiety emotions, whereas violating central values results in dejection emotions (sadness). (c) Anxiety-inducing emotions can trigger self-enhancement values.	(a) As values are emotion-based, SGs play that involves values should be emotionally salient. Also, as people often hold onto values without having supported them with cognitive reasoning, having players reason about their values may lead to value change. (b) Post-game surveys on affect can give insight to whether values were violated and whether they were central or peripheral to one's self-concept. (c) Beyond the game context, real-life consequences to one's safety and security may cause anxiety-related emotions that potentially lead to self-enhancement values. SGs designers could consider this as a potential constraint to learning transcendental values. SGs designers could also consider which emotions are involved in game play and test players' level of anxiety during the iterative design process.
Values are influenced by culture and cultural differences (a) Individual values are invariably consistent with those of their respective social groups. (b) Humans across cultures typically favor obedience to group norms and group fitness to individual maximization. (c) Values are best studied and experimentally manipulated by focusing on the social context in which they arise and are enacted.	To foster self-transcendental values, SGs can be designed to target the universal evolved biases (e.g., prestige bias, and in-group or out-group biases) through which individuals and social groups outsource their preferences by conceptualizing cultural groups as people united by shared expectations. Competing with a perceived out-group with an opposite set of values will also help reinforce a shift in value systems towards, for example, ecologically sustainable practices.
Values are abstract and do not always influence decisions or behavior (even though they guide overall behavior and decision-making across situations)	SGs design could help bridge the value–action gap by requiring the detection of value relevant features in a conflict and requiring cognitive value-elaboration in water management decisions.
Values can be learned from identification	SGs could consider value elaboration in group settings for the possibility of members to identify with role models that hold cooperative values. However, it would be difficult to ensure that positive role models are present and identified with. Alternatively, virtual role models could be introduced, however, it is unclear how virtual role models affect identification.
Values can be learned through cognitive reasoning	There is the possibility for SGs to design a space for reflection and a cognitive elaboration of reasoning behind values, which may lead to a change in values (because values are often experienced emotionally without cognitive reflection). Such a process also leads to an increased perception of how values are involved in water management decisions. This requires time and space for reflection on values involved in decision-making. To design games according to theories on values, time for reflection is ultimately necessary.
Available empirical tools for assessing value change in groups and individuals (self-report surveys)	During an iterative design process, SGs could monitor learning and values with tools available, such as the Schwartz's Value Survey (SVS). SGs designers would have to decide whether to test mean changes in groups, or rank order changes in individuals, and long-term versus short term value changes. SGs designers could develop a tool to test value change through actual decisions made during game play, as decisions reflect value prioritization. This likely would contribute both to the game design as well as the psychological understanding of values in decision-making behavior.

Table 3. Key constraints to consider in SGs design for developing transcendental values.

Key Features of Social Psychology	SGs Design Constraints
Values are central to self-concept and resist direct persuasion	SGs design should not demand value change or involve a facilitator that directly tries to persuade players to change their values. However, in group settings, direct persuasion from other players may occur. This represents a potential constraint to learning "transcendental" values.
Values compete with other psychological factors in influencing decisions and behaviors	SGs designers should manage expectations of learning outcomes as human behavior is complex. If SGs game design demands explicit value-congruent behavior during play, this may not always be translated outside the game context because other psychological factors may dominate. SGs designers could also consider other psychological factors at play that may compete with values in terms of decisions-making behaviors of players and explore ways to mitigate those factors.
Intergroup conflict (intergroup conflict is a considerable barrier to cooperation in water management issues and also arises when arbitrary groups are made)	SGs could consider ways to mitigate intergroup conflict in game play. However, the psychological mechanisms to mitigate intergroup conflict are well beyond the scope of this paper. Some interesting possibilities based on brief discussion are (1) increase the salience of individual versus group identity (which would likely require in person contact); and (2) the application of 'optimal distinctiveness theory' by assigning distinct roles/expertise towards a common goal. Importantly, real-life water management likely involves salient intergroup conflict. SGs design could alternatively decide to emphasize intergroup conflict, by making salient clues on the group identity of different stakeholders, or assigning arbitrary groups. In this way, SGs play could give insight to intergroup dynamics in real life water management problems and potentially offer insight into appropriate policy design to address this issue.
Individual differences in capacity for moral engagement (affective empathy and cognitive perspective taking are independent abilities in humans)	Perspective taking and role switching could induce cooperative values by only in individuals with capacity of affective empathy. Doing so for those who are incapable will only make them more manipulative. Again, this constraint should manage expectations of SGs design outcomes in teaching transcendental values. Further study could identify those with more or less capacity for empathy in managerial or decision-making roles.
Permanent value change correlates with high impact life events	The 'non-real' of a game may not be interpreted as an impactful event for game players. As game play is not a 'real life' event, this makes it difficult to be impactful for a long-term change of values. SGs could work around this constraint by (1) focusing on enhancing value perception water management decision-making; and (2) focusing on potential mechanisms of value change by delivering consistent cues through repeated game play and social interactions.

The list of research publications where SGs have been used and reviewed for IWRM is considerable and fast growing (e.g., [15,20,21,23,25,106]). A few studies stand out as providing experimental evidence of the ability and means through which SGs may facilitate behavioral changes. Vegt et al. [107], for example, demonstrate how, by changing the rules of a game, the players can be directed towards either competitive or collaborative behavior. Another example is a study by Kampf and Stolero [108], which highlights the value of serious games in enhancing players' empathy for the perspectives of others, encouraging players to critically reflect on their own position, and provide a forum where parties can develop a relationship by interacting in a safe environment. Tipton et al. [109] emphasize that learning happens during a serious game, while players observe the impacts of their decisions and receive feedback on their actions, but that a debriefing phase after the game is essential as a way to draw lessons and critically reflect on the game process and outcomes.

Despite a growing interest in serious games for water resources management, novel approaches towards serious game design from a sociological or psychological are still lacking [110]. Linehan et al. [110] stress the importance of better understanding the spectrum of relevant social and psychological processes acting on both the designer as well as a players of serious games. They state that to increase the potential benefits of SGs, a deeper understanding is required of the processes through which players are incentivized to behave in a productive and sustainable manner, answering questions about how to measure, understand, predict, and guide people's behavior.

Overall, several important recommendations can be derived from exploring psychological perspectives and the potential of value change for IWRM stakeholders through SGs play, namely:

- SGs design should target values systematically, as individual changes in values also result in changing non-targeted values.
- SGs design should involve many value conflicts that are faced in real life water management decisions, so that players are not only forced to prioritize values, but also learn to recognize how their values are involved across many water management situations.
- As values are often supported by emotion and water management decisions can be complex, having players engage in the conscious reasoning behind the values involved in decisions represent important potential learning mechanisms. Reasoning about values has been shown to cause values to change, and this could potentially be incorporated in SGs play. Importantly, this requires time for reflection about one's values, where time for reflection has already been recognized as a constraint in IWRM games [11]. Notably, a similar concept is being applied in SGs for clinical medical ethics, involving the conscious deliberation of deontological ethics (ethics based on principles rather than consequences) in clinical decision-making, to foster virtuous medical professionals [111] (p. 99).
- Intergroup conflict in water management represents an important barrier to value change, both within and beyond the game context. Intergroup conflict is a complex phenomenon between in-groups and out-groups, and SGs can address this issue either by enhancing group conflict to explore their dynamics, or by mitigating conflict to promote cooperation between out-groups. Moreover, the relationship between intergroup conflict and universal versus benevolent value prioritization in water management decisions warrants further study. Water management decisions may involve conflict between universal and benevolent values, even though they share the underlying motivation of self-transcendence. Such relationships can potentially be explored through SGs play. Notably, in order to explore such dynamics, it may be more interesting to have team players that can compete or collaborate in person (rather than interact online or with computer models).
- Whether or not players' values change (in the short or long-term) can be tested using SVS during the iterative SGs design process. Values are primarily based on emotion, and individual differences in the capacity for affective empathy constrain value change for certain individuals. In addition to considering how values are involved when designing SGs, Schwartz' value survey can be used to examine the prioritization of values for participants engaged in existing water related serious games. Interestingly, the researchers involved in this paper have been administering the SVS survey to subjects both before and after participating in a number of game simulation events. The results and insights from these events and the resulting surveys will be used to determine whether playing the SGs had any effect on participants' values.

All in all, this discussion attempts to address a significant challenge for SGs design, and whether it can have an impact on the underlying belief systems of water management stakeholders. The importance of this challenge is illustrated by some of the responses of the actual game players in this context. For instance, in her exploration of SGs for marine spatial planning and water management, Zhou found the following responses from interviews (interviewee No. 18), "Experience shows that it is very hard to change players' belief by playing games...they don't really relate the game to their real world problems or seriousness" [7] (p. 105), and (interviewee No. 20) "Policymakers look for excuses not to learn from the game. Gaming is not the thing to change the behaviors of individuals" [7] (p. 10). Despite this obvious challenge, the current understanding in psychology behind changing values provides insight into the mechanisms that can be incorporated into SGs design. Moreover, the empirical testing of values and value change, followed by improving SG design, could move beyond addressing SGs design challenges and provide some insight into important psychological phenomena. Further, studies for both psychology and SGs design could look into the impact of reflecting on values in groups compared to interaction with a computer or online game. Also, individual differences in the ability to embrace transcendental values could be studied. Finally, more effort could look into how to effectively communicate about values in groups or other mediums (such as ICT), and how SGs could provide a space for formulating policies aligned with values in complex water management decisions or dilemmas.

Author Contributions: D.M. conceived and conducted the literature review; W.M. and J.A. supervised this project; D.M. and W.M. wrote the paper; J.A., S.P.L.V., I.M. and A.E.J.W. provided ongoing reviews to the paper drafts.

Funding: This research was funded by the Social Sciences and Humanities Research Council of Canada (SSHRC) grant number SSHRC 890-2014-0056.

Acknowledgments: This study was funded by a SSHRC Partnership Development Grant held by Adamowski and contributed to by Nguyen from the McGill University Brace Centre for Water Resources Management.

Conflicts of Interest: The authors declare no conflict of interest.

References

1. Izac, A.-M.N.; Sanchez, P.A. Towards a natural resource management paradigm for international agriculture: The example of agroforestry research. *Agric. Syst.* **2001**, *69*, 5–25. [CrossRef]
2. Van Noordwijk, M.; Tomich, T.P.; Verbist, B. *Negotiation Support Models for Integrated Natural Resource Management in Tropical Forest Margins*; CABI Publication: Cambridge, MA, USA, 2003; pp. 87–108.
3. Adamowski, J.; Zyla, C.; Cuenca, E.; Medema, W.; Clamen, M.; Reig, P. *Integrated and Adaptive Water Resources Planning, Management, and Governance*; Water Resources Publications LLC: Littleton, CO, USA, 2013.
4. Clark, R.C. Applying cognitive strategies to instructional design. *Perform. Improv.* **2002**, *41*, 10–16. [CrossRef]
5. Medema, W.; Jeffrey, P.; McIntosh, B. From premise to practice: A critical assessment of integrated water resources management and adaptive management approaches in the water sector. *Ecol. Soc.* **2008**, *13*, 29–47. [CrossRef]
6. Pahl-Wostl, C.; Kranz, N. Water governance in times of change. *Environ. Sci. Policy* **2010**, *13*, 567–570. [CrossRef]
7. Zhou, Q. *The Princess in the Castle: Challenging Serious Game Play for Integrated Policy Analysis and Planning*; Delft University: Delft, The Netherlands, 2014.
8. Allouche, J. The birth and spread of iwrm-a case study of global policy diffusion and translation. *Water Alternatives* **2016**, *9*, 412–413.
9. Cherlet, J. Tracing the emergence and deployment of the 'integrated water resources management' paradigm. In Proceedings of the 12th EASA Biennial Conference, Ghent, Belgium, 10–13 July 2012.
10. Rahaman, M.M.; Varis, O. Integrated water resources management: Evolution, prospects and future challenges. *Sustain. Sci. Pract. Policy* **2005**, *1*, 15–21. [CrossRef]
11. Medema, W.; Adamowski, J.; Orr, C.; Wals, A.; Milot, N. Towards sustainable water governance: Examining water governance issues in québec through the lens of multi-loop social learning. *Can. Water Res. J.* **2015**, *40*, 373–391. [CrossRef]
12. Medema, W.; Furber, A.; Adamowski, J.; Zhou, Q.; Mayer, I. Exploring the potential of serious games on social learning and stakeholder collaborations for transboundary watershed management of the St. Lawrence river basin. *Water* **2016**, *8*, 175. [CrossRef]
13. Cervoni, L.; Biro, A.; Beazley, K. Implementing integrated water resources management: The importance of cross-scale considerations and local conditions in ontario and nova scotia. *Can. Water Res. J.* **2008**, *33*, 333–350. [CrossRef]
14. Medema, W.; Adamowski, J.; Orr, C.; Furber, A.; Wals, A.; Milot, N. Building a foundation for knowledge co-creation in collaborative water governance: Dimensions of stakeholder networks facilitated through bridging organizations. *Water* **2017**, *9*, 60. [CrossRef]
15. Chew, C.; Lloyd, G.J.; Knudsen, E. An Interactive Capacity Building Experience—An Approach with Serious Games. 2013. Available online: https://www.dhigroup.com/upload/publications/mikebasin/Chew_2013.pdf (accessed on 13 August 2018).
16. Termeer, C.; Bruinsma, A. Ict-enabled boundary spanning arrangements in collaborative sustainability governance. *Curr. Opin. Environ. Sustain.* **2016**, *18*, 91–98. [CrossRef]

17. Kleinhans, R.; Van Ham, M.; Evans-Cowley, J. Using social media and mobile technologies to foster engagement and self-organization in participatory urban planning and neighbourhood governance. *Plann. Pract. Res.* **2015**, *30*, 237–247. [CrossRef]
18. (IPCC), I.P.o.C.C. *Managing the Risks of Extreme Events and Disasters to Advance Climate Change Adaptation*; Cambridge University Press: New York, NY, USA, 2012; p. 582.
19. Zhou, Q.; Mayer, I. Models, simulations and games for water management: A comparative q-method study in The Netherlands and China. *Water* **2017**, *10*, 10. [CrossRef]
20. Aubert, A.H.; Bauer, R.; Lienert, J. A review of water-related serious games to specify use in environmental multi-criteria decision analysis. *Environ. Modell. Softw.* **2018**, *105*, 64–78. [CrossRef]
21. Hoekstra, A.Y. Computer-supported games and role plays in teaching water management. *Hydrol. Earth Syst. Sci.* **2012**, *16*, 2985–2994. [CrossRef]
22. Ewen, T.; Siebert, J. Learning about water resource sharing through game play. *Hydrol. Earth Syst. Sci.* **2016**, *10*, 4079–4091. [CrossRef]
23. Geurts, J.L.A.; Duke, R.D.; Vermeulen, P.A.M. Policy gaming for strategy and change. *Long Range Plann.* **2007**, *40*, 535–558. [CrossRef]
24. Douven, W.; Mul, M.L.; Son, L.; Bakker, N.; Radosevich, G.; Hendriks, A. Games to create awareness and design policies for transboundary cooperation in river basins: Lessons from the shariva game of the mekong river commission. *Water Resour. Manag.* **2014**, *28*, 1431–1447. [CrossRef]
25. Barreteau, O. The joint use of role-playing games and models regarding negotiation processes: Characterization of associations. *Artif. Soc. Soc. Simulat.* **2003**, *6*. Available online: http://jasss.soc.surrey.ac.uk/6/2/3.html (accessed on 17 August 2018).
26. Cleland, D.; Dray, A.; Perez, P.; Cruz-Trinidad, A.; Geronimo, R. Simulating the dynamics of subsistence fishing communities reefgame as a learning and data-gathering computer-assisted role-play game. *Simul. Gaming* **2012**, *43*, 102–117. [CrossRef]
27. van der Wal, M.M.; de Kraker, J.; Kroeze, C.; Kirschner, P.A.; Valkering, P. Can computer models be used for social learning? A serious game in water management. *Environ. Model. Softw.* **2016**, *75*, 119–132. [CrossRef]
28. Rumore, D.; Schenk, T.; Susskind, L. Role-play simulations for climate change adaptation education and engagement. *Nat. Clim. Chang.* **2016**, *6*, 745–750. [CrossRef]
29. Mayer, I.S. The gaming of policy and the politics of gaming: A review. *Simul. Gaming* **2009**, *40*, 825–862. [CrossRef]
30. Tanenbaum, J.G.; Antle, A.N.; Robinson, J. Three perspectives on behavior change for serious games. In Proceedings of the SIGCHI Conference on Human Factors in Computing Systems, Paris, France, 27 April–2 May 2013.
31. Young, M.F.; Slota, S.; Cutter, A.B.; Jalette, G.; Mullin, G.; Lai, B.; Simeoni, Z.; Tran, M.; Yukhymenko, M. Our princess is in another castle a review of trends in serious gaming for education. *Rev. Educ. Res.* **2012**, *82*, 61–89. [CrossRef]
32. Kangas, M. Creative and playful learning: Learning through game co-creation and games in a playful learning environment. *Think. Skills Creat.* **2010**, *5*, 1–15. [CrossRef]
33. Susi, T.; Johannesson, M.; Backlund, P. *Serious Games: An Overview*; University of Skövde: Skövde, Sweden, 2007.
34. Karpouzoglou, T.; Zulkafli, Z.; Grainger, S.; Dewulf, A.; Buytaert, W.; Hannah, D.M. Environmental virtual observatories (evos): Prospects for knowledge co-creation and resilience in the information age. *Curr. Opin. Environ. Sustain.* **2016**, *18*, 40–48. [CrossRef]
35. Duke, R.D. *Gaming: The Future's Language*; SAGE Publications: Thousand Oaks, CA, USA, 1974.
36. Savic, D.; Morley, M.S.; Khoury, M. Serious gaming for water systems planning and management. *Water* **2016**, *8*, 456. [CrossRef]
37. Schwartz, S.H. Universals in the content and structure of values: Theoretical advances and empirical tests in 20 countries. *Adv. Exp. Soc. Psychol.* **1992**, *25*, 1–65.
38. Bardi, A.; Goodwin, R. The dual route to value change: Individual processes and cultural moderators. *J. Cross-Cult. Psychol.* **2011**, *42*, 271–287. [CrossRef]
39. Forsyth, D.R. *Group Dynamics*; Wadsworth Cengage Learning: Belmont, CA, USA, 2014.
40. Kramer, M.R. Creating shared value. *Harvard Bus. Rev.* **2011**, *89*, 62–77.
41. Schultz, P.W. Changing behavior with normative feedback interventions: A field experiment on curbside recycling. *Basic Appl. Soc. Psychol.* **1999**, *21*, 25–36. [CrossRef]

42. Thøgersen, J.; Ölander, F. Spillover of environment-friendly consumer behaviour. *J. Environ. Psychol.* **2003**, *23*, 225–236. [CrossRef]
43. Dietz, T.; Fitzgerald, A.; Schwom, R. Environmental values. *Annu. Rev. Environ. Resour.* **2005**, *30*, 335–372. [CrossRef]
44. Leiserowitz, A. Climate change risk perception and policy preferences: The role of affect, imagery, and values. *Clim. Chang.* **2006**, *77*, 45–72. [CrossRef]
45. Sultana, F. Suffering for water, suffering from water: Emotional geographies of resource access, control and conflict. *Geoforum* **2011**, *42*, 163–172. [CrossRef]
46. Weber, E.U. Experience-based and description-based perceptions of long-term risk: Why global warming does not scare us (yet). *Clim. Chang.* **2006**, *77*, 103–120. [CrossRef]
47. Flanagan, M.; Nissenbaum, H. *Values at Play in Digital Games*; MIT Press: Cambridge, MA, USA, 2014.
48. Salen, K.; Zimmerman, E. *Rules of Play: Game Design Fundamentals*; MIT Press: Boston, MA, USA, 2004.
49. Brosch, T.; Sander, D. *Handbook of Value: Perspectives from Economics, Neuroscience, Philosophy, Psychology and Sociology*; Oxford University Press: New York, NY, USA, 2016; p. 424.
50. Bandura, A. Social cognitive theory: An agentic perspective. *Annu. Rev. Psychol.* **2001**, *52*, 1–26. [CrossRef] [PubMed]
51. Schwartz, S.H. An overview of the schwartz theory of basic values. *Online Read. Psychol. Cult.* **2012**. Available online: https://scholarworks.gvsu.edu/cgi/viewcontent.cgi?article=1116&context=orpc (accessed on 8 August 2018). [CrossRef]
52. Groenfeldt, D.; Schmidt, J.J. Ethics and water governance. *Ecol. Soc.* **2013**, *18*, 14. [CrossRef]
53. Flournoy, A.C. Building an environmental ethic from the ground up. *UC Davis Law Rev.* **2003**, *37*, 53.
54. Postel, S.; Richter, B. The missing piece: A water ethic. *Am. Prospect.* **2008**, *19*, 22–23.
55. Sandford, R.W. *Ethical Water: Learning to Value What Matters Most*; Rocky Mountain Books Ltd.: Victoria, BC, Canada, 2011.
56. Orlove, B.; Caton, S.C. Water sustainability: Anthropological approaches and prospects. *Annu. Rev. Psychol.* **2010**, *39*, 401–415. [CrossRef]
57. Hermans, L. *Stakeholder-Oriented Valuation to Support Water Resources Management Processes: Confronting Concepts with Local Practice*; Food and Agriculture Organization of the United Nations: Rome, Italy, 2006.
58. Ravesteijn, W.; Kroesen, O. Design for values in waterwater management. In *Handbook of Ethics, Values, and Technological Design*; Springer: Berlin/Heidelberg, Germany, 2015; pp. 847–867.
59. Webler, T.; Kastenholz, H.; Renn, O. Public participation in impact assessment: A social learning perspective. *Environ. Impact Assess. Rev.* **1995**, *15*, 443–463. [CrossRef]
60. Henrich, J. *The Secret of Our Success: HOW Culture is Driving Human Evolution, Domesticating Our Species, and Making us Smarter*; Princeton University Press: Princeton, NJ, USA, 2015.
61. Reed, M.S.; Evely, A.C.; Cundill, G.; Fazey, I.; Glass, J.; Laing, A.; Newig, J.; Parrish, B.; Prell, C.; Raymond, C.; et al. What is social learning? *Ecol. Soc.* **2010**, *15*, 4. [CrossRef]
62. Watson, J.B. Psychology as the behaviorist views it. *Psychol. Rev.* **1913**, *20*, 158. [CrossRef]
63. Gottfried, J.A.; O'Doherty, J.; Dolan, R.J. Encoding predictive reward value in human amygdala and orbitofrontal cortex. *Science* **2003**, *301*, 1104–1107. [CrossRef] [PubMed]
64. Skinner, B.F. *How to Teach Animals*; Freeman: San Francisco, CA, USA, 1951.
65. O'Reilly, R.C.; Frank, M.J.; Hazy, T.E.; Watz, B. Pvlv: The primary value and learned value pavlovian learning algorithm. *Behav. Neurosci.* **2007**, *121*, 31–49. [CrossRef] [PubMed]
66. Bandura, A. *Social Learning Theory*; General Learning Press: New York, NY, USA, 1971.
67. Driscoll, M.P. Psychological foundations of instructional design. In *Trends and Issues in Instructional Design and Technology*; Reiser, I.R.A., Dempsey, J.V., Eds.; Prentice Hall: Upper Saddle River, NJ, USA, 2002.
68. Maio, G.R. Mental representations of social values. *Adv. Exp. Soc. Psychol.* **2010**, *42*, 1–43.
69. Hitlin, S.; Piliavin, J.A. Values: Reviving a dormant concept. *Annu. Rev. Sociol.* **2004**, *30*, 359–393. [CrossRef]
70. Gregory, R.; Failing, L.; Harstone, M.; Long, G.; McDaniels, T.L.; Ohlson, D. *Structured Decision Making: A Practical Guide to Environmental Management Choices*; Wiley-Blackwell: West Sussex, UK, 2012.
71. Keeney, R.L. Value-focused thinking: Identifying decision opportunities and creating alternatives. *Eur. J. Oper. Res.* **1996**, *92*, 537–549. [CrossRef]
72. Hofstede, G.J. The cultural relativity of organizational practices and theories. *J. Int. Bus. Stud.* **1983**, *14*, 75–89. [CrossRef]

73. Hofstede, G. *Culture's Consequences: International Differences in Work-Related Values*; SAGE Publication: California, CA, USA, 1980.
74. Hofstede, G. National cultures in four dimensions: A research-based theory of cultural differences among nations. *Int. Stud. Manag. Organ.* **1983**, *13*, 46–74. [CrossRef]
75. Schwartz, S.H. Basic human values: Theory, measurement, and applications. *Revue Française de Sociologie* **2006**, *42*, 249–288.
76. Schwartz, S.H. Value orientations: Measurement, antecedents and consequences across nations. In *Measuring Attitudes Cross-Nationally-Lessons from the European Social Survey*; Jowell, R., Roberts, C., Fitzgerald, R., Eva, G., Eds.; SAGE Publications: London, UK, 2006.
77. Davidov, E.; Schmidt, P.; Schwartz, S.H. Bringing values back in: The adequacy of the European social survey to measure values in 20 countries. *Public Opin. Q.* **2008**, *72*, 420–445. [CrossRef]
78. Bilsky, W.; Janik, M.; Schwartz, S.H. The structural organization of human values – evidence from three rounds of the European social survey (ess). *J. Cross Cult. Psychol.* **2011**, *42*, 759–776. [CrossRef]
79. Schwartz, S.H.; Cieciuch, J.; Vecchione, M.; Davidov, E.; Fischer, R.; Beierlein, C.; Ramos, A.; Verkasalo, M.; Lönnqvist, J.-E.; Demirutku, K.; et al. Refining the theory of basic individual values. *J. Person. Soc. Psychol.* **2012**, *103*, 663–688. [CrossRef] [PubMed]
80. Kahneman, D. *Thinking, Fast and Slow*; Farrar, Straus and Giroux: New York, NY, USA, 2011.
81. Boyd, R.; Richerson, P.J. *Culture and The Evolutionary Process*; University of Chicago press: Chicago, IL, USA, 1988.
82. Gilovich, T.; Griffin, D.; Kahneman, D. *Heuristics and Biases: The Psychology of Intuitive Judgment*; Cambridge University Press: Cambridge, NJ, USA, 2002.
83. Haidt, J. The emotional dog and its rational tail: A social intuitionist approach to moral judgment. *Psychol. Rev.* **2001**, *108*, 814–834. [CrossRef] [PubMed]
84. Haidt, J.; Björklund, F. Social intuitionists answer six questions about moral psychology. *Moral Psychol.* **2008**, *2*, 181–217.
85. Schmidt, M.F.; Rakoczy, H.; Tomasello, M. Young children enforce social norms selectively depending on the violator's group affiliation. *Cognition* **2012**, *124*, 325–333. [CrossRef] [PubMed]
86. Henrich, J.; Gil-White, F.J. The evolution of prestige: Freely conferred deference as a mechanism for enhancing the benefits of cultural transmission. *Evol. Hum. Behav.* **2001**, *22*, 165–196. [CrossRef]
87. Henrich, J.; Boyd, R.; Bowles, S.; Camerer, C.; Fehr, E.; Gintis, H.; McElreath, R. In search of homo economicus: Behavioral experiments in 15 small-scale societies. *Am. Econ. Rev.* **2001**, *91*, 73–78. [CrossRef]
88. Henrich, J. Decision-making, cultural transmission and adaptation in economic anthropology. *Theory Econ. Anthropol.* **2002**, 251–295.
89. Mercier, H.; Sperber, D. *The Enigma of Reason*; Harvard University Press: New York, NY, USA, 2017.
90. Ramstead, M.J.; Veissière, S.P.; Kirmayer, L.J. Cultural affordances: Scaffolding local worlds through shared intentionality and regimes of attention. *Front. Psychol.* **2016**, *7*, 1090. [CrossRef] [PubMed]
91. Veissière, S. Cultural markov blankets? Mind the other minds gap! Comment on "answering Schrödinger's question: A free-energy formulation" by maxwell james désormeau ramstead et al. *Phys. Life Rev.* **2018**, *24*, 47–49. [CrossRef] [PubMed]
92. Norenzayan, A.; Shariff, A.F.; Gervais, W.M.; Willard, A.K.; McNamara, R.A.; Slingerland, E.; Henrich, J. The cultural evolution of prosocial religions. *Behav. Brain Sci.* **2016**, *39*. [CrossRef] [PubMed]
93. Kenter, J.O.; Bryce, R.; Christie, M.; Cooper, N.; Hockley, N.; Irvine, K.N.; Fazey, L.; O'Brien, L.; Orchard-Webb, J.; Ravenscroft, N.; et al. Shared values and deliberative valuation: Future directions. *Ecosyst. Serv.* **2016**, *21*, 358–371. [CrossRef]
94. Mostert, E.; Craps, M.; Pahl-Wostl, C. Social learning: The key to integrated water resources management? *Water Int.* **2008**, *33*, 293–304. [CrossRef]
95. Hitlin, S. Values as the core of personal identity: Drawing links between two theories of self. *Soc. Psychol. Q.* **2003**, *66*, 118–137. [CrossRef]
96. Bardi, A.; Lee, J.A.; Hofmann-Towfigh, N.; Soutar, G. The structure of intraindividual value change. *J. Pers. Soc. Psychol.* **2009**, *97*, 913–929. [CrossRef] [PubMed]
97. Kashima, Y.; Kobuko, T.; Kashima, E.S.; Boxall, D.; Yamaguchi, S.; Macrae, K. Culture and self: Are there within-culture differences in self between metropolitan areas and regional cities? *Pers. Soc. Psychol. Bull.* **2004**, *30*, 816–823. [CrossRef] [PubMed]

98. Inglehart, R. *The Silent Revolution: Changing Values and Political Styles Among Western Publics*; Princeton University Press: Princeton, NJ, USA, 1977.
99. Festinger, L.V. *A Theory of Cognitive Dissonance*; Stanford University Press: Stanford, CA, USA, 1962.
100. Baron-Cohen, S.; Leslie, A.M.; Frith, U. Does the autistic child have a "theory of mind"? *Cognition* **1985**, *21*, 37–46. [CrossRef]
101. Jones, A.P.; Happé, F.G.E.; Gilbert, F.; Burnett, S.; Viding, E. Feeling, caring, knowing: Different types of empathy deficit in boys with psychopathic tendencies and autism spectrum disorder. *J. Child Psychol. Psychiatry* **2010**, *51*, 1188–1197. [CrossRef] [PubMed]
102. Dadds, M.R.; Hawes, D.J.; Frost, A.D.; Vassallo, S.; Bunn, P.; Hunter, K.; Merz, S. Learning to 'talk the talk': The relationship of psychopathic traits to deficits in empathy across childhood. *J. Child Psychol. Psychiatry* **2009**, *50*, 599–606. [CrossRef] [PubMed]
103. Sherif, M. *Common Predicament: Social Psychology of Intergroup Conflict and Cooperation*; Houghton Mifflin: New York, NY, USA, 1966.
104. Tajfel, H. Social psychology of intergroup relations. *Annu. Rev. Psychol.* **1982**, *33*, 1–39. [CrossRef]
105. Brewer, M.B.; Gardner, W. Who is this "we" ? Levels of collective identity and self representations. *J. Pers. Soc. Psychol.* **1996**, *71*, 83–93. [CrossRef]
106. Jean, S.; Medema, W.; Adamowski, J.; Chew, C.; Delaney, P.; Wals, A. Serious games as a catalyst for boundary crossing, collaboration and knowledge co-creation in a watershed governance context. *J. Environ. Manag.* **2018**, *223*, 1010–1022. [CrossRef] [PubMed]
107. Vegt, N.; Visch, V.; Vermeeren, A.; de Ridder, H. Player experiences and behaviours in a multiplayer game: Designing game rules to change interdependent behaviour. *Int. J. Serious Games* **2016**, *3*, 69–85. [CrossRef]
108. Kampf, R.; Stolero, N. Computerized simulation of the israeli-palestinian conflict, knowledge gap, and news media use. *Inf. Commun. Soc.* **2014**, *18*, 644–658. [CrossRef]
109. Tipton, E.J.; Leigh, E.; Kritz, W.C.; Crookall, D. Debriefing: The real learning begins when the game stops. In *Simulation and Gaming in the Network Society*; Kaneda, T., Toyoda, Y., Rizzi, P., Eds.; Springer: Singapore, 2016; p. 473.
110. Linehan, C.; Kirman, B.; Roche, B. Gamification as behavioral psychology. In *The Gameful World: Approaches, Issues, Applications*; Walz, S.P., Deterding, S., Eds.; MIT Press: Cambridge, MA, USA, 2015; pp. 81–105.
111. Arnab, S.; Dunwell, I.; Debattista, K. *Serious Games for Healthcare: Applications and Implications*; IGI Global: Hershey, PA, USA, 2012.

water

MDPI

Concept Paper

Conceptualizing Serious Games as a Learning-Based Intervention in the Context of Natural Resources and Environmental Governance

Romina Rodela [1,2,*], Arend Ligtenberg [2] and Roel Bosma [2]

1. Technology and Environmental Studies, School of Natural Sciences, Södertörn University, SE-14189 Huddinge, Sweden
2. Wageningen University & Research, 6700 AH Wageningen, The Netherlands; arend.ligtenberg@wur.nl (A.L.); roel.bosma@xs4all.nl (R.B.)

* Correspondence: rominarodela@hotmail.com; Tel.: +39-334-9963-658

Received: 27 November 2018; Accepted: 14 January 2019; Published: 31 January 2019

Abstract: The use of serious games in the governance of natural resources and the environment is progressively increasing and includes games used for research and data collection, teaching and training, and fostering a change of practices. However, this diversity remains underexplored and underreported. In view of a growing interest in the use of serious games in natural resource and environmental governance, the absence of discussions about how differences in intended use and delivery influence the performance, assessment, and outcomes of games is problematic. Here we present an inventory, and a description, of such different uses then, by focusing on serious games used as interventions, we discuss when, and how, games could be used to generate learning and social learning. To that end we use a narrative review of selected literature, and insight from research on social learning, to develop an inventory of game use, and within that inventory we conceptualize the use of serious games as a social learning intervention. Also, by means of an illustrative case of a serious game (developed as part of the Assessing the Learning Effects of Games on Attitude of Stakeholders toward Sustainable Shrimp Farming – ALEGAMS research project) we reflect on a few key aspects of game use. We suggest that developing a serious game needs several iterations and, although the learning outcomes can be assessed, the impact of games aiming at changes in current practice and policy will likely fall beyond the timespan of usual project periods. This is something future research should consider as it has implications for the research design and methodology.

Keywords: serious games; role-playing games; learning-based intervention; transformative change; social learning; aquaculture; Mekong Delta; mangrove

1. Introduction

Using serious games in natural resource and environmental governance, inclusive of water governance, has been slowly increasing since mid-1990s [1–3]. For instance, Flood et al. [1] in their review note a trend of serious games being now developed and meant for use with stakeholders and policy makers. This trend seems to emerge as a response to the limitations ascribed to some participatory activities used in the research field of natural resources and environmental governance. For instance, while expert workshops might work well where participants have comparable levels of education and free speech is the norm, these do not suit contexts where participants might struggle with complex content and do not feel entitled to voice concerns due to cultural norms, power asymmetries, or other [4]. Aware of this, practitioners and researchers are now exploring alternative (participatory) tools, such as serious games [3].

Using games for serious purposes was first introduced for the training of military novices who participated in simulated situations of near-real-world conditions which were (often) modeled and replicated based on close-to-real data and parameters. This allows for players to safely experience and explore the likely implications certain decisions might have, and to learn from that. Today, serious games are used regularly for training in management, medicine, and policy [5,6]; they are gaining ground in education [7]; and furthermore are attracting the attention of researchers and practitioners working in natural resource and environmental governance. There, serious games are used for many, and different, purposes inclusive of games used as interventions intending to resolve or further understand environmental and/or resource issues and challenges. In such cases, when games are an intervention, this is commonly done under the assumption of games having the potential to engender learning and foster transformative change and/or more sustainable practices. Transformative change toward more sustainable alternatives is a very relevant and contemporary subject in which many researchers and practitioners of natural resource and environmental governance have an interest.

However, in a review of 25 serious games on environmental management done by Madani et al. [1] they note that literature on games often does not report on how assumptions about learning, or skill development, informed game development. Their review points at two major shortcomings across the serious games on environmental management that they reviewed: (i) evaluations of a game's effectiveness were missing; and (ii) the pedagogical foundations on which the games are based were not clear. Madani et al. [1] advance that many games do not seem well-suited for their intended use and conclude that evaluations of serious games' effectiveness is a critical research frontier. We agree that, in light of a growing interest for serious games in natural resources and environmental governance, more research is needed to meet its potential to perform as a tool for interventions. Except for a few studies [1–3], the diversity of serious games found in this field remains underreported and underexplored. Specifically, there is a need for research about the implications arising from differences of intended use e.g., data collection vs. education and training vs. transformative change. A growing interest the absence of discussions about how differences of intended use might influence the development, delivery, assessment, and outcomes of serious games, is problematic.

In this study, we focus on serious games in natural resource and environmental governance to conceptualize and discus differences of use, with a special interest in games as learning-based interventions. Therefore, we do not aim to classify all serious games available across the many scientific fields, but instead we focus on the research field of natural resource and environmental governance. This research field is a problem-solving oriented area of study and practice which we considered in developing our discussion of game use (Figure 1). Moreover, we focus on practitioners (i.e., adults) and so we are not looking at educational games for pupils and children, and we do not review literature on gamification. By extension, since we set out to review selected literature and map trends in a selected field, not reviewing literature on gamification is a further limitation since it relates mainly to enhancing existing education and decision processes with gaming elements.

In the next section, we first present an inventory of uses serious games have found in natural resource and environmental governance. Thereafter, we focus on games used as learning-based interventions in Section 3, and discuss and reflect in Section 4 on the work we did in the Assessing the Learning Effects of Games on Attitude of Stakeholders toward Sustainable Shrimp Farming (ALEGAMS) research project. Based on our current research and collectively accumulated experience, as well as social learning theory and practice, we present a "Development and evaluation framework for serious games used as learning-based interventions" in natural resource and environmental governance.

2. Differences across Serious Games: A Typology of Use in the Context of Natural Resource and Environmental Governance

Games come in many forms, and can be played in numerous ways, from dice, card and board games, computer games, and field games to role-playing (theater) games. However, setting aside genre differences, games are seen to share four defining traits: a *goal*, *rules*, a *feedback system*, and *participation* [8]. These all are important characteristics; for instance, when the game is compelling, participants/players will go on playing until completion. Much literature on game-play links participants/players' motivation to completion of the task at hand with the game being perceived as engaging, fun, and compelling. The way a game can trigger participants/players' interest and motivation will affect the likelihood for the game to perform its intended use at a high level. In environmental and natural resource management, serious games are used for many reasons, and these differences impact not only on how interest and motivation should be built into game-play, but differences of intended use have implications also on evaluating games and their outcomes [9].

Based on available literature, we now summarize and describe differences in using serious games found in environmental and natural resource management. In this we consider three aspects characterizing this field: (i) propensity to explore with, and use of, (participatory) tools that fit within local contexts; (ii) interest in engaging with and seeking input from different profiles, as locals, practitioners, and community members (all being adults); and (iii) a tradition of problem-solving in relation to resource and environmental issues. Therefore, assuming that games in this field are used to address given needs and issues, we focus on two core aspects (i) *purpose* and (ii) *expected outcomes* (Figure 1) to distinguish between different typologies of game use. We acknowledge that others before have looked at the purpose of games and offered different descriptions, for instance Djaouti et al. [10]. However, to the best of our knowledge, such an exercise has not been carried out yet for the field of natural resource and environmental governance.

Following McGonigal, [8] we suggest that while the four defining traits of a serious game she outlined (*goal*, *rules*, a *feedback system*, and *participation*) may stay constant, it is also important to acknowledge differences of intended game use (e.g., games for data collection *vs.* education; training vs. transformative change) which are likely to have implications for game development, test, use, and evaluation.

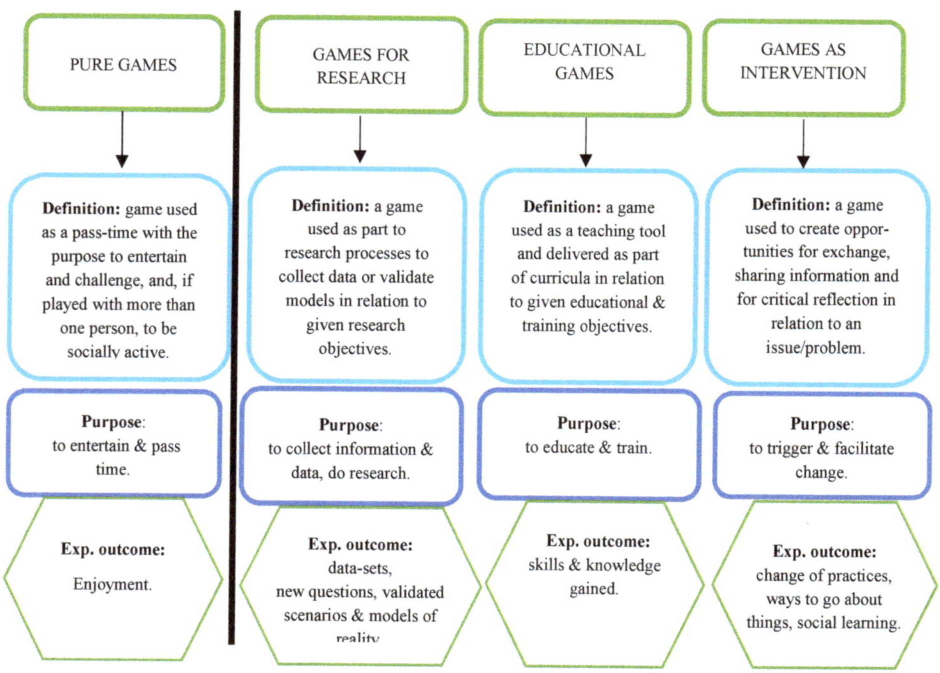

Figure 1. Differences in the use of serious games in natural resource and environmental governance.

2.1. Games Used for Research

The first group of literature we identify reports on games used as a tool to collect data in order to study aspects of interest. This literature often approaches games as an "experimental method" meant to reproduce near-real-world systems that participants/players engage in while researchers observe them and record/collect data produced during the game. Often these studies focus on *social interaction and other social processes* as there are group dynamics, decision-making, or behavior of interest. Several teams, as well as scholars working on commons and common-pool resources, have used games as an "experimental method" to study competition, collaboration and decision-making in resource use and management [11]. Janssen et al. [12] note how those early works were informed by experimental economics (e.g., public good, prisoner's dilemma, and bargaining games). A limitation of these studies was that by primarily focusing on decision-making, these did not include ecological dynamics. This aspect was picked up by later research that extended the use of games (type and number of parameters used), but also how games were administrated (on-campus experimental environment with students vs. field data collection with locals). This created new opportunities, such as the collection of data, that are contextualized (i.e., specific to a locality) and are more robust (i.e., better representing current resource use in the field). As a result, the literature reporting on games used to collect data in the field with local stakeholders (e.g., villagers, inhabitants, farmers) is increasing. For instance, Vieira Pak et al. [13] report on research where role-playing games were used to collect data about decision-making processes about land use in the Colombian Amazonian frontier. They reflect on how role-playing games allowed for discussions during which players shared information that the research team needed and that would not be effectively communicated in an interview. Lebel [14] used a serious game in a similar way—albeit combined with other methods of data collection—in a case of fish farming in Thailand. Meinzen-Dick et al. [15] used serious games in field experiments in India, while Bell et al. [16] used (a tablet-PC-based) serious games in field experiments about an irrigation system in Pakistan.

Furthermore, we identify an additional subgroup of literature that reports on serious games used in combination with computer models [17,18]. Often this use is called companion modeling (ComMOD). In contrast with the literature cited above, the main objective in companion modeling is not to collect data to answer research questions; rather, games are used to build and validate models that researchers have developed, or to provide additional information about scenarios these models reproduce. Étienne [19] notes that the model sometimes simulates the dynamics of a natural resource and provides a dynamic basis for a negotiating game, while in other cases the model is transposed into a game to validate and/or to explain the strategies stakeholders use. In this case a role-playing game facilitated a collective construction of the model and helps to discuss scenarios. For instance, Campo et al. [20] used a serious game with villagers in the Philippines to validate their model and report on how the information generated in this way revealed some model elements to be revised or corrected. Joffre et al. [21] have used a role-playing game to calibrate and validate a model representing different aquaculture scenarios, and they reported that farmers' involvement with the role-playing game helped them to fine-tune the model and identify key parameters and drivers in farmers' decision-making. In both cases, the game is instrumental to the research process and is one among several steps and tools used to achieve a much broader objective [18].

2.2. Educational Games

A second group of literature reports on games used as a teaching tool within formal curricula [3,22,23] targeting students undergoing formal education. Among the first, and perhaps best-known example of serious games used for natural resource management curricula in the 1970s is the Fish Banks game, which was used to educate players about dilemmas arising through the joint management of open access resources [3,24]. Using games as a teaching tool is a well-discussed topic, and as a teaching method serious games are generally preferred by students over more traditional lectures. These games put students into a more active role, allowing learning by doing and ownership of their role in ways that lectures do not [23]. Given the complexity of contemporary issues in resource management and environmental governance, which require the integration of multiple disciplinary knowledge, as well as good understanding of social and political processes, using games integrating scientific information with information about contemporary policy processes seems to hold promise for education.

Games are assumed to support a deeper understanding of contemporary environmental challenges [25]. For instance, Flood et al. [1] in their review of serious games on climate change, find that besides educational outcomes, often these games also lead to improved awareness. Flood et al. [1] comment that, during game-play, students' work with data and descriptions/simulations of real-world policy processes helps them in their learning processes, but also allows them to have a close-to-real-world experience. Chen and Martin [24] discuss the potential of role-playing as an educational tool for transformative learning. They anchor their assumptions in the way players engage with each other, and experience different situations during game-play. They use the Climate Change Policy course (for management students in the Community of European Management Schools, CEMS) network as reported by Paschall and Wüstenhagen [26] to provide an example of how such a (two-day intensive role-play) experience can (potentially) foster transformative learning. In addition to scientific facts, Chen and Martin [24] see value in participants having to choose a role to play (i.e., they need to take a stand/adopt a perspective), and then play and advocate for it. This, they assume, requires higher order thinking, reflection, and examination of both internal and external influences on environmental attitude and behavior.

Also, some literature, albeit much less, reports on serious games for training and education of stakeholders, farmers, and other practitioners. For instance, Pruksakorn et al. [27] report on using a role-playing game named "Soil analysis and appropriate fertilizer use game" (SAAFU) designed for farmers with a low formal education to learn about soil analysis before planting, and, by extension, to improve the appropriate chemical fertilizer usage. Ferrero et al. [28] report on using a serious game for water professionals to complement existing training materials for them to better understand the problem at hand, to explore with strategic decisions financial investments, and to see how stakeholder collaboration impacts water supply safety management. Complementing this report, Erisman et al. [29] report on a game named NitroGenious, developed and used to improve understanding among policy makers about the nitrogen pollution problems in a selected area in The Netherlands.

However, we note that literature reporting on games used for training of practitioners, and other stakeholders, is less frequent compared to literature on games used for the teaching of students undergoing education. This could be linked to differences of context within which these two groups are located, and to differences of opportunities educators have to assess and reply of the needs learners have. Teachers/educators have daily contact and a good overview of student performance in class. That knowledge might allow them to better grasp what needs students have and search /choose among alternative pedagogical methods/tools. While the extensions agents, and related agencies, know well what information and content should be delivered on a subject matter, they might not have an overview of practitioners' learning needs and preferences. Nevertheless, practitioners' are a more heterogeneous group (e.g., age, education, interests) compared to students. These are, however, tentative assumptions given that, to the best of our knowledge, no overview of used serious games for training practitioners / stakeholders is available.

In contrast, literature reporting on serious games used as part of formal education on natural resource and environmental governance is thoroughly reviewed [2,3,30]. These reviews, however, find that studies often fail to establish links between the learning theory used to inform game development, expectations in terms of outcomes, and frameworks used to establish the game effectively and deliver learning outcomes [2]. Much of the literature on educational games seems to agree that the core objective of these games is to educate, train and help students acquire knowledge. However, these aspects often go unreported. For instance, Chen and Martin [24] discuss the potential of role-playing games, and in so doing use as an example the game developed by Paschall and Wüstenhagen [26] which conceptualizing it as a tool for transformative learning. The discussion by Chen and Martin [24] is relevant and points to several aspects of interest pertaining to game-play. However, that they remain silent on the hypothesis and objectives, in terms of transformative change, as applied by those who developed the game. Albeit, of interest is insight and suggestions they offer on how different aspects of game-play (e.g., discussion, confrontation with opponents, etc.) provide space for participants' transformative learning and the resulting outcomes. Mezirow [31] assumed that transformative learning results in lasting change processes, yet Chen and Martin [24] and Paschall and Wüstenhagen [26] report on this not being evident for the stated educational game on climate change.

2.3. Games as Interventions

A third group of literature reports on games used as an intervention under the assumption that these contribute to change processes that go beyond the individual (often referred to by current literature as social learning processes). Such use of games might have emerged as a response to the limitations ascribed to participatory interventions e.g., workshops and/or focus groups, that some research critiques do not fit well in all local contexts [1,3]. Serious games are described by some as participatory tools able to engage people (e.g., problem owners) in debate and activity that can accommodate their needs and create certain opportunities in different ways. For instance, Flood et al. [1], while focusing on climate change, propose serious games as tools that can trigger learning about the subject at hand, and assume this could give way to change of practices and behaviors. Interestingly, several other papers assume games to have potential in terms of social learning and to bring about transformative change [15,32–36]. This literature advances specific assumptions, as summarized below, about what a serious game can do, and what its outcomes can be, which contrasts within the literature outlined above describing educational games, or games for research (Figure 1).

First, educational games are broader in scope and often are about a general issue, at a larger geographical and/or administrative scale, e.g., climate change [1], and over-fishing [25]. However, games meant to perform as interventions often target a rather narrowly identified issue in natural resource and environmental governance, and this much often at a local scale [15,28,32,37–40]. Serious games meant to perform as interventions often integrate contextual features and have finer levels of detail, to ensure that local stakeholders, or problem owners, get motivated and engaged in game-play and become part of the change processes the research team envisaged.

The second difference is notable in the assumptions and expectations about what a game-play can trigger. Educational games, while targeting individual "learners" seek to inform, deliver information, and share knowledge on a subject, whereas research games seek to gather or validate data. In contrast, games used as interventions place the attention on elements/aspects regarded (also from a theoretical view point) to have an important role in fostering the transformative change of interest that goes beyond the individual player. Players are then often understood and described in their role of "stakeholders" rather than "learners" (e.g., educational games), or "testers" or "data points" (e.g., research games). Souchère et al. [41] reported a role-playing game created with groups of local stakeholders, meant to facilitate negotiations on the future management of erosive runoff in the Pays de Caux (France). In this, the objective was to create opportunities for co-operation between stakeholders, and this was done based on theoretically grounded assumptions about participatory and collaborative activities. Souchère et al. [41] defined the game as a tool for mediation and collective thinking that allowed players to see the collective consequences of their individual decisions, and to share knowledge, ideas and their vision of the problem, in order to test different strategies. Rebolledo-Mendez et al. [42] reported on research in which they sought to explore the effectiveness of serious games in improving the awareness on flooding among the general public in the UK; they considered the game as a tool helping participants acquire information and a better understanding of the system.

Third, when serious games are used as interventions, substantial effort is also put into integrating updated scientific data, contextual features, and local circumstances which together can best simulate the current state of affairs. This is often done with the ambition to recreate near-real-world circumstances for the game-play to be a meaningful activity to the problem owners. The recreation of a near-real, meaningful experience is assumed to help players to maintain high motivation, complete the activity, and be influenced by the experience to which they were exposed.

However, for games used as interventions with the intention of pursuing and/or fostering change processes, a question that emerges is to what extent such assumptions are validated in practice. The latter question, however, is poorly understood and underreported. With very few exceptions [42], only few studies provide robust evidence of game-play-induced change. Future research should try to bridge that gap and include methods that assess the effectiveness of serious games when used as interventions. Assessments need to go beyond verifying whether the game is an accurate representation of the reality it represents, and beyond reports of a player's subjective experience. Rather, assessments need to explore and test the assumptions made about whether, how, and when the serious game has delivered on the expected outcomes, i.e., change processes and actions. For instance, the assessment of the FloodSim serious game offers useful input here. Rebolledo-Mendez et al. [42] note that FloodSim was designed with a stated aim to raise awareness about flooding policy, however, their assessment shows that most players who were interviewed gained only a superficial level of awareness, and not of the kind that would allow them to engage with and understand the policy debate on flood and government expenditure. They conclude for FloodSim that game design and development should have been more interdisciplinary and should have better integrated well-established frameworks and knowledge about the human behavior that is assumed to change. Establishing a conceptual and methodological link between the *expected outcomes* with the different traits of game-play [8] early in the process is a recommendation from which future projects, interested to use serious games as intervention, could benefit.

To conclude this section we want to clarify that we do not suggest the above being a rigid categorization, and/or mutually exclusive, of game use in the governance of natural resources and the environment. Rather we suggest the typology (Figure 1), and corresponding descriptions, as a support to those needing to plan and deliver games in a local context. Attention to *purpose* and *expected outcomes* may guide researchers and practitioners over choices of game development, test, use, and evaluation—inclusive of theoretical and analytical frameworks. In the next chapter, we report and reflect on the work done by our team on developing a serious game meant to perform as a learning-based intervention. In building on this experience, in Section 4 we then consider social learning research and practice and suggest a "Development and evaluation framework for serious games used as learning-based interventions".

3. Developing a Serious Game within the Assessing the Learning Effects of Games on Attitude of Stakeholders toward Sustainable Shrimp Farming (ALEGAMS) Research Project

The development and use of serious games meant as interventions is often done as part of larger projects needing to achieve broader objectives. Some projects might focus on the development and prototyping of a serious game only—then feature it as an output for others to use—while other projects might develop a game as an intervention/tool and use it to achieve pre-defined *outcomes e.g., change processes.* The ALEGAMS project (Box 1) aimed for both: the design and development of a serious game meant as a learning-based intervention seeking to generate social learning across local communities in three pilot sites. Below, we summarize our experience and work, including the challenges and constraints encountered during the project.

Box 1. The "Assessing the Learning Effects of Games on Attitude of Stakeholders toward Sustainable Shrimp Farming" research project (acronym: ALEGAMS).

Background: Over recent years mangrove forest, along the Vietnamese Mekong Delta, is under pressure from the expansion of aquaculture, shrimp farming more specifically [30]. This is a problem recognized by policy and practice, among whom most agree that loss of mangrove is an issue since these forests perform important functions in regards to climate change adaptation and local economy. Dedicated efforts to restore and maintain mangrove have been implemented in Vietnam, inclusive of a policy program supporting integrated mangrove shrimp farming (IMS). IMS can be economically viable and allow for sufficient ecological quality [43], particularly compared to extensive shrimp systems [30]. However, it has been observed that farmers who practice extensive farming do not adopt IMS voluntarily.
Objectives: ALEGAMS aimed to design, test, and use a serious game (i.e., board role-play game) as an intervention. The game reproduces, through game-play, the dynamics of the main different shrimp farming systems present in the region. This has the aim to help farmers/players to understand the implications of the different shrimp farming systems, and of their mix, applied in the Vietnamese Mekong Delta.
Output: A serious game (board game) to be handed over to the International Union for the Conservation of Nature - IUCN Asia (Figure S2).
Outcome: Exchange and learning about risk and IMS by shrimp farmers across selected cases in the Mekong Delta.
Impact: Social learning manifested in increased knowledge and change of attitude on shrimp farming systems. Increased awareness on IMS which will support farmers in decisions to shift toward IMS. Eventually, an increase of IMS farms in two provinces is envisaged.

3.1. Summary of the Work Done and Steps Taken for Developing the ALEGAMS Learning-Based Game Intervention

In the following section, we summarize and reflect on the work done during the development of the ALEGAMS serious game in order to extract input useful for future research. Methodologically speaking, ours is a reflective exercise and in hindsight we see our work articulated in four phases, which we name and use to guide the summary presented below.

3.1.1. Conceptualization in Context

ALEGAMS focused on documented issues and challenges in the fast-growing intensive shrimp aquaculture in the Mekong Delta. This game intervention targets shrimp farmers and, to a degree, also other decision-makers active in this sector. This idea emerged during an earlier project, when an earlier version of the serious game was used for research to collect data for the calibration of an agent-based simulation model [21].

In ALEGAMS, we built on that experience and designed a game that would help shrimp farmers to learn about the risk and financial performance of three shrimp farming systems, a game that would allow them to explore and experience the ups and downs of each in the face of real-life events, with a focus on shrimp disease outbreaks. We assumed that learning about the risk and financial performance of shrimp farming systems could help farmers to recognize the risk of monocultural, intensive shrimp farming and the value of integrated mangrove shrimp farming, and subsequently increase their willingness to maintain, or shift to more sustainable practice. We conceptualized a game where players, by starting at either an integrated mangrove or extensive farm on plots with different area and value (considering infrastructure) could experience (mixes of) different farming systems and evaluate which turned out to be the most profitable. This, we hypothesized, would allow players/farmers to weigh the (high) risks and rewards of intensive farming versus the (medium) risks and rewards of integrated farming which is bringing about other advantages (timber, ecosystem services, etc).

A board game was considered a suitable tool to bring farmers together for learning within a mimicked local context. Computer games would not suit this context.

3.1.2. Iterative Game Design and Testing

After having formulated assumptions about what the ALEGAMS game should achieve, and locating these in a socio-economic but also biophysical context, we expanded on the work reported by Joffre et al. [21] while focusing on the impact of shrimp disease outbreaks. A first test of the game-play was done with students and researchers to check on clarity of the instructions, ease of application, and financial aspects regarding cost and benefits of farming. Feedback on game-play was collected during several iterations with different players (both Vietnamese and European) and then integrated into a prototype. A group of six shrimp farmers were invited to join us at a project progress meeting held in September 2016 in Can Tho, where they played the game and then reported their feedback orally to the ALEGAMS project team. This included also their views about non-technical matters. For instance, farmers reported not finding the game enjoyable and lacking motivation to complete all the rounds planned to deliver the information about differences between the farming systems. Farmers commented on what they identified as a mismatch with reality: the game did not allow farmers to reduce the risk of shrimp disease outbreaks by investing in technology and/or by collaborating. The increased cost of the former would be outweighed by the extra revenues, and the latter was assumed to reduce cost and increase margins.

This input was considered with care, and our team agreed to develop "risk cards" intended to be introduced during game-play in the form of a card with written text about real-life circumstances relating to risks. One set of cards reflected communal risks, such as flooding, storms, and new or more virulent shrimp diseases, and one card was drawn in each round and applied to all players. The second set of cards was drawn individually at each turn and primarily concerned household matters. Both sets of cards were carefully balanced in numbers having positive and negative impact on the player's game result. Moreover, the option to buy technology and become a member of a cluster ('cooperative') was added, both are decisions players can make to reduce risk.

The background data was updated and contextualized to make the identified parameters reflect reality (i.e., cost and income, and risk, for each farming system). We assumed these changes would make the game-play feel closer to real circumstances since farmers pointed to these as important elements to maintain their motivation, while also making the game more dynamic and interesting.

We tested the new version of the game multiple times: first, with students and researchers familiar with the shrimp sector in Vietnam; then, in The Netherlands with a group of researchers to check the game flow, and the new instructions; then, also with farmers in Vietnam. We integrated all input emerging from these sessions. Then we repeated the tests and ran a pilot with shrimp farmers, i.e., profiles matching our target group, who also offered valuable feedback.

3.1.3. Game Use in the Field

When a project aims to develop and use a game in the field, certain questions need to be addressed. For instance, what is the minimum number of rounds and how to acquaint the players with the rules and to engage them in learning, respectively. The timespan for concentration and availability to complete the gameplay is to be considered. In our case, the numbers of rounds was related to the number of playing rounds/cycles needed before an effect could be observed on the results (bank account) and of the new technologies, to all hazards cards had been turned, and players felt to have had a fair chance to obtain one good dice score. The total time available to farmers was limited to a morning or afternoon since we wanted them to play at least two sessions. For the ALEGAMS game, we saw the need for a first session to make farmers acquainted with all the rules and a second to have them play independently of the moderator.

3.1.4. Evaluation of the Game Intervention

The final prototype of the game was produced by a professional company in Vietnam (Figures S1 and S2). We agreed on a longitudinal design early in the process and developed an evaluation strategy meant to collect data about game-play and learning, along with the post-game outcomes at different points of time. ALEGAMS intended to trigger change in a semi-experimental design to collect data about learning and attitude, and repeated exposure (one or two times) to the play. The game was used with local shrimp farmers and practitioners on two occasions at several months' interval in three coastal locations (Figure S1). This sample excluded those who had played the prototype version.

Data on the learning outcomes are being analyzed and will be published later. The intended impact, i.e., an increase in applying integrated mangrove shrimp systems, was not aimed to be assessed within the three-year project period.

4. Development and Evaluation Framework for Games as Learning-Based Interventions

In a narrative review of literature in natural resource and environmental governance, we identified eight studies reporting on games used as learning-based interventions seeking to promote social learning and so contributing at changing in how things are done [32–36,44,45]. These studies differ in the way social learning is defined and operationalized, but share assumptions about game-play offering opportunities for (i) debate and interaction with other players/participants, (ii) critical reflection, either alone or collectively, (iii) development of a shared perception on the problem domain, (iv) development of a shared stock of knowledge, and (v) seeing the issue from a different perspective. All are process and outcome features discussed in the literature on social learning [46,47] that partially agree with the operationalization of social learning within three dimensions: relational, cognitive, and moral/normative [47–50]. Based on this compilation of papers, on broader literature on social learning, and most important, informed by work done on the ALEGAMS project, we now propose a "*Development and evaluation framework for serious games meant as learning-based interventions*".

4.1. Conceptualization in Context

When serious games are utilized as interventions, a first step is the identification and description of the problem, or issue, games are expected to contribute at resolving and these issues likely are about social matters. Games are about the engagement of people, and it is important that the theoretical basis operationalizing social processes under study (i.e., learning and transformative change) is chosen, described, and made operational early on. Literature on social learning assumes that social learning can be operationalized within cognitive, behavioral, normative, and relational aspects [46,47,50]. Drawing links between these and ways in which game-play exerts an influence on them are relevant. Two recent reviews of the learning literature may help to reflect upon, identify and select process and outcome features assumed to be most relevant to a project [46,47].

Also, the assumed underlying mechanism and linkages between these (social) processes and the problem issue at hand should be explicated, evidence for these collected and considered, and the expected short and longer effects of game-play listed. This may build on well-established theories about learning and literature in psychology (e.g., when attitudes are under focus) and on social learning literature. One can reach out also to more informal evidence/local expertise regarding pressures, or local beliefs, assumed to promote or inhibit (behavioral) change. Current empirical literature on this subject has acknowledged benefiting from informal input as well [35,36].

This work, inclusive of the assessment and choice of a best-suited theory, will help preliminarily to identify a suitable game typology (computer-based vs. board game; group game-play vs. solo-play: role-playing etc.) and study designs (case study; quasi-experiment; controlled study; ethnographic research; etc.) [51–53]. The choice of game typology should fit well to local circumstances inclusive of its needs, expectations, and practices and socio-economic characteristics, all of which must be identified at an early stage. This will affect choices on whether to use a computer-based vs. board game; group game-play vs. solo-play, or others. It is useful to note that the choice of a game type will have implications for the (social) processes the game-play will bring about. De Vente et al. [4] addressed the question of how the context and design of participatory processes affect their outcomes more broadly and that discussion can offer some useful input on that aspects. Furthermore, literature focused on social learning offers input and informs us that activities fostering *collaboration* between the participants overall seem to perform differently on learning compared to activities fostering *competition* [54]. Studies have, for instance, reported on activities undertaken as part of group work where deliberation, and talking over different viewpoints, resulted in developing a shared understanding [48,49].

4.2. Iterative Game Design and Testing

After having addressed theoretical questions and developed a sharper perspective on the intervention needed, the components the game intervention is expected to have are listed (i.e., opportunity for interaction, for reflection), and how these are inter-related and will lead to desired outcomes outlined. Input on these aspects can be sought in studies reporting on and discussing implications about how choices of design (of participatory processes) impact on learning and interaction. Studies have also discussed how that experience may affect (enhance or hinder) social learning after the activity also [52,53,55]. At this stage, a paper-and-pencil plan can describe the idea and assumptions behind the intervention, or this can be done with a preliminary computer simulations, or modeling. Where needed, and possible, testing of assumptions regarded as not well-supported by available scientific evidence can be done in focus groups, surveys, or shorter observational studies. This work will result in a game-play conceptual idea.

Serious games used as interventions should represent the reality these seek to mimic. When this is not the case, research shows that incentives and motivation drop. Players (particularly when well-informed on the topic) struggle to find meaning, become disengaged, or drop out before completing the rounds. The team needs to populate the conceptual idea of the game intervention with current data and information in order to safeguard meaning and purpose. Games are expected to be an engaging experience also; thus, adequate incentives, matching well with the target group profile, should be considered and integrated into the game (e.g., suitable degree of difficulty, challenge, entertainment features). This step may result in a rudimentary prototype.

Testing the game intervention allows for checking how different components are received and how these are appreciated. The way the prototype is perceived and first assessed by players can be established with short post-test questionnaires, interviews, and/or observation of the game-play. The data collected will help to identify areas to be adjusted/strengthened, or features to be added. For instance, identify issues with the representation/recognition of the natural environment, of the problem(s) at hand and the goal(s) to accomplish, and identification of processes/events present in reality but missing in the game. Several re-tests are needed, and suitable testers should recruited [56]. Where needed, control groups and outcome measures could be used. As the prototype is being refined and tested, there are opportunities to verify alternative ideas. After several iterations, when the majority of the testers no longer signal significant shortcomings, and the research team is satisfied with the learning outcomes observed, the game can regarded as fit for purpose.

4.3. Game Use in the Field

Projects aiming to develop a game intervention to be used in the field and to foster change processes are advised to consider constrains and opportunities for learning and social learning present in a local context. In this there is a need to acknowledge differences about what is to be looked at more precisely, and evaluated (the game vs. long terms change processes). For instance, evaluation of the *game* as an output, developed as part to a project, can be done within the project lifespan and can involve self-reported assessments the players fill out after completing game-play. However, assessments aiming to assess the *impact* in terms of change at a more meta level might to be based on objective measures. Also, since change processes are best observed over longer time periods, the questions of when is a good time to collect evidence, and for how long, become relevant.

4.4. Evaluation of the Game Intervention

A game intervention is developed with the intent to be used and put to work. When implementation is scheduled as part of the core project, that might also include evaluation. Choice of a suitable evaluation method is a challenge, given that change of practice and transformative processes require longer observational periods to be detected and recorded [53]. A suitable evaluation method, as some critical papers noted, needs to be chosen at an early stage, and revised when needed [54]. Evaluating learning and other processes at the level of individual players is recommended, and can be done with different post-test data collection tools and approaches. The broader applicability of the game intervention outside the context where it was developed may also be considered early in the processes. Appropriate arrangements are needed for handover and implementation when this is done by a different organization than that developing it.

To conclude, we want to note this framework should not be taken as a rigid progression, and not all research teams need to move hierarchically, and some projects might proceed fast, skip steps, or reiterate. But we expect this overview can offer some support to project teams in embedding game development and its delivery within social learning theory and practice, and within learning-led change process.

5. Conclusions

The inventory of literature conducted suggests that serious games are mainly used and studied for educational purposes, while their use to support research and interventions has emerged more recently. We noted that several studies report on games used for more than one purpose, e.g., to collect data or as an educational tool. However, the assessment of the outcomes games were expected to deliver is not always undertaken. In this paper, we aimed to conceptualize serous games as a learning-based intervention and then have used the illustrative case to consider a few aspects, including the assessment of expected outcomes.

In our case, developing a serious game with the objective to serve as a learning-based intervention within the ALEGAMS research project needed the articulation of several steps. The game was developed in several iterations; each version of the game, updated with a main new element, was tested with the target group (farmers). We observed that having adequate research design that allows flexibility, and complementary revision, is paramount. Based on literature, we developed a framework that included assumptions about learning outcomes the game would deliver. In ALEGAMS the degree to which the learning outcomes were reached was assessed within the project timeline. The latter helped for work to be framed within a period useful to make timely recommendations and thus contribute to innovation of serious games as a participatory tool. However, we recognize this might not always be possible for all projects. Projects aiming at a change of practice, or policy, should consider that this change can only be assessed over a longer while, which most likely goes beyond short projects.

Future research could try to add to this debate by discussing the challenges and issues faced with serious games in natural resource management. Active debate, based on empirical studies and experience, about the extent to which serious games can foster learning and social learning and then bring about change, should form the basis of this field of research.

Supplementary Materials: The following are available online at http://www.mdpi.com/2073-4441/11/2/245/s1. Figure S1: The players, Figure S2: The game.

Author Contributions: The idea for this conceptual paper originates in the work done as part to the ALEGAMS project for the "Research Protocol about the Assessment of the Potential for Role-play Games to Foster Learning" prepared by the first author with other authors. The conceptualization of the paper and review of selected literature has been done by R.R. Preparation of the illustrative case has been done by R.R., R.B. and A.L.

Funding: The ALEGAMS project was funded by IUCN and the WOTRO founding scheme of the Dutch Research Agency (NWO).

Acknowledgments: The authors wish to acknowledge the contributions during the game development of the project director Arnold Bregt, and the Vietnamese team members Tran Thi Phung Ha, Nguyen Huynh Thi Phuong and Tong Quoc Hiep.

Conflicts of Interest: The authors declare no conflict of interest. The funders had no role in the design of the study; in the collection, analyses, or interpretation of data; in the writing of the manuscript; or in the decision to publish the results.

Abbreviations

ALEGAMS	Assessing the Learning Effects of Games on Attitude of Stakeholders toward Sustainable Shrimp Farming
IMS	Integrated mangrove shrimp farming
IUCN	International Union for the Conservation of Nature

References

1. Flood, S.; Cradock-Henry, N.A.; Blackett, P.; Edwards, P. Adaptive and interactive climate futures: Systematic review of 'serious games' for engagement and decision-making. *Environ. Res. Lett.* **2018**, *13*, 063005. [CrossRef]
2. Madani, K.; Pierce, T.W.; Mirchi, A. Serious games on environmental management. *Sustain. Cities Soc.* **2017**, *29*, 1–11. [CrossRef]
3. Wesselow, M.; Stoll-Kleemann, S. Role-playing games in natural resource management and research: Lessons learned from theory and practice. *Geogr. J.* **2018**, *184*, 298–309. [CrossRef]
4. de Vente, J.; Reed, M.S.; Stringer, L.C.; Valente, S.; Newig, J. How does the context and design of participatory decision making processes affect their outcomes? Evidence from sustainable land management in global drylands. *Ecol. Soc.* **2016**, *21*, 24. [CrossRef]
5. Mayer, I.; Bekebrede, G.; Harteveld, C.; Warmelink, H.; Zhou, Q.; van Ruijven, T.; Lo, J.; Kortmann, R.; Wenzler, I. The research and evaluation of serious games: Toward a comprehensive methodology. *Br. J. Educ. Technol.* **2014**, *45*, 502–527. [CrossRef]
6. Neys, J.; Jansz, J. Political Internet games: Engaging an audience. *Eur. J. Commun.* **2010**, *25*, 227–241. [CrossRef]
7. Cai, Y.; Goei, S.L.; Trooster, W. *Simulation and Serious Games for Education*; Springer: Berlin, Germany, 2016.
8. McGonigal, J. *Reality Is Broken, Why Games Make Us Better and How They Can Change the World*; The Penguin Press: New York, NY, USA, 2011.
9. Wilson, K.A.; Bedwell, W.; Lazzara, E.; Salas, E.; Burke, S.; Estock, J.; Orvis, K.; Conkey, C. Relationships Between Game Attributes and Learning Outcomes: Review and Research Proposals. *Simul. Gaming* **2009**, *40*, 217–266. [CrossRef]
10. Djaouti, D.; Alvarez, J.; Jessel, J.P. *Classifying Serious Games: The G/P/S Model*; IGI Global: Hershey, PA, USA, 2011.
11. Baggio, J.A.; Rollins, N.D.; Pérez, I.; Janssen, M.A. Irrigation experiments in the lab: Trust, environmental variability, and collective action. *Ecol. Soc.* **2015**, *20*, 12. [CrossRef]
12. Janssen, M.A.; Lee, A.; Waring, T.M. Experimental platforms for behavioral experiments on social-ecological systems. *Ecol. Soc.* **2014**, *19*, 20. [CrossRef]
13. Vieira Pak, M.; Castillo Brieva, D. Designing and implementing a Role-Playing Game: A tool to explain factors, decision making and landscape transformation. *Environ. Model. Softw.* **2010**, *25*, 1322–1333. [CrossRef]
14. Lebel, P.; Sriyasak, P.; Kallayanamitra, C.; Duangsuwan, C.; Lebel, L. Learning about climate-related risks: Decisions of Northern Thailand fish farmers in a role-playing simulation game. *Reg. Environ. Chang.* **2016**, *16*, 1481–1494. [CrossRef]
15. Meinzen-Dick, R.; Janssen, M.A.; Kandikuppa, S.; Chaturvedi, R.; Rao, K.; Theis, S. Playing games to save water: Collective action games for groundwater management in Andhra Pradesh, India. *World Dev.* **2018**, *107*, 40–53. [CrossRef]
16. Bell, A.R.; Ward, P.S.; Shah, M.A.A. Increased water charges improve efficiency and equity in an irrigation system. *Ecol. Soc.* **2016**, *21*, 23. [CrossRef]

17. D'aquino, P.; Le Page, C.; Bousquet, F.; Bah, A. Using self-designed role-playing games and a multi-agent system to empower a local decision-making process for land use management: The SelfCormas experiment in Senegal. *J. Artif. Soc. Soc. Simul.* **2013**, *6*, 5.
18. Ligtenberg, A.; van Lammeren, R.J.A.; Bregt, A.K.; Beulens, A.J.M. Validation of an agent-based model for spatial planning: A role-playing approach. *Comput. Environ. Urban Syst.* **2010**, *34*, 424–434. [CrossRef]
19. Étienne, M. Sylvopast: A multiple target role-playing game to assess negotiation processes in sylvopastoral management planning. *J. Artif. Soc. Soc. Simul.* **2003**, *6*, 5.
20. Campo, P.; Bousquet, F.; Villanueva, T. Modelling with stakeholders within a development project. *Environ. Model. Softw.* **2010**, *25*, 1302–1321. [CrossRef]
21. Joffre, O.M.; Bosma, R.H.; Ligtenberg, A.; Tri, V.P.D.; Ha, T.T.P.; Bregt, A.K. Combining participatory approaches and an agent-based model for better planning shrimp aquaculture. *Agric. Syst.* **2015**, *141*, 149–159. [CrossRef]
22. Gordon, S.; Thomas, I. 'The learning sticks': Reflections on a case study of role-playing for sustainability. *Environ. Educ. Res.* **2018**, *24*, 172–190. [CrossRef]
23. Hoekstra, A.Y. Computer-supported games and role plays in teaching water management. *Hydrol. Earth Syst. Sci.* **2012**, *16*, 2985–2994. [CrossRef]
24. Chen, J.C.; Martin, A.R. Role-Play Simulations as a Transformative Methodology in Environmental Education. *J. Transform. Educ.* **2015**, *13*, 85–102. [CrossRef]
25. Meadows, D.L. Learning to Be Simple: My Odyssey with Games. *Simul. Gaming* **1999**, *30*, 342–351. [CrossRef]
26. Paschall, M.; Wüstenhagen, R. More Than a Game: Learning About Climate Change Through Role-Play. *J. Manag. Educ.* **2011**, *36*, 510–543. [CrossRef]
27. Pruksakorn, S.; Kiratiprayoon, S.; Uttaranakorn, S.; Sukreeyapongse, O.; Dumrongrojwatthana, P. Game for Low-Formal Education Farmers to Learn How to Improve Soil Quality. *Simul. Gaming* **2018**, *49*, 146–167. [CrossRef]
28. Ferrero, G.; Bichai, F.; Rusca, M. Experiential Learning through Role-Playing: Enhancing Stakeholder Collaboration in Water Safety Plans. *Water* **2018**, *10*, 227. [CrossRef]
29. Erisman, J.W.; Hensen, A.; de Vries, W.; Kros, H.; van de Wal, T.; de Winter, W.; Wien, J.E.; Elswijk, M.V.; Maat, M.; Sanders, K. NitroGenius: A Nitrogen Decision Support System. *J. Hum. Environ.* **2002**, *31*, 190–196. [CrossRef]
30. Tran, T.P.H. Resilience and Livelihoods Dynamics of Shrimp Farmer and Fishers in the Mekong Delta, Vietnam. Ph.D. Thesis, Wageningen University, Wageningen, The Netherlands, 2012; pp. 117–138.
31. Mezirow, J. Transformative Learning as Discourse. *J. Transform. Educ.* **2003**, *1*, 58–63. [CrossRef]
32. Becu, N.; Amalric, M.; Anselme, B.; Beck, E.; Bertin, X.; Delay, E.; Long, N.; Marilleau, N.; Pignon-Mussaud, C.; Rousseaux, F. Participatory simulation to foster social learning on coastal flooding prevention. *Environ. Model. Softw.* **2017**, *98*, 1–11. [CrossRef]
33. Medema, W.; Furber, A.; Adamowski, J.; Zhou, Q.; Mayer, I. Exploring the Potential Impact of Serious Games on Social Learning and Stakeholder Collaborations for Transboundary Watershed Management of the St. Lawrence River Basin. *Water* **2016**, *8*, 175. [CrossRef]
34. Magnuszewski, P.; Królikowska, K.; Koch, A.; Pająk, M.; Allen, C.; Chraibi, V.; Giri, A.; Haak, D.; Hart, N.; Hellman, M.; et al. Exploring the Role of Relational Practices in Water Governance Using a Game-Based Approach. *Water* **2018**, *10*, 346. [CrossRef]
35. Salvini, G.; van Paassen, A.; Ligtenberg, A.; Carrero, G.C.; Bregt, A.K. A role-playing game as a tool to facilitate social learning and collective action towards Climate Smart Agriculture: Lessons learned from Apuí, Brazil. *Environ. Sci. Policy* **2016**, *63*, 113–121. [CrossRef]
36. Villamor, G.B.; Badmos, B.K. Grazing game: A learning tool for adaptive management in response to climate variability in semiarid areas of Ghana. *Ecol. Soc.* **2016**, *21*, 39. [CrossRef]
37. Bourgoin, J.; Castella, J.-C. "PLUP FICTION": Landscape Simulation for Participatory Land Use Planning in Northern Lao PDR. *Mt. Res. Dev.* **2011**, *31*, 78–88. [CrossRef]
38. Craven, J.; Angarita, H.; Corzo Perez, G.A.; Vasquez, D. Development and testing of a river basin management simulation game for integrated management of the Magdalena-Cauca river basin. *Environ. Model. Softw.* **2017**, *90*, 78–88. [CrossRef]

39. Ducrot, R.; van Paassen, A.; Barban, V.; Daré, W.S.; Gramaglia, C. Learning integrative negotiation to manage complex environmental issues: Example of a gaming approach in the peri-urban catchment of São Paulo, Brazil. *Reg. Environ. Chang.* **2015**, *15*, 67–78. [CrossRef]
40. Gurung, T.R.; Bousquet, F.; Trébuil, G. Companion modeling, conflict resolution, and institution building: Sharing irrigation water in the Lingmuteychu Watershed, Bhutan. *Ecol. Soc.* **2006**, *11*, 36. [CrossRef]
41. Souchère, V.; Millair, L.; Echeverria, J.; Bousquet, F.; Le Page, C.; Étienne, M. Co-constructing with stakeholders a role-playing game to initiate collective management of erosive runoff risks at the watershed scale. *Environ. Model. Softw.* **2010**, *25*, 1359–1370. [CrossRef]
42. Rebolledo-Mendez, G.; Avramides, K.; De Freitas, S.; Memarzia, K. Societal impact of a serious game on raising public awareness: The case of FloodSim. *Sandbox* **2009**, *9*, 15–22.
43. Bosma, R.H.; Nguyen, T.H.; Siahainenia, A.J.; Tran, H.T.; Tran, H.N. Shrimp-based livelihoods in mangrove silvo-aquaculture farming systems. *Rev. Aquac.* **2014**, *6*, 43–60. [CrossRef]
44. Carson, A.; Windsor, M.; Hill, H.; Haigh, T.; Wall, N.; Smith, J.; Olsen, R.; Bathke, D.; Demir, I.; Muste, M. Serious gaming for participatory planning of multi-hazard mitigation. *Int. J. River Basin Manag.* **2018**, *16*, 379–391. [CrossRef]
45. Van der Wal, M.M.; de Kraker, J.; Kroeze, C.; Kirschner, P.A.; Valkering, P. Can computer models be used for social learning? A serious game in water management. *Environ. Model. Softw.* **2016**, *75*, 119–132. [CrossRef]
46. Gerlak, A.K.; Heikkila, T.; Smolinski, S.L.; Huitema, D.; Armitage, D. Learning our way out of environmental policy problems: A review of the scholarship. *Policy Sci.* **2017**, *51*, 335–371. [CrossRef]
47. Suškevičs, M.; Hahn, T.; Rodela, R.; Macura, B.; Pahl-Wostl, C. Learning for social-ecological change: A qualitative review of outcomes across empirical literature in natural resource management. *J. Environ. Plan. Manag.* **2018**, *1*, 1085–1112. [CrossRef]
48. Baird, J.; Plummer, R.; Haug, C.; Huitema, D. Learning effects of interactive decision-making processes for climate change adaptation. *Glob. Environ. Chang.* **2014**, *27*, 51–63. [CrossRef]
49. Ensor, J.; Harvey, B. Social learning and climate change adaptation: Evidence for international development practice. *Wiley Interdiscip. Rev. Clim. Chang.* **2015**, *6*, 509–522.
50. Rodela, R. Social Learning, Natural Resource Management, and Participatory Activities: A reflection on construct development and testing. *Wagening. J. Life Sci.* **2014**, *69*, 15–22.
51. Rodela, R.; Cundill, G.; Wals, A.E.J. An analysis of the methodological underpinnings of social learning research in natural resource management. *Ecol. Econ.* **2012**, *77*, 16–26. [CrossRef]
52. Pahl-Wostl, C.; Hare, M. Processes of Social Learning in Integrated Resources Management. *J. Community Appl. Soc. Psychol.* **2004**, *14*, 193–206. [CrossRef]
53. Shackleton, C.M.; Cundill, G.; Knight, A.T. Beyond Just Research: Experiences from Southern Africa in Developing Social Learning Partnerships for Resource Conservation Initiatives. *Biotropica* **2009**, *41*, 563–570. [CrossRef]
54. Cundill, G.; Rodela, R. A search for coherence: Social learning in natural resource management. *Rev. Soc. Learn. Lit.* **2012**, *5*, 31–38.
55. Vinke-de Kruijf, J.; Bressers, H.; Augustijn, D.C.M. How social learning influences further collaboration: Experiences from an international collaborative water project. *Ecol. Soc.* **2014**, *19*, 61. [CrossRef]
56. Rodela, R.; Pérez-Soba, M.; Bregt, A.; Verweij, P. Spatial decision support systems: Exploring differences in pilot-testing with students vs. professionals. *Comput. Environ. Urban Syst.* **2018**, *72*, 204–211. [CrossRef]

water

MDPI

Article

A Serious Game Designed to Explore and Understand the Complexities of Flood Mitigation Options in Urban–Rural Catchments

Mehdi Khoury [1,*], Michael J. Gibson [1], Dragan Savic [1,2], Albert S. Chen [1], Lydia Vamvakeridou-Lyroudia [1], Harry Langford [3] and Sarah Wigley [4]

1 Centre for Water Systems, University of Exeter, North Park Road, Exeter EX4 4QF, UK; m.gibson@exeter.ac.uk (M.J.G.); d.savic@exeter.ac.uk (D.S.); a.s.chen@exeter.ac.uk (A.S.C.); l.s.vamvakeridou-lyroudia@exeter.ac.uk (L.V.-L.)

2 KWR Watercycle Research Institute, 3433 PE Nieuwegein, The Netherlands; dragan.savic@kwrwater.nl

3 Department of Geography, University of Sheffield, Winter Street, Sheffield S10 2TN, UK; h.langford@sheffield.ac.uk

4 Westcountry Rivers Trust Head Office, Rain-Charm House, Kyl Cober Parc, Stoke Climsland, Callington, Cornwall PL17 8PH, UK; sarahw@wrt.org.uk

* Correspondence: m.khoury@exeter.ac.uk; Tel.: +44-753-907-4931

Received: 13 September 2018; Accepted: 11 December 2018; Published: 19 December 2018

Abstract: Flood prevention in mixed urban–rural environments has become a greater concern due to climate change. It is a complex task requiring both efficient management of resources and the involvement of multiple stakeholders from diverse backgrounds. As Serious Games (games used for purposes other than mere entertainment) have emerged as an effective means of engaging stakeholders, this work proposes a new Serious Game applied to flood mitigation in the village of Millbrook in the UK. Results show that the game has both an informative and a transformative effect (statistical significance levels from 0.01 to 0.05), improving participants' understanding of the problem, and helping them to find a new and improved approach to flood risk management in Millbrook, with the potential to improve resilience significantly. Furthermore, the game successfully transformed participants into "citizen scientists" in the purest sense of the term—it led them to use inductive reasoning from data produced by the game to correctly confirm or reject hypotheses and resulted in more than 70% of the participants revising their initial assumptions. Interestingly, the game instigated the formation of new local partnerships and helped to prioritize the discussion of natural flood management measures in Millbrook Parish Council meetings.

Keywords: serious gaming; flood; urban; rural; infrastructure; decision making

1. Introduction

Climate change is currently causing a substantial increase in flood risk across the UK [1,2]. High residual risk of flooding will remain, even with the current investment plans on flood defense [3], and hard engineering solutions will not be affordable to protect all areas. In this context, finding other ways to improve resilience to flooding by preventing or mitigating flood impacts becomes a necessity. The complexity of the problem is significant because flood prevention in an urban–rural environment requires both the management of limited resources and the involvement of multiple stakeholders with different perspectives such as residents, farmers, business owners, utilities companies and policy makers.

On the evening of the 30th of September 2016, a workshop involving comprehensive communication and engagement of stakeholders by the Westcountry Rivers Trust took place in the in the center of Millbrook, a small village in Cornwall, UK. Millbrook has had many recent problems caused by large pluvial flood events. The goal was to present causes to the Millbrook flood events as mostly pluvial, and to emphasize land use and natural flood management measures as potential solutions recommended by hydrology reports. Much of the participants' attention ended up focusing on other relevant topics such as the presence of the reservoir in the village and the drainage infrastructure. In that context, the limitations of traditional methods as an effective conduit for translating complex data, modelled outputs and academic research findings, into clear and concise messages that can be understood by a non-technical audience became apparent. As an action research to respond to that problem, the idea of designing a Serious Game that could potentially fulfill that role was then born. This is also in line with the concept of Shared Vision Planning [4], which engages stakeholders in developing and experimenting with interactive simulation models for building consensus [5].

Serious Games (games used for purposes other than mere entertainment), as an effective means of engaging stakeholders, have been previously used for water system planning and management [5,6] with common aims of raising awareness [7–11], facilitating dialogue [11–16], or training for crisis response [17,18]. The Serious Game presented in this work aims to fulfill the research action goal in three ways. First, the software developed introduces novel technical elements to make it an effective Virtual Reality tool for visualizing results of simulations of 3D animated floods—a category of the Serious Game visualization system that is missing from the present body of work [6]. While existing work tends to emphasize "step-based" 3D flood visualization [19–23], the tool developed in this work focuses on the real-time deformation of 3D meshes representing volumes of water, with animated shaders allowing the mapping of color gradients to the mesh surface in order to emphasize flood depth. Second, to the best of our knowledge, this is the very first time that a Serious Game combining hydraulic simulation results, expert knowledge about farming systems and soil science, drainage infrastructure management, as well as representative damage and cost estimates have been put together to help players find an efficient (and previously unexplored) solution to the flood resilience problem for a real-world case study. Finally we propose a novel Serious Game intervention design inspired by the Socratic Method [23] and aim to prove that it can (a) enable participants to use inductive reasoning based upon game-derived data (hydraulic and costs) to confirm or reject hypotheses in a rational manner, and (b) encourage the players to question their initial assumptions (perception of the truth) and assist them in shifting towards a more reasoned, scientifically sound, deductive solution, i.e., closer to the evidence ("scientific truth").

From an experimental design point of view, the Serious Game combines pre- and post-game questionnaires with a traditional interpersonal group-based intervention. The intervention is based on asking participants to collaborate and answer questions in a specific order while playing the digital game developed using the Unity® [24] game engine. It must be emphasized that the approach does not consist solely of the digital game, but also integrates the questionnaires, intervention, and software all together. In this paper, the following materials and methods section presents all the elements used to design the Serious Game. The results are then outlined in Section 3 and their implications are discussed in Section 4.

2. Materials and Methods

First, we present the design of the Serious Game as an intervention specifically focused on enabling non-scientists to understand complex scientific models. Second, we present the choices made during the game development to produce a Virtual Reality tool suitable for visualizing results of simulations of 3D animated floods. Third, we present the experimental design composed of pre- and post-game questionnaires and used to assess the intervention based on this Serious Game.

2.1. The Overall Design of the Intervention

Using criteria extracted from [5], the game design can be characterized as follows. The primary aim or outcome of the game is for the user to explore different flood mitigation actions and corresponding outcomes in order to answer five questions. Finding the answers to these questions leads to the discovery of the optimal flood outcome to the given problem while minimizing costs. During the initialization of the game, a facilitator presents the Millbrook flood problem and provides contextual information regarding the history of flooding in the village, explanations regarding how to use the interface as well as technical concepts such as terrain roughness, and the different farming systems that the players can choose from. Although the game is built as a single player experience, several participants can play it in parallel in the same room. Communication between participants is not compulsory, but is encouraged so that they can freely discuss their choice with their neighbors when answering questions. Players do not have a limited and differentiated role, and they are each encouraged to explore the whole range of possible flood mitigation choices during playtime. The user interface makes use of sophisticated 3D video technology comparable to that already available in the videos game industry so that the user is able to visualize results of simulations via a Virtual Reality tool. The simulation model used relies on cellular automata and a series of conditional statements. The level of realism of the game can be seen as moderate when considering its individual aspects, such as costs and hydraulics models, in isolation. When looking at all of them combined, the level of sophistication of the socio-technical-environmental system presented is significant. Performance feedback during gameplay is instant as the player is immediately informed of the consequences of their actions on the flood and cost outcomes. Progress monitoring, such as the "capability of saving intermediate game results for follow up analysis" [5] is ignored as the user can switch at any time the combination of measures they wish to explore. Game portability is good because it can be played off-line on most Windows computers, though we have limited this portability by delivering the Serious Game intervention in a supervised environment. The game is designed to be a challenging, utilitarian, and demanding exercise.

To fulfill our aim to help participants to explore and understand complex scientific models, the Serious Game intervention is designed using three steps as shown in Figure 1.

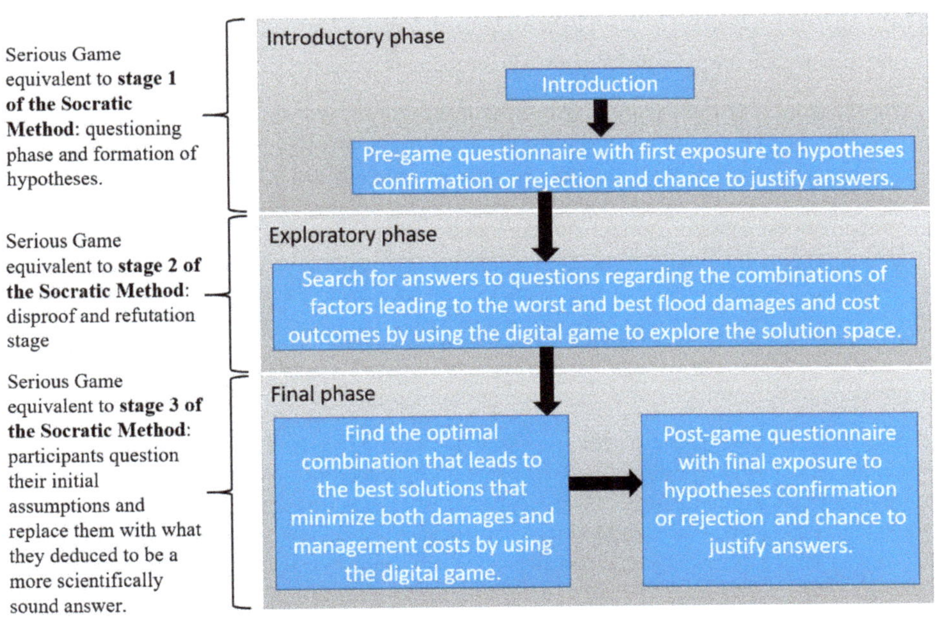

Figure 1. Three parts of the Serious Game process.

2.1.1. The Introductory Phase

Participants are given a general introduction (see link to additional materials at the end) on the Millbrook case study, and then asked to fill in a pre-game questionnaire (Appendix A) in which they need to confirm or validate different hypotheses, for the following reasons:

- The evaluation of any eventual cognitive change brought about by the Serious Game. This allows us to understand the pre-game state of the initial participants' views on the Millbrook flood problem. This includes possible pre-formed opinions (perception of the truth) on what would be the best possible decisions to achieve the balance between flood prevention and cost.
- It alerts participants to the crucial questions that they need to keep in mind while they improve their understanding of the problem, and therefore helps them to consolidate memories on technical issues, and to validate or reject hypotheses at the end of their investigative work.
- It creates a Serious Game equivalent of the questioning phase of the Socratic Method. The only difference is that instead of it emerging naturally through dialogue, the hypotheses are artificially created by the mix of introductory materials and the pre-game questionnaire.

2.1.2. The Exploratory Phase

During the exploratory phase of the game, players must interrogate the possible solutions and answer questions (Appendix B). The first two questions focus on finding the combinations of factors leading to the worst and best flood damage outcomes, e.g., "What is the combination of measures that leads to the worst flood outcome? Also note the damage/cost and the number of flooded houses." The next two questions focus on finding combinations leading to outcomes with the lowest and highest associated management costs. The last question is about finding an optimal solution that minimizes both flood and cost. During this phase, the digital game provides the players with the freedom to apply their own management decisions. It gives them feedback in the form of simulation results, which they can use to check the validity of the given hypotheses.

The use of compulsory ordered questions is essential because:

- It reduces an overwhelmingly large combination of solutions into a manageable list of five tasks that are essentially guiding participants' explorations of the different solution scenarios.
- It forces participants to answer questions where the solutions shown by the game might contradict what they initially thought were the best decisions, hence targeting preconceptions. For example, the exploration of answers to question 2: "What is the combination of measures that leads to the best flood outcome?" leads to the discovery that changing the farming systems around the village has a much greater impact than changing the level of investment in the drainage infrastructure. This corresponds to the Serious Game equivalent of the disproof and refutation stage of the Socratic method.

2.1.3. The Final Phase

Finally, in the third phase, players are asked to capitalize on the previous explorative work by finding the solutions that minimize both damages and management costs. They then have to fill in the post-game questionnaire (Appendix C) and apply inductive reasoning from the data revealed by their exploratory work to reject or confirm the presented hypotheses. Note that players can justify each answer in their questionnaire, giving them the opportunity to reach the Serious Game equivalent of the third step of the Socratic Method, namely to question their initial assumptions ("perception of the truth") and replace them with what they deduced to be a more scientifically sound answer ("the truth").

2.2. Game Development Choices

Millbrook has had many recent problems caused by large pluvial flood events. A 2 km square terrain around the village was selected, and uniform rain input of 45 mm/h for the first hour of simulation was applied, which equates to a 1 in 100-year event. In the game, players can simply change four parameters: the level of drainage infrastructure investment and the type of farming system applied in three instances (one for each sub-catchment in the village). They are then able to see flood impacts and associated management costs in numerous resulting combinations.

A 3D virtual table that comprises different informative documents and post-it notes introducing the case has been constructed as shown in Figure 2 to provide additional information to the users.

Figure 2. Overview of the 3D virtual table with narrative and interactive elements.

This innovative and original setting for flood visualization is quite powerful and expandable as follows:

- This allows a wealth of information to be represented as a spatial construct of 3D meshes on a virtual table of practically infinite size (Figure 2), where the geographic location of each item on the virtual table leads to better recall, in a similar way to mnemonic "visual palaces" [25].
- It can structure this information visually so that the user can see how components are related to each other in the case study, and allows the user to move, rotate, and zoom around components, and finally potentially display geographically distant elements as interconnected infrastructure nodes. For example, a distant power plant that influences the electricity supply of a village can be represented near the village on the virtual plan due to the cause–effect relationship between the two.
- Furthermore, this setting is easily translatable as a Virtual Reality headset-based Serious Game without major code modification.

The Unity® rendering engine has been improved in our previous work [26] to allow the terrain/flood height to be rendered with a higher degree of precision. The height data are provided by standard images in .png format where each pixel's red, green, blue, and alpha values encode in 32 bits the height of the terrain. From a practical point of view, these floating-point height data are rescaled into large integer numbers, between 0 and 2^{32} (allowing the system to display the difference in height with sufficient precision). Furthermore, by using .png rather than Unity® default .raw file format, more compact images are produced that can easily be sent through the internet. Although, so far, the data are stored locally with the game, whole sequences of images could potentially be sent from a "game" server hosting many test cases to provide multiple animations steps through a flood sequence.

Terrain and flood surfaces can now be deformed in real-time using advanced Shaders [27] software that exploits the computational capabilities of graphic cards. Shaders—programs implemented in the OpenGL Shading Language to display vertices and visual fragmentation using the accelerated computational power of the graphic card—are adopted to output in real time a 3D geometric linear interpolation of the flood "height-map" between the starting and the end state by looking at the "counter" input. Flood animation is implemented by providing a series of images and "counter" float numbers between 0 and 1, expressing how far the animations are from the end stage.

The Shaders are engineered not only to deform 3D meshes efficiently, but also to change the flood mesh color gradient from clear blue to dark blue with increasing flood depth (Figure 3). This works seamlessly even on fairly modest machines or small laptops, animating meshes with more than 1,440,000 triangles in real time with ease, as long as they have a DirectX 11 compatible graphic card.

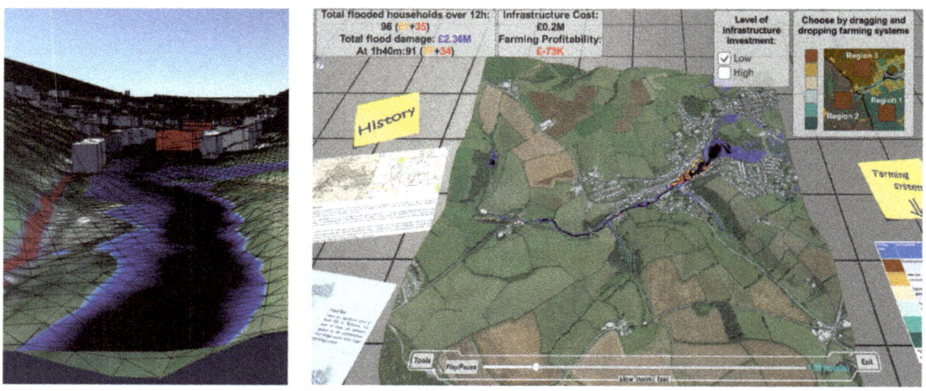

(a) (b)

Figure 3. Shaders: (**a**) is a close up with the vertices of the mesh made visible while (**b**) is an overview of the final result.

The Serious Game allows players to change (i.e., make decision about): (a) the level of investment in the drainage infrastructure, and (b) the type of farming system applied in each of the three sub-catchments in the village. They can then see the resulting flood impact and associated management costs. These inputs are integrated in the user interface as shown in Figure 4. The simplicity of the interface is the product of a compromise between computability, accessibility, and realism. Allowing the player to change the type of farming system in every individual field in Millbrook would provide a more realistic setting, but it could lead to a usability problem (manually changing each of the hundred fields before computing the consequences would detract from the games effectiveness because participants would lose sight of the problem while doing so). Furthermore, this would have been computationally impractical because it would require 2×6^{100} hydraulic modeling results to reflect the flood risk of different management scenarios. The present choice of grouping the fields into three cultivated areas allows the user to set farming systems quickly and explore different consequences of their choices, while also keeping the overall problem in mind. While the spatial resolution for targeting measures is quite low in the current model, it still provides a useful indication of which parts of the catchment provide the greatest constraints or opportunities for land use change and flood risk alleviation. It reduces the number of hydraulic computations to a manageable 432 combinations. Although it might initially look more complicated, the user interface and search space structure is quite simple because it closely follows the infiltration gradient of the farming systems as they are presented to the participants, from dark brown (system with the lowest infiltration rate) to dark green (system with the highest infiltration rate). Participants are therefore expected to find the best solution available and master the problem of minimizing both the flood damage and management cost in under an hour without having to explore all possible choices. Each farming system, as described in Table 1, comes with an associated infiltration rate, surface roughness value, sedimentation rate leading to a percentage of pipe blockage in the village downstream, and profitability (ranging from "– –" for higher loss, "–" for average loss, "0" for null, "+" for average profit, and "++" for higher profit).

Table 1. Farming systems.

Type	Infiltration Rate (mm/h) *	Surface Roughness (Manning Coefficient)	Drainage Pipe Blockage (%)	Profitability (– –, –, 0, +, ++)
Heavily grazed permanent pasture	10	0.3	40	–
Intensive arable under tillage and standard rotation	15	0.4	90	0
Young grassland with moderate grazing and mechanical improvement	30	0.35	10	– –
Arable with mixed rotation and increased soil amendment	38	0.45	70	+
Arable with mixed rotation, cover crops and minimum tillage	70	0.50	50	++
Arable with no-till, cover crops, mixed rotation and conservation agriculture	90	0.50	20	+

* Underlying soil is Soilscape 6 [28] (freely draining, slightly acid loam); as such, a bare soil infiltration rate of between 10–20 mm/h is likely.

These farming systems are used as model values in the Serious Game in accordance with the latest available agri-environmental data for the Millbrook catchment and from a thorough review of relevant literature detailed in Appendix D.

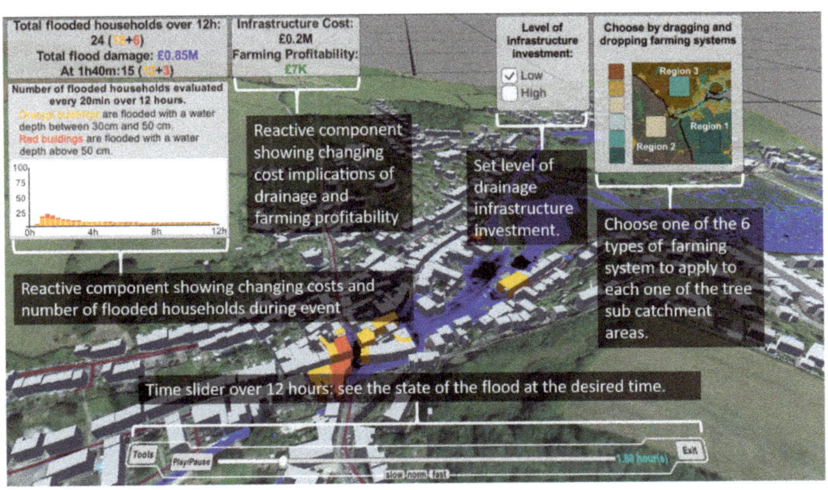

Figure 4. The user inputs as seen in the Serious Game interface.

To communicate effectively the consequences of the decisions made by users, three elements have been included on the interface of the Serious Game as shown in Figure 4.

- A temporal slider at the bottom allows a user to change the timing during a selected event to see the progression of the flood. Note that the overall speed of the animation can also be paused, slowed, or accelerated.
- A reactive flood damage information window in the top left corner shows the total number of households flooded, split by the severity of flood inundation, and tracks the costs associated with this. A further indicator shows the number of households flooded at a given point in time. When the mouse hovers on the window, a graph pops up showing the distribution of the number of flooded buildings over time to highlight the peak of the flood event as well as compare the number of occurrences of major versus minor flooding of houses.
- A window at the top center-left of the screen shows the total cost of the selected drainage infrastructure management policy, as well as the total farming profitability of all three cultivated areas around the village.

The 3D map of Millbrook is 2000 m × 2000 m. Houses colored in orange have a water depth between 15 and 30 cm inside the building, while houses colored in red have a water depth beyond 50 cm. A player can choose between two levels of investment in the 16 km of the urban drainage infrastructure of Millbrook. The infiltration rates are shown in the context of a one in a hundred-year event of 45 mm of rainfall in one hour. The investment levels indicate the capacity of the drainage network to manage surface runoff under different maintenance conditions (additional data about drainage management costs are available in Appendix E).

- A high-maintenance (high cost, well maintained) drainage infrastructure that can cope with surface runoff up to 26 $mm/h/m^2$ in urbanized areas. At the high level of investment scenario, the yearly management cost of the urban drainage infrastructure is evaluated at 406k GBP and includes cleaning 10% of the length of all drainage pipes per year, repairing 3% of total length of pipes, and replacing 1%.
- A low-maintenance (low cost, not well maintained) drainage infrastructure that can discharge surface runoff up to 13 $mm/h/m^2$ in urbanized areas. At this low level of investment scenario, the yearly management cost of the urban drainage infrastructure is evaluated at 203k GBP and includes cleaning 5% of the length of all drainage pipes per year, repairing 1.5% of total length of pipes, and replacing 0.5%.

Flood depths encoded in images are produced using the WCA 2D flood model [29], a part of the CADDIES [30,31] modelling framework. CADDIES is an open-source framework designed to facilitate the design and deployment of Cellular Automata (CA), specifically for 2D urban flood modelling. It is ideal for fast evaluations of floods [31], and therefore allows the swift computation of multiple simulation results corresponding to various users' choices in the context of serious gaming. In the present work, the flood models were pre-run to produce output images used to make the 3D animated mesh of the flood (more details about computation times in Appendix F).

The Millbrook example presents the users with a high-speed model, where a cell size of 2.5 m was selected for the flood modelling with WCA 2D as a compromise between accuracy and computability to produce the 432 different flood outcomes in a few days. The speed of execution makes it suitable in the context of serious gaming for quickly displaying numerous flood outputs resulting from different combination of parameters.

The existing farming situation on the ground in Millbrook (a mixture of "heavily grazed permanent pasture" and "intensive arable cultivated areas under tillage with standard rotation") leads to the worst possible combination in the game. It has a maximum damage outcome (108 houses flooded for 2.6 million GBP of damage), independently of the management cost (high or low maintenance infrastructures both lead to the same result), and a poor overall farming profitability (−30k GBP/year).

The optimal outcomes found in the game maximize farming profitability (+109k GBP/year) and reduce the number of flooded houses by approximately 80% (20 to 22 houses for 0.68 to 0.8 million GBP of damage depending on the choice for either a high or a low maintenance drainage infrastructure). It corresponds to a mixture of the "Arable with no-till, cover crops, mixed rotation and conservation agriculture" and the "Arable with mixed rotation, cover crops and minimum tillage". It is notable, that despite the high cost, the high maintenance drainage infrastructure seems to have a negligible influence on the flood impact. Internal testing was carried out prior to the first game event on a beta version by a group of six staff members from the Westcountry Rivers Trusts, comprising farm advisors, fisheries and river management specialists and data/communication officers. The feedback included some suggestions for improving the visual aspects or 'usability' of the game, and some comments about improvements to the accuracy or completeness of the model, mainly regarding the type of farming systems available.

2.3. The Experimental Design Used to Assess the Intervention

The Serious Game combines pre- and post-game questionnaires with an intervention where participants are encouraged to collaborate and asked to answer questions in a specific order while using a digital game.

The initial idea was to measure the impact of the Serious Game on the participants from the Millbrook community. However, after the first gaming session took place at the Westcountry Rivers Trusts headquarters in Cornwall in the UK, invitations were sent via email but attendance was poor (2 of 20 residents). We were left with only six valid contributions (with correctly filled forms). Only two of the participants were residents from Millbrook (non-specialists), and the other four were environmental and land management advisors. Other contributors submitted incomplete forms and therefore their participation was disregarded. To overcome this problem we had to get more people to play the game even if not linked to Millbrook, and therefore focus more on the cognitive impact of the game on participants rather than its direct effects on the Millbrook community. The second gaming session took place at the IHE Delft Institute for Water Education in the Netherlands as part of a compulsory introduction to Serious Gaming and had nine valid contributions (eight of the participants were MSc students in Hydrology, and one was a non-specialist). The last gaming session took place in Southampton in the UK, wherein neighbors from the same street were invited to participate using "door-to-door" presentations. This session had seven valid contributions from non-specialists from various backgrounds. The total number of valid participations was 22 of a total of 31 (ten were non-specialists).

Thus, our primary aim was to measure whether there was a statistically significant difference between the pre-game and post-game players' views on the best solutions to the Millbrook flood problem. More specifically, we aimed to prove that playing the Serious Game induced a statistically significant directional change in the way participants confirmed or rejected seven different "Millbrook flood" hypotheses towards what the hydraulic and cost models of the game point to as the correct rational answers. We created seven different hypotheses (see Table 2) in order to test the limits of the Serious Game, regarding how effective it can be at making such problems accessible to the general public (non-specialists). The first hypothesis set the context to describing the extent of the flood damages caused by a one in a hundred-year event of 45 mm of rainfall in one hour in the village of Millbrook. The next hypotheses explore whether investing in drainage infrastructures or land use are the best strategies to limit risks of flood damage. Finally, the last group of hypotheses use a more specific technical language and focus on characterizing the impact of different farming systems on both flood impact and profitability.

Using a statistical directional test, we aimed to find whether playing the game would induce a change in the way players confirmed or rejected the seven different "Millbrook flood" hypotheses in the pre- and post-game questionnaires. Levels of significance were computed using a Wilcoxon Signed-Ranks One-Tailed Test [32]—see Appendix G for more details—where data are paired (between pre- and post-game answers) and answers are set on an ordinal scale (-2 = strongly disagree, -1 = disagree, 0 = neutral, 1 = agree, 2 = strongly agree). With a total of 22 participants, the standard

Z-value based version of the test could not be used, and was replaced by the W-value suitable for sample sizes where $N < 30$. The null hypothesis was that playing the Serious Game does not change the players' views. Levels of significance considered were: "Not Significant" (for $p > 0.05$), "Significant" (for $p < 0.05$), and "Highly Significant" (for $p < 0.01$).

We also intended to show to a lesser extent that playing the Serious Game was informative and significantly changed the participants' views on the solutions to the Millbrook flood mitigation problem. In order to achieve this we sought to demonstrate that (a) there was an increase in the number of correct or relevant numerical values used in the justification of the answers, (b) there was a gain in technical vocabulary, (c) there was an increase in the use of comparative, superlative, and quantitative words that translates into a lexical field more oriented toward showing the degree of influences of factors over the flood and cost outcomes, and (d) there possibly was a growth in confidence and accuracy in the participants' answers.

In response to the shift in the aim of the paper, showing a real impact on the Millbrook community itself became less important than demonstrating that the Serious Game could induce cognitive change in the participants and therefore be a suitable tool developed in the context of the action research.

3. Results

The differences measured between answers in the pre- and the post-game questionnaires (Appendices A and C) are used to infer the results as described in the following sections.

3.1. *Transformative Nature of the Millbrook Serious Game*

Table 2 shows, with the associated statistical level of significance, how playing the game induced a directional change in the way participants confirmed or rejected the seven hypotheses, while Figures 5 and 6 provide radar charts showing, for all participants, where answers to hypotheses (excluding Q5 that is a different type of question) were changed to stronger agreement or disagreement respectively after playing the game.

Table 2. Playing the game induces a directional change in the way participants confirm or reject the seven hypotheses.

Hypothesis	Correct Answer	Directional Change [1]	Level of Significance [2] of Directional Change
Q1—A one in-a-hundred years rainfall event of 45 mm of water in one hour could cause serious flood problems in Millbrook (valid answer: true)	True	Increased Agreement	+ ($p = 0.05$)
Q2—Improving the village sewer infrastructure that directly drains excess water could result in neutralizing up to 90% of the flood resulting from such an event	False	Increased Disagreement	++ ($p = 0.01$)
Q3—Improving farming practices that change the ground cover of the cultivated areas around the village could result in neutralizing up to 90% of the flood resulting from such an event	True	Increased Agreement	++ ($p = 0.01$)
Q4—Investing money in the sewer infrastructure gives a better result for your money when fighting against flood than trying to support different farming practice	False	Increased Disagreement	++ ($p = 0.01$)
Q6—Heavily grazed permanent pasture increases greatly the capacity of the soil to absorb water	False	Increased Disagreement	++ ($p = 0.01$)

Table 2. *Cont.*

Hypothesis	Correct Answer	Directional Change [1]	Level of Significance [2] of Directional Change
Q7—Growing crops while minimizing tillage increases greatly the capacity of the soil to absorb water	True	Increased Agreement	++ ($p = 0.01$)
Q8—Intensive farming that compacts the ground and lowers the capacity of the soil to absorb water, remains the most profitable way to exploit a cultivated area	False	Increased Disagreement	+ ($p = 0.05$)

[1] Difference between pre- and post-game questionnaire answers (−2 = strongly disagree, −1 = disagree, 0 = neutral, 1 = agree, 2 = strongly agree). [2] Significance using a Wilcoxon signed rank one-tailed test for paired ordinal data. + means medium significance, and ++ means higher significance.

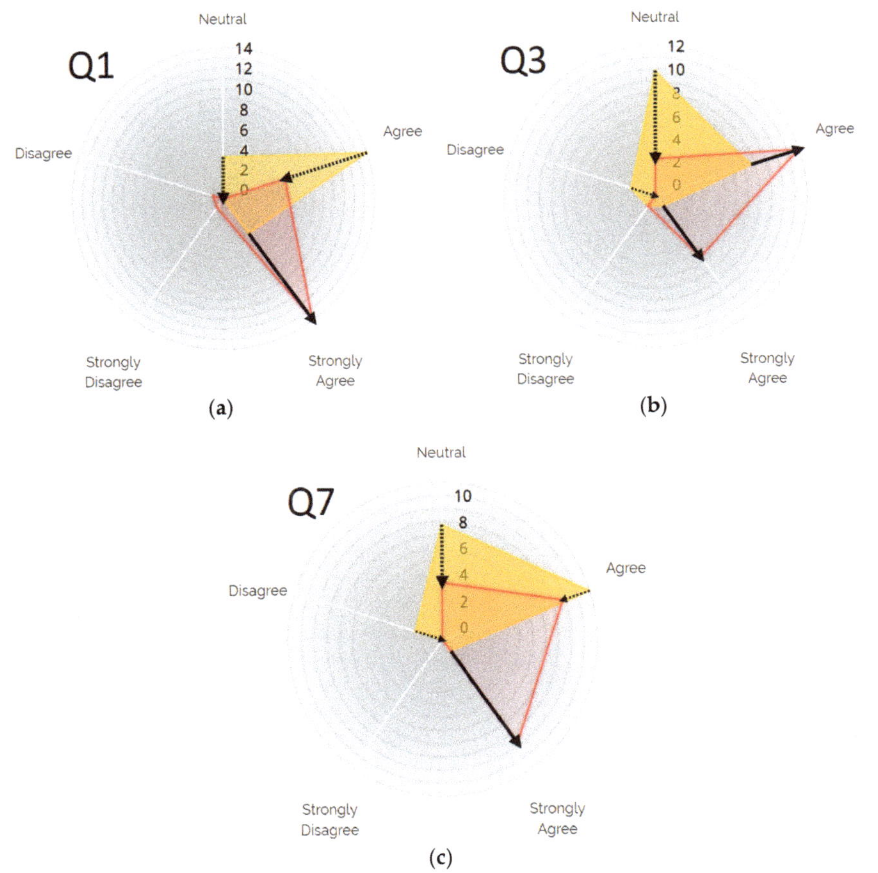

Figure 5. Radar charts showing for all participants, where the changes induced by the games for answers to hypothesis reflects a stronger agreement. Each participant answer (−2 = strongly disagree, −1 = disagree, 0 = neutral, 1 = agree, 2 = strongly agree) is rated on one axis in yellow (pre-game) and red (post-game) areas. The questions answered are (**a**) Q1; (**b**) Q3; (**c**) Q7.

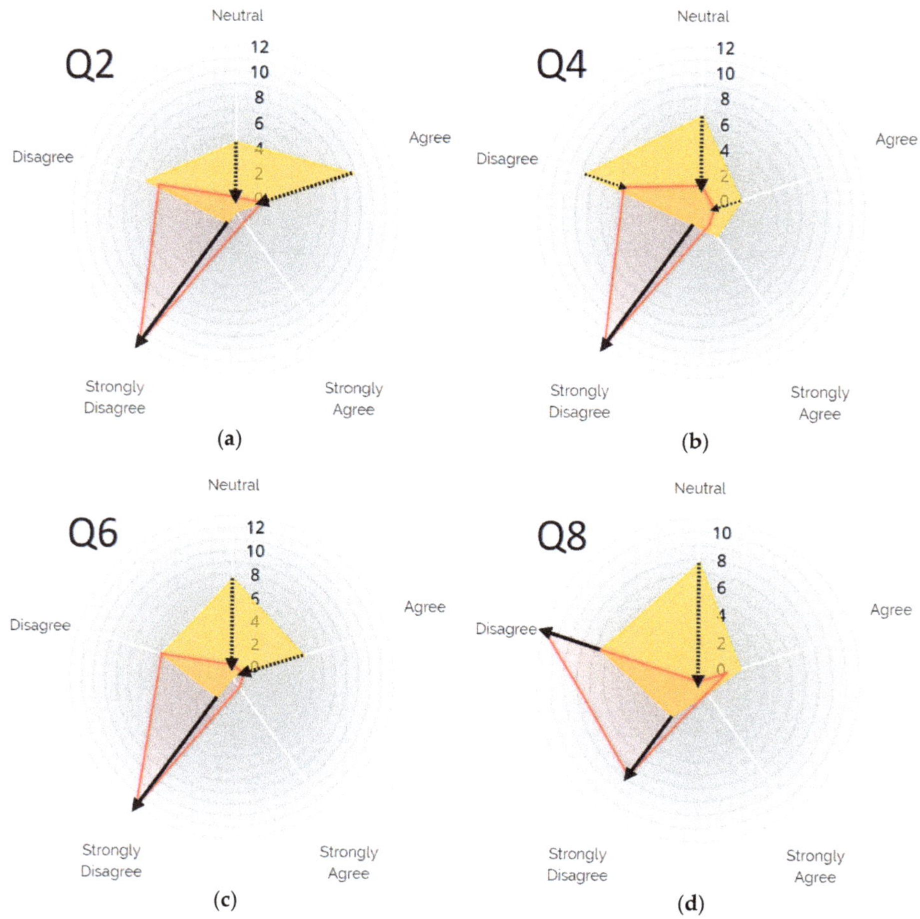

Figure 6. Radar charts showing for all participants, where the changes induced by the games for answers to hypothesis should reflect a stronger disagreement. Each participant answer (−2 = strongly disagree, −1 = disagree, 0 = neutral, 1 = agree, 2 = strongly agree) is rated on one axis in yellow (pre-game) and red (post-game) areas. The questions answered are (**a**) Q2; (**b**) Q4; (**c**) Q6; (**d**) Q8.

Figure 7 shows how participants perceived in 70% of cases that playing the game made them revise their views and assumptions on what they initially thought were the causes and best solutions to the Millbrook Flood problem (see supplementary data Table S1 in "tables" table). Note that the "perceived" change seems to match the "measured" changes of views shown in Table 3.

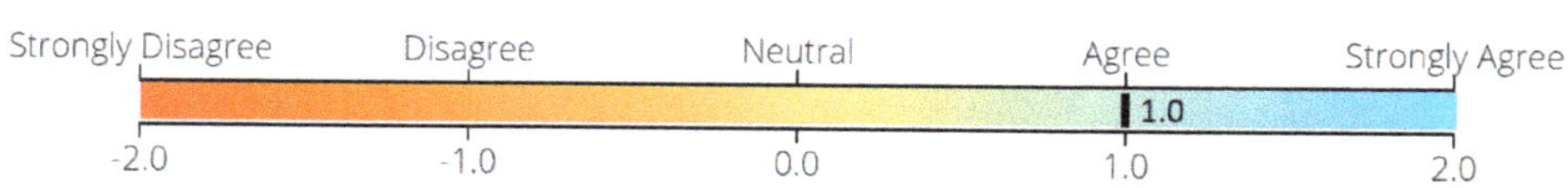

Figure 7. Shows the average score of answers from participants when asked if the game made them revise their views on the problem—range [0,2] and standard deviation 0.71.

We can report inconclusive effects when trying to select the most important flood factors in Q5, which asked: "what are the two most important factors in this list of three that influence flooding when looking at the ground cover of cultivated areas?" Participants were required to select two of the three following factors in the pre-game and the post-game questionnaire.

- The capacity of the ground to absorb water
- The amount of washed sediments that can block drainage pipes in the village downstream
- The roughness of the terrain that slows the travelling speed of the water

There does not seem to be any significant difference between pre- and post-game questionnaires amongst the participants. After playing the Serious Game, the majority of participants were able to establish that the drainage system was having a limited effect on the flood level for the modeled pluvial scenarios. Using deductive reasoning, participants would then have been expected to decrease the relative importance of "the sediments and drain blockage" factor to stay consistent with their other conclusions. This did not happen. One possible explanation is that the influence of terrain roughness and sediment induced drain blockage are shown in a much subtler way than water absorption capability in the Serious Game model. As these factors were perceived as less salient, the difference in ranking between them was not the subject of the analysis. The participants did not have a prior opinion on this matter, thus they did not have pre-conceptions to revise in relation to it.

Table 3. Shows the shifting of average scores as arrows between all the participants' answers to a hypothesis from being a non-specialist (starting point of the rectangle) to being a domain expert in either environmental science or hydrology (pointing end).

Question	Shifting of Average Scores from Non-specialist to Domain Expert
Q1	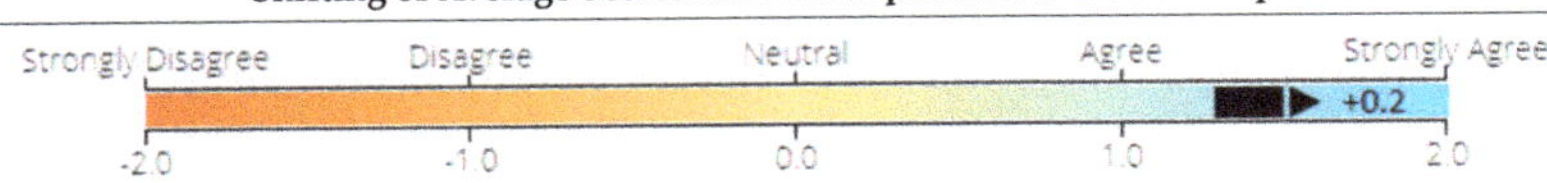
Q2	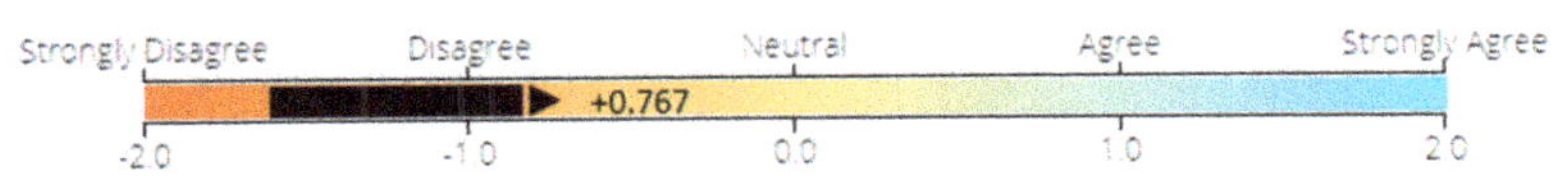
Q3	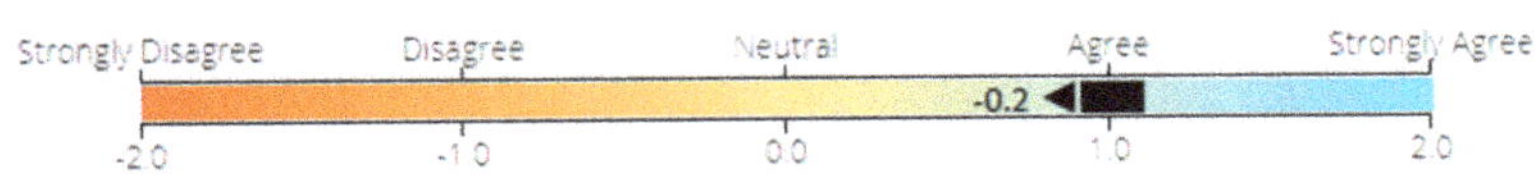
Q4	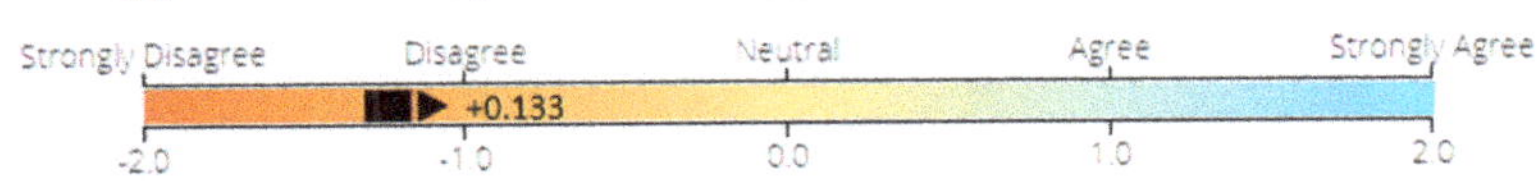
Q6	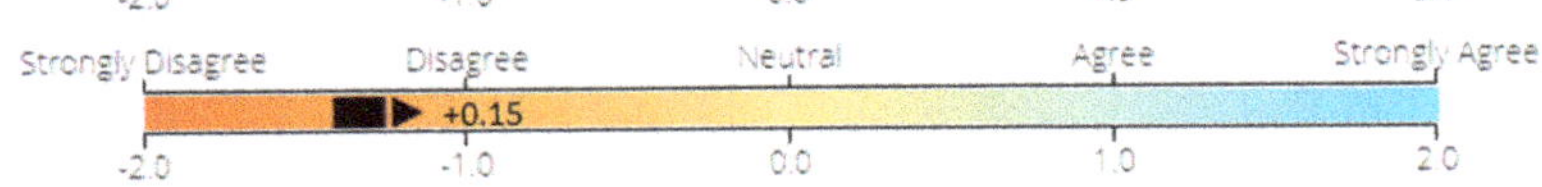
Q7	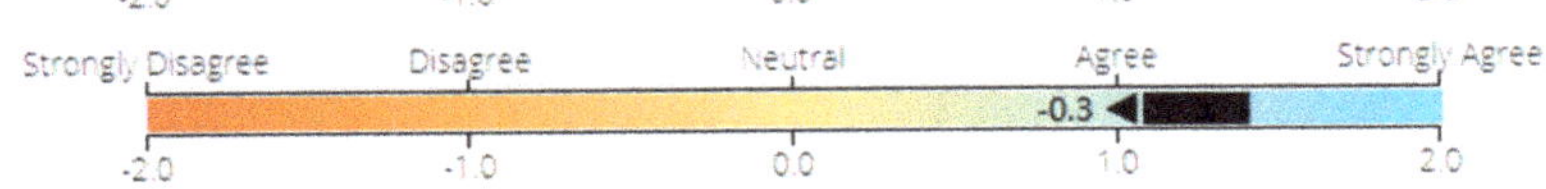
Q8	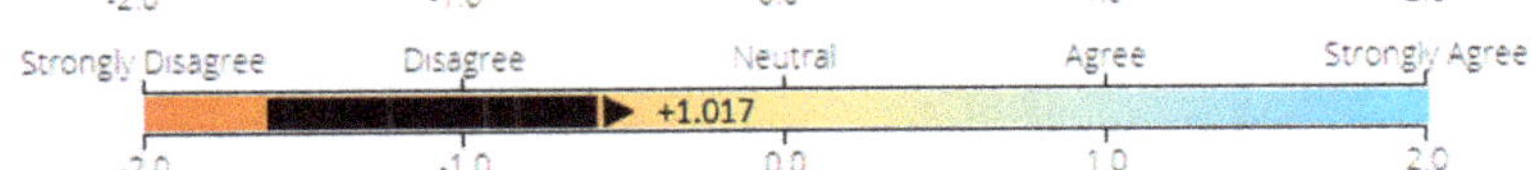

Slightly more than half the participants (12 of 22) were experts in either environmental matters or hydrology. Results seem to show that the presence of experts boosts the accuracy of the answers in the very first hypothesis only, and in fact decreases the accuracy of the answers towards neutrality in all other hypotheses, as shown in Table 3. We cannot divide our sample into two groups of size 12 and 10 and apply a Mann-Whitney U Test in order to confirm this hypothesis because the sizes of these groups are too small to achieve statistical significance. Still, simply by showing the change in mean (as opposed to a change in median, which is less meaningful) it allows us to speculate that experts act as dampeners because overall they tend to be more reserved and "neutral" in their attitude. As such, the presence of "moderating" experts amongst our participants suggests that we should obtain even stronger results with a non-specialist audience.

3.2. Informative Nature of the Millbrook Serious Game

Changes induced by the game in terms of use of numerical values, vocabulary, and confidence and accountability are listed here.

3.2.1. The Accuracy and Increasing Use of Numerical Values in Participants' Answers

Overall, the words written by participants to justify their answers underwent a qualitative change from speculative in tone and unfocused to a greater focus on numerical values learned from the game. In Q1, the word used with the highest frequency in the pre-game responses is "Unknown". By contrast, in the post-game responses, the word used with the highest frequency becomes "108" which is the exact maximum number of houses flooded in the worst scenario. This shows a better understanding and a more precise quantification of the magnitude of the flood issues.

From Q2 to Q7 the degree of numeracy of the answers—the number of correct or relevant figures used in the justification of the answer—is nil in the pre-game questionnaire (none of the very few numbers written in justifications of these answer were relevant as mostly they were just copied from the questions) and higher in the post-game questionnaire (10 words in 293, approximately 3.4% of words written in justifications of these answer are accurate numbers).

3.2.2. The Acquisition of a Domain Specific Vocabulary and the Increased Utilization of Technically Accurate Words.

Although the size of the lexical field (list of all the words) does not change significantly (pre-game questionnaire uses 278 words to justify their answers while the post-game questionnaire uses 293)—in the post-game questionnaire there are many more words used appropriately, with a high frequency (underlying data can be found in supplementary file in "grid data" tab and in Tables S1–S4 in the "table" tab). In the pre-game answers, "Unknown" (17 times), "neutral" (15 times), "runoff" (7 times), and "reduce" (7 times) are the four most used words. In the post-game answers, "Farming" (17 times), "Impact" (14 times), "Soil" (12 times) and "Infrastructure" (11 times) are the four most encountered words. This indicates that players have acquired a more specialized vocabulary from playing the game and that they more often tend to use domain specific and technical words to justify their answers.

Furthermore, the proportion of words expressing a comparative relationship between two elements (for example: "bigger", "reduces", etc.) or a relative importance ("high", "greatly", etc.) also changes from 18.3% (51 words out of 278) in the pre-game answers to 29.3% (86 words out of 293) in the post-game answers. This shows a better grasp of the influences of different factors associated with the flood and damages and their relative importance.

3.2.3. The Growth in Confidence and Answer Accountability Induced by the Game.

Due to the fact that justifying their answer was not compulsory during the exercise, 14 participants sometimes chose answers without justifying them in the forms (resulting in one-third of the answers lacking an explanation). In these instances, we cannot say if the participants were able to justify their answers or not. That was only possible if they stayed "neutral" on their agreement/disagreement scale—in which case we concluded that they were not able to justify their answer. Most of the participants that were not able to justify their answers before playing the game were then found able to do so after playing it. This was done in an explicit and technically detailed manner—only 2 of the 28 answers were either wrong or not explicit enough (exact details can be found in supplementary data in "grid data" tab).

When looking specifically at the users that were not able to justify the answers to the questions before playing the game (33.33% of "Unknown"), the overwhelming majority of these players were able to justify their answers after playing the Serious Game (only 1.8% of "Unknown") (see Figure S1 in "grid data" table in data file for more details). There is only one exception, in questions 5 and 7, where the same user could justify his/her answer in the pre-game questionnaire, but failed to justify his/her answer in the post-game questionnaire—that particular player admitted having lost sight of that question while playing the game.

When directly asked if they learned anything new after playing the game, on average, the participants' answers were mostly positive, as shown in Figure 8 (Table S2 in "tables" tab in supplementary data shows the complete list of associated answers). A couple of notable exceptions came from a land management advisor and a hydrology MSc student who felt that the game simply confirmed their pre-existing understanding.

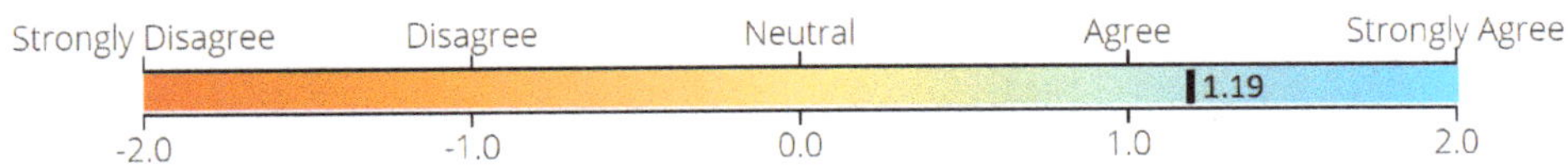

Figure 8. Shows the average score of answers from participants when asked if they have acquired new knowledge from playing the game—range [−1,2] and standard deviation 0.74.

Finally, feedback from the participants regarding the suitability of the Serious Game for helping people to understand the Millbrook flood issues was positive overall, as shown in Figure 9 (Table S3 in "tables" tab in supplementary data shows the complete list of associated answers).

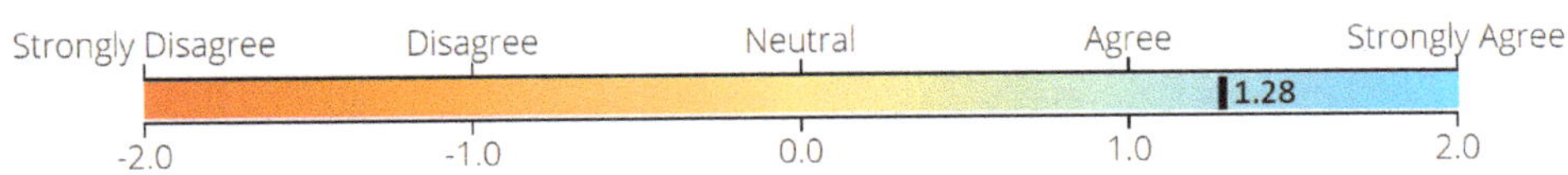

Figure 9. The average score from all participants' answers when asked if they would recommend this type of Serious Game for stakeholders—range [1,2] and standard deviation 0.46.

When asked why they would recommend the Serious Game, participants' answers can be divided into three broad categories:

- It is a good informative tool that allows the user to view and understand the effects of decisions for many different scenarios.
- It encourages thought and discussion.
- It is good at changing attitude and promoting positive changes.

Regarding the impact of the Serious Game on the Millbrook community, eleven months after playing the Serious Game, a number of participants have communicated the outputs with other local stakeholders and have initiated the formation of an ongoing natural flood management partnership for action in Millbrook.

4. Discussion and Conclusions

In this section we discuss the cognitive changes induced by the Serious Game on players, the general impact on the Millbrook community, the limitations of this paper, and future work.

4.1. Cognitive Impact of the Serious Game

In this work we have shown that playing the Millbrook Serious Game significantly changed the participants' views on the Millbrook flood problem. Primarily, we have demonstrated that the game was informative (Section 3.2) because it improved participants' knowledge of the quantitative effects of flood damage, increased their use of technical vocabulary, improved their understanding of the influences and relative importance of multiple factors affecting the flood and cost outcomes, and caused a growth in confidence and accuracy in their answers. In Section 3.1 we established our most important aim whereby statistically significant changes induced by playing the game were directional and lead the participants to strengthen their confirmation and rejection of specific hypotheses. This also encouraged them to deduce rational and correct conclusions regarding fairly technical and non-trivial decisions for the Millbrook flood problem. In effect, the game briefly and successfully transformed the participants into "citizen scientists" in the purest sense of the term—it led them to use inductive reasoning from data produced during the game (hydraulic and costs) to correctly confirm or reject hypotheses in a rational manner. Feedback from participants even strengthened this claim when they admitted that the game allowed them to revise their views and assumptions on what they initially thought were the causes and best solutions to the Millbrook flood problem.

4.2. Impacts of the Serious Game on the Millbrook Community

Surprisingly, considering that we only had two residents from the Millbrook community playing the game, there nonetheless was a significant impact. These two participants did communicate with other local stakeholders and together have initiated the formation of an ongoing natural flood management partnership for action in Millbrook. The formation of partnerships is hugely beneficial for instigating action on the ground [33] because they encourage collaborative working, the sharing of resources, funds, knowledge and skills, and they can provide local groups with a clearer vision and drive for change. Westcountry Rivers Trust have stated that the Serious Game has helped to strengthen the relationship between Westcountry Rivers Trust and Millbrook residents, parish councilors, academic researchers, and local experts. They also added that this game will be a valuable tool, enabling them to communicate natural flood management concepts to a wide range of individuals across the south west of England. They concluded that the combined effect of the aforementioned benefits has increased the likelihood that natural flood management measures will be implemented in the Millbrook catchment area.

4.3. Limitations and Future Work

Possible limitations of this work regarding the efficiency of the Serious Game as a teaching and decision support tool were also found. The first concerns the barrier of technology as a limiting external factor. One participant's unfamiliarity with computer technology interfered with the consistency of game play and thus their answers, and so had to be discarded. This experience leads us to suspect that for a mixed audience, another kind of game interface or type may be more appropriate (possibly board games or numeric games with a comprehensive training component).

A second limitation is that some results indicated limits in the understanding of the participants. As shown in Figure S2 in the supplementary data file ("grid data" tab) there were inconclusive effects when participants tested the relative importance of more subtle factors such as "the sediments and drain blockage", rather than terrain roughness. This is likely a limitation in the depth of the cognitive effects of the Serious Game, possibly due to its relatively short duration. The game has a duration of one hour only, which is probably not enough time for participants to thoroughly investigate these subtler effects among the 432 combinations of flood management strategies. Each of these strategies results in outcomes that probably require more in-depth investigation. Eight of the participants were MSc students in hydrology, and they themselves seemed to have conflicting views on the prevalence of either roughness or sediment-induced drain blockage as the second most important factor. Potentially this could be improved by increasing the duration of the game, introducing more questions, and extending the duration of the exploration phase over multiple sessions.

A third *perceived* limitation was that the presence of too many participants with a high level of technical expertise might limit the generalization of the results. The Serious Game was originally built for a non-specialist audience—and more specifically for the Millbrook residents. As such, having slightly more than half the participants being experts in either environmental matters or hydrology might be seen as a limitation. At first, it would seem that this would: (a) lessen our ability to demonstrate that the Serious Game improves participants' knowledge as they are already experts, (b) "boost" the accuracy of answers unnecessarily and, therefore, skew the results towards correct hypothesis confirmation or rejection. As far as we can see, by carefully examining the data, it seems that these reservations are unfounded. First, the fact that we can demonstrate a strong informative value in Section 3.2 despite the presence of experts suggests that the game will work even better with a non-specialist audience. Second, results in Table 3 in Section 3.1 seem to show that the presence of experts only boosts the accuracy of the answers in the very first hypothesis and in fact decreases the accuracy of the answers towards neutrality in all other six hypotheses. As such, the presence of "moderating" experts in the midst of our participants suggests that we should obtain even stronger results with a non-specialist audience. Future work will emphasize testing with a targeted audience sample.

Another limitation, from a flood modelling point of view, is that the model is relatively coarse as we only consider three large agricultural areas and two infrastructure interventions. Future development of the model could incorporate finer spatial units and the ability to model a wider range of measures at specific sites across the catchment (e.g., storage ponds or leaky dams) so as to better support land management and potentially enable farmers participation. This of course would have to be done carefully so as to limit the resulting increase in complexity by selectively emulating key features of water and the possible actions of players in catchments [34].

Whilst the small size of our study limits our ability to generalize these results, the positive outcomes we have observed so far lead us to consider this work as a template applicable to other urban–rural flood case studies throughout the world. Potential new case studies are under consideration and should add to the available information to provide a more complete picture of this novel type of teaching tool and its suitability for improving the understanding of complex issues. We also intend to add control treatment groups in future experiments. Four possible control groups come to mind: a standard workshop without a serious game; a workshop without a serious game but with an intervention that includes a dialogue with participants based on the Socratic method and the validation or rejection of hypotheses; a facilitated workshop with a serious game without the Socratic method-based intervention; and finally, a treatment using another common exiting flood visualization tool.

Finally, the design of the Serious Game presented in this work could scale very well to complex problems with thousands of variables that would require weeks of practice to master—an area of serious gaming that is presently attracting increased interest. In the context of the video game industry, modern strategy games used for pure entertainment have presently reached a staggering level of complexity. Players usually need to practice for days or even weeks to understand how to simultaneously balance dozens of variables and their combined influences on the dynamic of the systems they play. In this respect, the Serious Game described in this paper has been engineered to be a "challenging/utilitarian" exercise that scales well with complex problems that could take long periods of playing time to master. A solution could be to extend the "exploration phase" of the Socratic Method inspired process described in Section 2.1.2 to several weeks if necessary, by asking participants to answer different groups of questions per session, focusing on a specific sub-problem at a time and this way avoid the kind of problems resulting from a broad-scale approach [35]. Although useful, this design does not provide an answer to the open question of motivational affordance, and additional work is needed in this area. We intend to pursue this avenue in the context of the ongoing (2016–2020) SIM4NEXUS research project [36], with a Serious Game that has possibly the most complex underlying model ever attempted.

Supplementary Materials: The following are available online at http://www.mdpi.com/2073-4441/10/12/1885/s1. Table S1: Average response to Q10: Has this Serious Game changed any previous view you had on the Millbrook problem? Table S2: Average response to Q9: Did you learn anything new today? Table S3: Average response to Q11: Would you recommend this type of Serious Game for other people involved in Millbrook flood problem? Table S4: Average response to Q1: A one in-a-hundred years rainfall event of 45mm of water in one hour could cause serious flood problems in Millbrook, Figure S1: Justifying answers before and after playing the game, Figure S2: Prioritization of flood factors.

Author Contributions: Conceptualization, M.K. and D.S.; Investigation, M.K., H.L. and S.W.; Methodology, M.K. and D.S.; Software, M.K., M.J.G. and A.S.C.; Supervision, D.S. and L.V.-L.; Validation, M.K., M.J.G., A.S.C., L.V.-L., H.L. and S.W.; Visualisation, M.K.; Writing—original draft, M.K., M.J.G., D.S., H.L. and S.W.; Writing—review and editing, M.K., D.S., A.S.C., L.V.-L., H.L. and S.W.

Funding: The authors would like to acknowledge the funding provided by the UK Engineering and Physical Sciences Research Council, grant EP/M018865/1 (The Nexus Game), the ongoing EC H2020 EU-CIRCLE (GA 653824) project, and the ongoing e European Union's Horizon 2020 research and innovation programme under Grant Agreement No. 689,150 SIM4NEXUS. Harry Langford acknowledges his knowledge exchange funding from the N8 AgriFood Resilience programme.

Acknowledgments: Many thanks to the Westcountry Rivers Trust for sharing data and organizing the first Serious Game meeting with Millbrook stakeholders.

Conflicts of Interest: The authors declare no conflict of interest. The founding sponsors had no role in the design of the study; in the collection, analyses, or interpretation of data; in the writing of the manuscript, and in the decision to publish the results.

Appendix A

Millbrook Serious Game Questionnaire 1

University of Exeter—Westcountry Rivers Trust

Privacy Statement: Please complete this questionnaire. Any concerns can be communicated to Mehdi Khoury (m.khoury@exeter.ac.uk) at University of Exeter. Thank you for your time and cooperation.

Answer the following questions by circling the most appropriate answer.

1. **A one in-a-hundred years rainfall event of 45 mm of water in one hour could cause serious flood problems in Millbrook.**
   Strongly Disagree *Disagree* *Neutral* *Agree* *Strongly Agree*
   How many residences could potentially be flooded? Any worst-case cost approximation?

   ---

2. **Improving the village sewer infrastructure that directly drains excess water could result in neutralising up to 90% of the flood resulting from such an event.**
   Strongly Disagree *Disagree* *Neutral* *Agree* *Strongly Agree*
   Can you justify your answer?

   ---

3. **Improving farming practices that change the ground cover of the cultivated areas around the village could result in neutralising up to 90% of the flood resulting from such an event.**
   Strongly Disagree *Disagree* *Neutral* *Agree* *Strongly Agree*
   Can you justify your answer?

   ---

4. **Investing money in the sewer infrastructure gives a better result for your money when fighting against flood than trying to support different farming practice.**
   Strongly Disagree *Disagree* *Neutral* *Agree* *Strongly Agree*
   Can you justify your answer?

   ---

5. **What are the two most important factors in this list of three that influence flooding when looking at the ground cover of cultivated areas?**
   - *the capacity of the ground to absorb water*
   - *the amount of washed sediments that can block drainage pipes in the village downstream*
   - *the roughness of the terrain the slows the travelling speed of the water*

   Can you justify your answer?

   ---

6. **Heavily grazed permanent pasture increases greatly the capacity of the soil to absorb water.**
Strongly Disagree *Disagree* *Neutral* *Agree* *Strongly Agree*
Can you justify your answer?

---

7. **Growing crops while minimising tillage increases greatly the capacity of the soil to absorb water.**
Strongly Disagree *Disagree* *Neutral* *Agree* *Strongly Agree*
Can you justify your answer?

---

8. **Intensive farming that compacts the ground and lowers the capacity of the soil to absorb water, remains the most profiTable way to exploit a cultivated area.**
Strongly Disagree *Disagree* *Neutral* *Agree* *Strongly Agree*
Can you justify your answer?

Appendix B

Questions

1- **What is the combination of measures that leads to the worst flood outcome? Also note the damage/cost and number of flood houses.**

2- **What is the combination of measures that leads to the best flood outcome? Also note the damage/cost and number of flood houses.**

3- **What is the combination of measures that leads to the worst financial outcome from both the infrastructure cost and the farming profitability?**

4- **What is the combination of measures that leads to the best financial outcome from both the infrastructure cost and the farming profitability?**

5- **Is there a satisfying solution that manages to minimise the flood and bring an acceptable financial outcome?**

Appendix C

Millbrook Serious Game Questionnaire 2

University of Exeter—Westcountry Rivers Trust

Privacy Statement

Please complete this questionnaire. Any concerns can be communicated to Mehdi Khoury (m.khoury@exeter.ac.uk) at University of Exeter. Thank you for your time and cooperation.

Answer the following questions by circling the most appropriate answer.

1. **A one in-a-hundred years rainfall event of 45 mm of water in one hour could cause serious flood problems in Millbrook.**
Strongly Disagree *Disagree* *Neutral* *Agree* *Strongly Agree*
How many residences could potentially be flooded? Any worst-case cost approximation?

---

2. **Improving the village sewer infrastructure that directly drains excess water could result in neutralising up to 90% of the flood resulting from such an event.**
Strongly Disagree *Disagree* *Neutral* *Agree* *Strongly Agree*
Can you justify your answer?

---

3. **Improving farming practices that change the ground cover of the cultivated areas around the village could result in neutralising up to 90% of the flood resulting from such an event.**
Strongly Disagree Disagree Neutral Agree Strongly Agree
Can you justify your answer?

4. **Investing money in the sewer infrastructure gives a better result for your money when fighting against flood than trying to support different farming practice.**
Strongly Disagree Disagree Neutral Agree Strongly Agree
Can you justify your answer?

5. **What are the two most important factors in this list of three that influence flooding when looking at the ground cover of cultivated areas?**
- *the amount of washed sediments that can block drainage pipes in the village downstream*
- *the capacity of the ground to absorb water*
- *the roughness of the terrain the slows the travelling speed of the water*

Can you justify your answer?

6. **Heavily grazed permanent pasture increases greatly the capacity of the soil to absorb water.**
Strongly Disagree Disagree Neutral Agree Strongly Agree
Can you justify your answer?

7. **Growing crops while minimising tillage increases greatly the capacity of the soil to absorb water.**
Strongly Disagree Disagree Neutral Agree Strongly Agree
Can you justify your answer?

8. **Intensive farming that compacts the ground and lowers the capacity of the soil to absorb water, remains the most profiTable way to exploit a cultivated area.**
Strongly Disagree Disagree Neutral Agree Strongly Agree
Can you justify your answer?

9. **Did you learn anything new today?**
Strongly Disagree Disagree Neutral Agree Strongly Agree
If so, what was news to you?

10. **Has this Serious Game changed any previous view you had on the Millbrook problem?**
Strongly Disagree Disagree Neutral Agree Strongly Agree
If so, what was the view that was changed?

11. **Would you recommend this type of Serious Game for other people involved in Millbrook flood problem?**
Strongly Disagree Disagree Neutral Agree Strongly Agree
If so, why?

Appendix D. Soil Science Review behind Chosen Farming Systems Values and Parameters in the Game

1: Overview of infiltration and overland flows across different farming practice scenarios

Soil infiltration rates are fundamentally affected by soil type and specifically the texture of the soil (i.e., the ratio of sand, silt and clay) [37]. Beyond this, there are several positive impactors that can contribute to improved infiltration rate and capacity, including: crop diversity, residue maintenance, cover cropping, managed traffic (minimizing area compacted by heavy vehicles), increased organic matter content, conservation agriculture, contour farming and increased earthworm content. Conversely, some practices and outcomes have negative effects on infiltration rate and capacity, including: surface crusting, low organic matter content, compaction, high soil moisture (waterlogging) and winter break rotations.

Infiltration rate is often estimated theoretically via such standards as the Richards equation [32] or the Green and Ampt [38] method. In these methods, the existing saturation of the soil and the porosity allow for such estimation, though they often assume a homogeneous soil. Empirically, infiltration rate data is often measured in isolation, or for simple scientific comparison between two system types (e.g., organic versus conventional); there is a paucity of data contrasting the variable farm practices across a region or watershed, measuring infiltration dynamics over time, and accounting for differing soil profile compositions. As such, to compile representative values, it is necessary to interrogate the literature for suitable data with which to set generalized values appropriate for a Serious Game.

The soils in the Millbrook watershed belong to Soils cape 6 [39] (a freely draining, slightly acid loam). These shale-derived sandy loam soils are of below average fertility, with neutral to acid pastureland dominating their use. Their carbon stocks are typically low and they are prone to both soil erosion and compaction.

In conventional tilled fields in the UK, infiltration rates on sandy loam soils typically vary between 15–45 mm/h, though compaction by agricultural machinery increases bulk density, reduces porosity and rapidly reduces infiltration rates to 3 mm/h [40–44]. With these values in mind, and considering the likely bare infiltration rates of Soils cape 6 soils (10–20 mm/h), it is considered unlikely that arable soils in the Millbrook watershed will have an infiltration rate greater than 30 mm/h, unless conservation agriculture techniques are employed. Similarly, for grazed pasture soils, infiltration rates typically fall between 3–36 mm/h [45,46], with a mean of 9 mm/h for permanent pasture, as low as 0.1 mm/h with heavy grazing pressure. As such, it is considered unlikely that permanent pasture in the Millbrook watershed, unless regular leys or mechanical improvement have been incorporated into the management strategy, will have an infiltration rate greater than 30 mm/h.

With the above summarized data in mind, the following values (mm/h) were chosen: red (heavily grazed permanent pasture) = 10, orange (tilled arable with standard rotation and traffic) = 15, yellow (young pasture with mechanical improvement and moderate grazing) = 30, aqua (tilled arable with diverse rotation and increased organic soil amendment) = 38.

Cover crops and crop residues increase infiltration rates by breaking the kinetic energy of rainfall, increasing residence time on the soil surface, and increasing the prevalence of root holes and thus heterogeneous porosity. Data on the effects of these vegetation types is sparse, though relevant literature suggests that cover crops and crop residues can increase infiltration rates up to at least 70 mm/h [42,47]. As such, a value for farming systems (blue) in which cover crops and minimum-tillage contribute to permanent cover was set at 70 mm/h. Beyond this, a wholesale shift to a no-till, conservation agriculture approach that built soil carbon, maintained soil cover, and minimized traffic could conceivably increase infiltration rates in a similar vain to a lightly, trafficked organic field, i.e., 6–10 times greater [40].

As such, for the purpose of this Serious Game, we have settled on 90 mm/h (6× that of conventional tillage) as the fastest infiltration rate possible for this farm system.

2: Sediment fluxes

The principal factors affecting soil erosion are: (a) the amount and intensity of rainfall and wind, (b) topography, (c) soil properties, and (d) natural and managed vegetation resulting from farming practice. Positive impactors, such as the presence of year-round ground cover, high organic matter contents and good aggregate stability, protect against soil erosion, whilst negative impactors, such as surface crusting, compaction, and winter break crop rotations, promote soil erosion [40]. The largest factors contributing to erosion were found in [48] to be crop cover and valley features, with erosive rainfall events typically at >10 mm d^{-1} and erosion dominating bare soil environments. Beyond 15% crop cover, erosion was more typically associated with tramlines and wheel tracks and exacerbated by landscape features. Cropland soils often erode at a higher rate than comparable grassland soils, due to their reduced ground cover, with [48] finding cropland erosion rates to be between 0.5–4× greater than grassland erosion rates. The median rate of soil erosion in the UK is approximately 1.3 tones/ha, with higher mean rates (>2.6 tones/ha) being associated with sandy or silty soils ([49] and references therein). Sediment losses vary substantially with crop type and agricultural management practice, though research does suggest that traffic and tillage both increase sediment losses [47,49].

As such, for the Serious Game, greater sediment loss rates were forecast from arable fields, with changes in management practice towards less intense, conservation agriculture approaches with greater ground cover yielding lower sediment losses. We present these as a percentage of the sediment loss needed to block the local drainage systems, from 10% to 90%. Note that at particular times, i.e., just after tillage in a bare arable field, the sediment loss has the potential to be both very substantial and enough to block 100% of the local drainage system.

3: Farm cost estimates

Table A1 presents a summary of the profitability for each of the three sub-regions of the Millbrook catchment, and estimated for a pre-Brexit scenario for the year 2016–2017 using the online calculator in DEFRA's Farm Business Survey 2016/17 [50]. Profitability for a given area is computed by multiplying either the high or the average net profitability per hectare—respectively 628.7 GBP/Hectare and 283.9 GBP/Hectare—by 85% of each region surface so as to take into account the fact that a small portion of the said "cultivated areas" around the village is not being exploited. The survey provided recent average business incomes for different types of farming.

Table A1. Profitability (in GBP) of each region in the catchment around Millbrook.

Profitability Class	Region 1 (75 Hectares)	Region 2 (123 Hectares)	Region 3 (101 Hectares)
high profit (++)	40,502	65,961	54,476
average profit (+)	18,289	29,785	24,599
null	0	0	0
average loss (−)	−18,289	−29,785	−24,599
high loss (−−)	−40,502	−65,961	−54,476

When defining profitability for the Serious Game, the fact that the average cereal farm profitability was more than double the average profitability of a livestock farm determined the basic profitability scores. The average mixed farm was somewhere in between, with 'general cropping' (i.e., diversifying the rotation) usually leading to improved profitability. These incomes are inclusive of single farm and agri-environment payments, with the majority of farms making a loss on agricultural income alone. Compounding factors, such as the potential for longer 'overwintering' periods in certain livestock farming approaches, the profitability of grass leys, reduced fuel costs in minimum tillage approaches, and the profitability of diversifying the rotation, were all considered in making subtle adjustments to the in-game profitability scores.

Appendix E. The Level of Investment in the Drainage Infrastructure

We have excluded lining (involving the installation of an internal resin layer that bonds to the inside of the pipe, essentially creating a pipe within a pipe), as data on this specific procedure were not readily available. The formula applied for estimating the management costs is confidential (South West Water). However, the distribution of pipe diameters by total length used to compute these costs is presented in Table A1. It should be pointed out that any pipes of unknown diameter were assigned an estimated diameter of 150 mm for the calculations.

Table A1. Total length of pipes under 225 mm, 225 to 450 mm, and larger than 450 mm.

Pipe diameter (mm)	100	125	150	225	300	350	380	400	450	600	unknown
Total Length (m)	312	817	4548	3224	2427	150	54	571	822	138	3328
Category	<225 mm			225–450 mm					≥450 mm		
Total Length (m)	5677			6426					960		3328

Appendix F. Computational Complexity of the Hydraulic Model and Compromises Made

A 1 m resolution cell size took approximately two hours to complete, while a 2 m resolution model took approximately 1 h. Reducing the resolution to 5 m still yield a processing time of approximately 1.5–5 min, and the 10 m resolution was completed in less than 10 s with a recent PC with a Nvidia GTX 1070 graphic card. A compromise of spatial quality over processing time is sought by selecting a resolution of 2.5 m, as the loss of precision between 2 m and 2.5 m is not really noticeable from a player point of view.

The simulation was run for twelve hours of simulation time to allow the flow to propagate after the rain events. The open boundary condition was set to allow flow to escape the terrain. The velocity between two neighboring cells with a water level difference less than 0.1 mm was assumed as zero to speed up the calculation for insignificant quantities of flows.

Different types of surfaces shown in in were associated with an appropriate infiltration rate and roughness as shown Table A1.

Table A1. Types of terrain surfaces with associated infiltration rate and roughness values.

Surface Type	**Forest**	**Cultivated Area**	**Urban**	**Urban Drainage**	**Road**
Infiltration rate (mm/h)	120 mm/h	*From 10 to 90 mm/h—see farming systems Table 2*	20.25 mm/h	*See Equations (1) and (2) in Section 2.1.1*	4.5 mm/h
Roughness (Manning coef)	0.75	*From 0.3 to 0.5 Manning coef—See farming systems Table 2*	0.25	0.25	0.05

The flood damage evaluation uses the residential sector average damage curve for short duration floods (less than 12 h) from the Multi-Colored Manual [51,52]. We simplify calculations as followed to compute damages from flood depth. Due to the average slope of the terrain and history of floods, house ground floors were slightly elevated. Assuming that the floor of each house starts 15 cm above the ground level, we subtract 15 cm from the flood depth to get the water depth measured inside the house to use it as a proxy for residential damage as plotted on the curve. Then we consider two possible levels of flood damage per building. Any flood depth between 0.3 m and 0.5 m will be considered inside a building as a 0.15 m water depth damage of 314 GBP/m^2 and any flood beyond 0.5 m will be considered inside a building as a 0.35 m water depth damage of 476 GBP/m^2. This simplification allows us to limit the complexity of the game visualization to two levels of flood damage while also the over-estimation that would result from not taking into account ground floor elevation. The data regarding the surface and location of each building were provided by The Westcountry Rivers Trust.

Appendix G. Detail of Wilcoxon Signed-Ranks One-Tailed Test computations

To prove that the Serious Game induces a directional change in the way participants confirm or reject each hypothesis, we calculate the level of significance as follows. We rank the differences for each pair, and affix a sign to each rank. From the sum of positive ranks W+ and the sum of negative ranks W− we compute the W test value such that W = min {W+, W−}. The number of instances where the subject's difference score is zero is then deducted from the sample size N. The critical value T at $p = 0.05$ and $p = 0.01$ for a one-tail test is then given by the test as provided in [31], and if $W \leq T$, the null hypothesis (being that that the median difference is zero) can be rejected. Details of the results are shown below for all seven hypotheses.

- For Question 1, the W-value is 29. The critical value of W for N = 15 at $p \leq 0.05$ is 30. Therefore, the result is significant at $p \leq 0.05$.
- For Question 2, the W-value is 16. The critical value of W for N = 18 at $p \leq 0.01$ is 32. Therefore, the result is significant at $p \leq 0.01$.
- For Question 3, the W-value is 31. The critical value of W for N = 18 at $p \leq 0.01$ is 32. Therefore, the result is significant at $p \leq 0.01$.
- For Question 4, the W-value is 6.5. The critical value of W for N = 17 at $p \leq 0.01$ is 27. Therefore, the result is significant at $p \leq 0.01$.
- For Question 6, the W-value is 16.5. The critical value of W for N = 19 at $p \leq 0.01$ is 37. Therefore, the result is significant at $p \leq 0.01$.
- For Question 7, the W-value is 0. The critical value of W for N = 11 at $p \leq 0.01$ is 7. Therefore, the result is significant at $p \leq 0.01$.
- For Question 8, the W-value is 12.5. The critical value of W for N = 13 at $p \leq 0.05$ is 21. Therefore, the result is significant at $p \leq 0.05$.

References

1. Alfieri, L.; Dottori, F.; Betts, R.; Salamon, P.; Feyen, L. Multi-Model Projections of River Flood Risk in Europe under Global Warming. *Climate* **2018**, *6*, 6. [CrossRef]
2. Kerbs, J.; Hall, J.; Johnson, A. *UK Climate Change Risk Assessment 2017-Synthesis Report: Priorities for the next Five Years*; UK Committee on Climate Change: London, UK, 2016.
3. UK Environment Agency. *Flood and Coastal Erosion Risk Management Long-term Investment Scenarios (LTIS) 2014*; UK Environment Agency: Bristol, UK, 2014.
4. Palmer, R.N.; Cardwell, H.E.; Lorie, M.A.; Werick, W. Disciplined planning, structured participation, and collaborative modeling—Applying shared vision planning to water resources. *J. Am. Water Resour. Assoc.* **2013**, *49*, 614–628. [CrossRef]
5. Savic, D.A.; Morley, M.S.; Khoury, M. Serious Gaming for water system planning and management. *Water* **2016**, *8*, 456. [CrossRef]
6. Aubert, A.H.; Bauer, R.; Lienert, J. A review of water-related serious games to specify use in environmental Multi-Criteria Decision Analysis. *Environ. Modell. Softw.* **2018**, *105*, 64–78. [CrossRef]
7. Rebolledo-Mendez, G.; Avramides, K.; De Freitas, S.; Memarzia, K. Societal impact of a serious game on raising public awareness: The case of FloodSim. In Proceedings of the 2009 ACM SIGGRAPH Symposium on Video Games, New Orleans, LA, USA, 4–6 August 2009; pp. 15–22.
8. Aqua Republica. Available online: http://aquarepublica.com/ (accessed on 10 October 2016).
9. Meera, P.; McLain, M.L.; Bijlani, K.; Jayakrishnan, R.; Rao, B.R. Serious game on flood risk management. In *Emerging Research in Computing, Information, Communication and Applications*; Springer: New York, NY, USA, 2016; pp. 197–206.
10. Douven, W.; Mul, M.L.; Son, L.; Bakker, N.; Radosevich, G.; Hendriks, A. Games to create awareness and design policies for transboundary cooperation in river basins: Lessons from the shariva game of the Mekong river commission. *Water Resour. Manag.* **2014**, *28*, 1431–1447. [CrossRef]

11. Valkering, P.; van der Brugge, R.; Offermans, A.; Haasnoot, M.; Vreugdenhil, H. A perspective-based simulation game to explore future pathways of a water-society system under climate change. *Simul. Gaming* **2013**, *44*, 366–390. [CrossRef]
12. Van der Wal, M.M.; de Kraker, J.; Kroeze, C.; Kirschner, P.A.; Valkering, P. Can computer models be used for social learning? A Serious Game in water management. *Environ. Model. Softw.* **2016**, *75*, 119–132. [CrossRef]
13. Haasnoot, M.; Middelkoop, H.; van Beek, E.; van Deursen, W.P.A. A method to develop sustainable water management strategies for an uncertain future. *Sustain. Dev.* **2011**, *19*, 369–381. [CrossRef]
14. Stefanska, J.; Magnuszewski, P.; Sendzimir, J.; Romaniuk, P.; Taillieu, T.; Dubel, A.; Flachner, Z.; Balogh, P. A gaming exercise to explore problem-solving versus relational activities for river floodplain management. *Environ. Policy Gov.* **2011**, *21*, 454–471. [CrossRef]
15. Seibert, J.; Vis, M.J.P. Irrigania—A web-based game about sharing water resources. *Hydrol. Earth Syst. Sci.* **2012**, *16*, 2523–2530. [CrossRef]
16. Bassi, A.M.; de Rego, F.; Harrisson, J.; Lombardi, N. WATERSTORY ILE: A systemic approach to solve a long-lasting and far-reaching problem. *Simul. Gaming* **2015**, *46*, 404–429. [CrossRef]
17. De Kleermaeker, S.; Zijderveld, A.; Thonus, B. Training for crisis response with serious games based on early warning systems. In Proceedings of the 8th International ISCRAM Conference, Lisbon, Portugal, 8–11 May 2011.
18. De Kleermaeker, S.; Arentz, L. Serious gaming in training for crisis response. In Proceedings of the 9th International ISCRAM Conference, Vancouver, BC, Canada, 22–25 April 2012.
19. Leskens, J.; Kehl, C.; Tutenel, T.; Kol, T.; Haan, G.; Stelling, G.; Eisemann, E. An interactive simulation and visualization tool for flood analysis usable for practitioners. *Mitig. Adapt. Strateg. Glob. Chang.* **2015**, *22*, 307–324. [CrossRef] [PubMed]
20. Kemec, S.; Zlatanova, S.; Duzgun, H. A Framework for Defining a 3D Model in Support of Risk Management. *Geogr. Inf. Cartogr. Risk Cris. Manag.* **2010**, 69–82. [CrossRef]
21. Mioc, D.; Anton, F.; Nickerson, B.; Santos, M.; Adda, P.; Tienaah, T.; Ahmad, A.; Mezouagh, M.; MacGillivray, M. Flood progression modelling and impact analysis. 2011. Available online: http://orbit.dtu.dk/files/51177341/InTech1.pdf (accessed on 13 June 2018).
22. Swiss Federal Geoportal. Available online: https://map.geo.admin.ch (accessed on 14 June 2018).
23. Hunnicutt, B.K. Leisure and play in Plato's teaching and philosophy of learning. *Leisure Sci.* **1990**, *12*, 211–227. [CrossRef]
24. "Unity®" (or Unity3d) Software. Available online: https://unity3d.com (accessed on 11 March 2018).
25. Qureshi, A.; Rizvi, F.; Syed, A.; Shahid, A.; Manzoor, H. The method of loci as a mnemonic device to facilitate learning in endocrinology leads to improvement in student performance as measured by assessments. *Adv. Physiol. Educ.* **2014**, *38*, 140–144. [CrossRef] [PubMed]
26. Khoury, M.; Chen, A.; Gibson, M.; Vamvakeridou-Lyroudia, L.; Stewart, D.; Wood, M.; Savic, D.; Djordjevic, S. A Serious Game to explore different flooding scenarios and their respective effects on infrastructures. In Proceedings of the 13th International Conference on Hydroinformatics, Palermo, Italy, 1–6 July 2018.
27. Lammers, K. *Unity Shaders and Effects Cookbook*; Packt Publishing: Birmingham, UK, 2013.
28. Guidolin, M.; Chen, A.S.; Ghimire, B.; Keedwell, E.C.; Djordjević, S.; Savić, D.A. A weighted cellular automata 2D inundation model for rapid flood analysis. *Environ. Modell. Softw.* **2016**, *84*, 378–394. [CrossRef]
29. CADDIES Framework University of Exeter. 2016. Available online: emps.exeter.ac.uk/engineering/research/cws/resources/caddies-framework/ (accessed on 25 August 2018).
30. Gibson, M.; Savic, D.; Djordjevic, S.; Chen, A.; Fraser, S.; Watson, T. Accuracy and computational efficiency of 2D urban surface flood modelling based on cellular automata. *Procedia Eng.* **2016**, *154*, 801–810. [CrossRef]
31. Online Calculator Used for the Wilcoxon Signed-Ranks One-Tailed Test: "Social Science Statistics". Available online: https://www.socscistatistics.com/tests/signedranks/Default2.aspx (accessed on 25 August 2018).
32. Lankford, B.; Watson, D. Metaphor in natural resource gaming: Insights from the river basin game. *Simul. Gaming* **2007**, *38*, 421–442. [CrossRef]
33. Van Huijstee, M.; Francken, M.; Leroy, P. Partnerships for sustainable development: A review of current literature. *Environ. Sci.* **2007**, *4*, 75–89. [CrossRef]
34. Harteveld, C. *Triadic Game Design: Balancing Reality, Meaning and Play*; Springer: London, UK, 2011.

35. Sušnik, J.; Chew, C.; Domingo, X.; Mereu, S.; Trabucco, A.; Evans, B.; Vamvakeridou-Lyroudia, L.; Savić, D.A.; Laspidou, C.; Brouwer, F. Multi-Stakeholder Development of a Serious Game to Explore the Water-Energy-Food-Land-Climate Nexus: The SIM4NEXUS Approach. *Water* **2018**, *10*, 139. [CrossRef]
36. Foth, H.D. *Fundamentals of Soil Science*, 8th ed.; Wiley: New York, NY, USA, 1991; ISBN 978-0-471-52279-9.
37. Richards, L.A. Capillary conduction of liquids through porous mediums. *Physics* **1931**, *1*, 318–333. [CrossRef]
38. Green, W.H.; Ampt, G. Studies of soil physics. *J. Agric. Sci.* **1911**, *4*, 1–24. [CrossRef]
39. Soilscape Soils Types Map. Available online: http://www.landis.org.uk/soilscapes/index.cfm (accessed on 8 September 2018).
40. Kooistra, M.J.; Bouma, J.; Boersma, O.H.; Jager, A. Physical and morphological characterization of undisturbed and disturbed ploughpans in a sandy loam soil. *Soil Tillage Res.* **1984**, *4*, 405–417. [CrossRef]
41. Meek, B.D.; Rechel, E.R.; Carter, L.M.; DeTar, W.R.; Urie, A.L. Infiltration rate of a sandy loam soil: Effects of traffic, tillage, and plant roots. *Soil Sci. Soc. Am. J.* **1992**, *56*, 908–913. [CrossRef]
42. Stobart, R.M.; Morris, N.L. The impact of cover crops on yield and soils in the New Farming Systems programme. *Asp. Appl. Biol.* **2014**, *127*, 223–232.
43. Chyba, J.; Kroulik, M.; Kristof, K.; Misiewicz, P.A. The influence of agricultural traffic on soil infiltration rates. *Agron. Res.* **2017**, *15*, 664–673.
44. Fullen, M.A. Compaction, hydrological processes and soil erosion on loamy sands in east Shropshire, England. *Soil Tillage Res.* **1985**, *6*, 17–29. [CrossRef]
45. Hathaway-Jenkins, L.J.; Sakrabani, R.; Pearce, B.; Whitmore, A.P.; Godwin, R.J. A comparison of soil and water properties in organic and conventional farming systems in England. *Soil Use Manag.* **2011**, *27*, 133–142. [CrossRef]
46. Slaymaker, O.; Spencer, T.; Embleton-Hamann, C. *Geomorphology and Global Environmental Change*; Cambridge University Press: New York, NY, USA, 2009.
47. Wang, X.; Gao, H.; Tullberg, J.N.; Li, H.; Kuhn, N.; McHugh, A.D.; Li, Y. Traffic and tillage effects on runoff and soil loss on the Loess Plateau of northern China. *Soil Res.* **2008**, *46*, 667–675. [CrossRef]
48. Evans, R.; Collins, A.L.; Foster, I.D.; Rickson, R.J.; Anthony, S.G.; Brewer, T.; Deeks, L.; Newell-Price, J.P.; Truckell, I.G.; Zhang, Y. Extent, frequency and rate of water erosion of arable land in Britain–benefits and challenges for modelling. *Soil Use Manag.* **2016**, *32*, 149–161. [CrossRef]
49. Rohde, K.; Yule, D. Soil compaction and controlled traffic farming research in Central Queensland. *ISTRO* **2003**, *16*, 1020–1027.
50. Online Farming Profitability Indicators: "Farm Business Survey". Available online: http://www.farmbusinesssurvey.co.uk/benchmarking/Default.aspx?Module=ProfitResults_ (accessed on 25 August 2017).
51. Penning-Rowsell, E.; Priest, S.; Parker, D.; Morris, J.; Tunstall, S.; Viavattene, C.; Chatterton, J.; Owen, D. *Flood and Coastal Erosion Risk Management: A Manual for Economic Appraisal*; Routledge: London, UK, 2013; ISBN 9780415815154.
52. Multi-Colored_Manual Online, Middlesex University. May 2018. Available online: https://www.mcm-online.co.uk/ (accessed on 20 August 2018).

water

MDPI

Article

Serious Games as Planning Support Systems: Learning from Playing Maritime Spatial Planning Challenge 2050

Steven Jean [1,*], Laura Gilbert [2,*], Wietske Medema [1], Xander Keijser [3,4], Igor Mayer [5], Azhar Inam [1] and Jan Adamowski [1]

1 Department of Bioresource Engineering, McGill University, 21 111 Lakeshore, Ste Anne de Bellevue, QC H9X3V9, Canada; wietske.medema@mcgill.ca (W.M.); muhammad.baig@mcgill.ca (A.I.); jan.adamowski@mcgill.ca (J.A.)

2 Department of Natural Resource Sciences, McGill University, 21 111 Lakeshore, Ste Anne de Bellevue, QC H9X3V9, Canada

3 Environmental Policy Group, Wageningen University, Hollandseweg 1, 6706 KN Wageningen, The Netherlands; xander.keijser@rws.nl

4 Rijkswaterstaat, Zuiderwagenplein 2, Lelystad, P.O. Box 2232, 3500 GE Utrecht, The Netherlands

5 Academy for Digital Entertainment, Breda University of Applied Sciences, Monseigneur Hopmansstraat 1, 4817 JT Breda, The Netherlands; i.s.mayer@hotmail.com

* Correspondence: steven.jean@mail.mcgill.ca (S.J.); laura.gilbert2@mail.mcgill.ca (L.G.)

Received: 14 August 2018; Accepted: 30 November 2018; Published: 5 December 2018

Abstract: The inherent complexity of planning at sea, called maritime spatial planning (MSP), requires a planning approach where science (data and evidence) and stakeholders (their engagement and involvement) are integrated throughout the planning process. An increasing number of innovative planning support systems (PSS) in terrestrial planning incorporate scientific models and data into multi-player digital game platforms with an element of role-play. However, maritime PSS are still early in their innovation curve, and the use and usefulness of existing tools still needs to be demonstrated. Therefore, the authors investigate the serious game, MSP Challenge 2050, for its potential use as an innovative maritime PSS and present the results of three case studies on participant learning in sessions of game events held in Newfoundland, Venice, and Copenhagen. This paper focusses on the added values of MSP Challenge 2050, specifically at the individual, group, and outcome levels, through the promotion of the knowledge co-creation cycle. During the three game events, data was collected through participant surveys. Additionally, participants of the Newfoundland event were audiovisually recorded to perform an interaction analysis. Results from survey answers and the interaction analysis provide evidence that MSP Challenge 2050 succeeds at the promotion of group and individual learning by translating complex information to players and creating a forum wherein participants can share their thoughts and perspectives all the while (co-) creating new types of knowledge. Overall, MSP Challenge and serious games in general represent promising tools that can be used to facilitate the MSP process.

Keywords: serious games; planning support systems; knowledge co-creation; sustainability; maritime spatial planning

---

1. Introduction

Human activities are increasingly affecting ocean spaces across the globe [1,2]. Traditionally used for fishing and transportation, today's oceans are a hub of human activities that are constantly diversifying [1–5]. Although many ocean activities occur in the same space, these activities are traditionally managed separately, leading to a fragmented system of ocean management. Fragmented management fails to address or resolve user conflicts and trade-offs, leaving oceans susceptible to the cumulative effects of these multiple uses [3]. Given the increased demand for maritime goods and resources, the main obstacle for human activities at sea is competition for maritime space [6]. Maritime spaces are prime examples of socio-technical complexity; i.e. complexity stemming from the overlap between the natural-technical-physical realm and the socio-political world [2]. Addressing this complexity requires new methods of ocean management that coordinate the different sectors to maintain ocean health. Maritime spatial planning (MSP) refers to the inherent complexity of planning at sea and requires balancing both social and economic objectives with ecological conservation by properly allocating human ocean activities spatially and temporally [3]. Tools are required to manage the complexity and to pre-empt conflicts that may arise throughout the planning and decision-making stages. Not all conflicts can be fully resolved, but tools that allow planners to identify and predict areas of conflict are useful to the planning process [7].

The term, MSP, first appeared on the European Union legal stage in July 2014 with the MSP Directive [6], and, although it borrows many of the same methods and applications as terrestrial spatial planning, it lags in some respects [8]. Terrestrial planning has a long history of planning support systems (PSS) that MSP lacks. PSS are tools and technologies that support the planning process [9]. For the purposes of this research, serious games (SG) will be isolated as a potential PSS for use in MSP. First coined in 1989 by Harris [10], PSS are informational frameworks that integrate the information technologies for planning and (ideally) provide interactive, integrated, and participatory procedures. According to Kuller et al. [11], these systems are meant to (1) provide users with a deeper understanding of the task, and (2) the formulation and communication of ideas, values, and preferences of stakeholders. Modern PSS help synthesize information and models for users. Information is transferred to users via a user-friendly interface that simplifies the technological complexities operating behind the scenes, allowing for a simplistic representation of complex spatial data [12,13]. PSS focus on the planning process as a whole and although they may facilitate decision-making, this is not their primary purpose [12].

PSS have been developing rapidly due to advances in technology that allow for portability and rapid computation [14]. It is important that the focus of a PSS is not the technology it uses, but rather the problems it was meant to solve. This ensures the system remains user-friendly and not overly complex [15]. An effective PSS must emphasize communication and collaboration in order to support collective design, a form of interaction and communication that seeks to achieve collective goals while dealing with common concerns [14]. An effective PSS allows for what Pelzer, Geertman, van der Heijden, and Rouwette [14] refer to as added value. This added value can be divided into three levels: (1) The individual level, (2) the group level, and (3) the outcome level. At the individual level, PSS have the added value of teaching the users about specific issues [14,16,17]. Pelzer, Geertman, van der Heijden, and Rouwette [14] classify this individual learning as two-fold: (1) Learning about the object of planning, and (2) learning about other stakeholders' perspectives and views [14]. At the group level, added value takes the form of collaboration, communication, consensus building, and efficiency [14,18]. These group level outcomes seek to bring users and stakeholders together by fostering environments that facilitate communication. This allows for users to hone their collaboration skills, facilitate consensus, and therefore increase the efficiency of the planning process [14,18]. The outcome level, Pelzer, Geertman, van der Heijden, and Rouwette [14] explain, is much more difficult to gauge, though it generally refers to whether or not a PSS improves the validity of the plans created.

Looking at these individual and group learning goals, a link to SG can be made. SG have long been used for planning in the healthcare, defense, and education sectors [19]. They are a subset of game-based learning, a field of research that uses gamification principles to help players achieve a specific goal [20]. These games can take the form of either board games or computer games; the format depends, among other factors, on the goal of the game, the complexity of the issue being modeled, and the resources available to the target audience [21,22]. Given the multiple uses of SG, their potential to act as PSS must be investigated.

1.1. Knowledge Co-creation and Stakeholder Engagement

PSS place a focus on communication and collaboration between users. Although certain SG can stimulate competition between players, the trend in SG for natural resource management is to promote collaboration and communication, a goal similar to that of a PSS [23]. This focus on collaboration is based on a constructivist view of learning. Constructivist learning theories view learning as both a process and a by-product [24]. The process of learning involves acquiring information and assimilating it while potentially resulting in changes to one's values [24]. The by-product of learning constitutes a change in the subject's knowledge, skills, values, or worldview [24,25]. The constructivist view emphasizes the importance of one's past experiences and how those experiences can be used to inform others [15]. As such, the constructivist view places an emphasis on both tacit and explicit knowledge, concepts developed by Polanyi [26]. Explicit knowledge is readily transferred in formal and systematic ways (e.g., knowledge of certain legislation) [27,28], while tacit knowledge is more experiential and personal, taking the form of intuitions and hunches [29] (e.g., ease of use with computers/technology). Tacit knowledge is less transferable than explicit knowledge, yet a significant portion of knowledge comes in the tacit form [29,30]. Tacit and explicit knowledge, however, are not mutually exclusive, rather they are mutually complementary and intermingle with each other through interactions between diverse individuals and groups [30–34].

These concepts link to the knowledge co-creation cycle developed by Nonaka et al. [35]. Knowledge co-creation refers to an institutional mechanism that enables learning and collaboration within a governance setting. The four stages of the knowledge co-creation process are described as follows:

- Socialization: From tacit to tacit knowledge, which involves the sharing and transferring of tacit knowledge between individuals and groups through physical proximity and direct interactions;
- externalization: From tacit to explicit knowledge, which requires tacit knowledge to be articulated and translated into comprehensible forms that can be understood by others;
- combination: From explicit to more complex sets of explicit knowledge, which requires communication and diffusion processes and the systematization of knowledge; and
- internalization: From explicit to tacit knowledge.

Preliminary research [36] involving the SG AquaRepublica has shown that SG have the potential to stimulate the four stages of the knowledge co-creation process, and Barreteau, Le Page, and Perez [23] add that SG and gaming simulations offer a unique opportunity to facilitate decision-making processes. One type of SG used in environmental management combines computer simulations powered by environmental models with role-play as a method to address the complexity inherent to environmental problems, while promoting collaboration and learning among stakeholders [37–39]. By gamifying the experience, stakeholders are given the space to exchange perspectives, knowledge, and ideas in a low-stakes environment, allowing them to better understand the boundaries that separate them. This allows for the opportunity to co-create new forms of knowledge that can help mitigate these boundaries. Many different types of boundaries can exist between stakeholder groups, for example, these boundaries can be administrative, jurisdictional, socio-cultural, and cognitive. Boundaries often carry a negative connotation denoting a divide or a separation that can lead to discontinuity or inaction [36,37,40]. However, the discontinuities created by boundaries can trigger learning by a process of collaborative reframing between stakeholders, resulting in the creation of new and different types of knowledge [16,22]. In this way, boundaries can be viewed as connecting rather than separating. By crossing boundaries, a variety of learning mechanisms may occur [41]: (1) Identification, referring to the understanding of how different practices on the boundary relate to each other; (2) coordination, referring to learning about how to work on the boundary with others; (3) reflection, referring to the expansion of perspectives gleamed from working along the boundary; and (4) transformation, referring to the co-development of new knowledge and practices. This last transformative mechanism refers back to knowledge co-creation and the constructivist view of learning that states that learning can simply be a by-product of experience [24].

An important application of SG is their use as a tool to facilitate the socialization process among players. This process is important to the phenomena of boundary crossing by fostering relationships between participants and allowing for the sharing and creation of new types of tacit and explicit knowledge [41]. Although the game artefacts themselves generally provide some information on the planning and management processes, players generally learn from each other through the exchange of information that occurs throughout gameplay [42]. The game environment lowers the pressure associated with planning and decision-making and allows the players to discuss different management strategies, which promotes externalization of knowledge [42]. Additionally, Van Bilsen, Bekebrede, and Mayer [42] found that as game-play progresses, trust builds between participants, which helps them articulate a common understanding of issues. One of the assumptions of game-based learning is that the learning that occurs via gameplay can sometimes be transferred to similar situations outside of the game [39]. To ensure that participants understand how they can use what they have learned during gameplay in a real-world setting, a debriefing session following a game event is particularly important [43,44]. The role of the facilitator during the debriefing allows for the combination of tacit and explicit knowledge through the communication of lessons learnt by certain players to the rest of the group [36,43].

1.2. Research Objective and Questions

For this study, the SG MSP Challenge 2050 was used. MSP Challenge 2050 uses ecosystem models to simulate the spatial dimension of MSP and adds a game layer atop the embedded models. This game layer allows for a greater focus on communication and collaboration, which has been isolated as lacking with current PSS that focus more on the technology behind them rather than the outcomes [14]. Ultimately, this paper seeks to investigate whether MSP Challenge 2050 fosters the added values of PSS as defined by Pelzer, Geertman, van der Heijden, and Rouwette [14]. The following research questions focus on each of the three levels of added value. This objective will be achieved by addressing the following research questions:

1. Individual level: Does MSP Challenge 2050 offer a platform for participants to learn about MSP by helping them understand information derived from data, analyses, and models?
2. Group level: Does playing MSP Challenge 2050 promote quality interactions and cooperation between participants facilitating the knowledge co-creation cycle? and
3. Outcome level: What are the characteristics of the plans developed while playing the game and how do the plans differ from team-to-team?

Research question 1 will determine whether MSP Challenge 2050 fulfills one of the main criteria of a PSS by determining whether participants are able to more easily understand the complex information that the game is transmitting to participants via the user-interface. This first question also helps to determine whether or not MSP Challenge 2050 promotes individual added value referring to explicit knowledge transfer to an individual either via the game artefact itself or through team discussions. Research question 2 will explore the group added value by taking a critical look at the interactions of players throughout the planning stage of MSP Challenge 2050. Research question 3 will explore the outcome level added value by analyzing the plans produced by the teams playing the game and how those relate to the interactions investigated in question 2. Together, the three questions will help to determine if MSP Challenge 2050 can provide an effective PSS.

2. Materials and Methods

2.1. MSP Challenge 2050

This research used MSP Challenge 2050, which was developed in 2013 in the Netherlands by the Ministry of Infrastructure and Water Management and its executive agency called Rijkswaterstaat [45]. It follows the previous version of the game, MSP Challenge 2011, and introduces more complex data sets from the North Sea using the Ecopath food chain model to calculate the cumulative impacts of players' decisions [45]. It was developed to mimic four concepts inherent to MSP: (1) Vague system boundaries, (2) ambiguity, (3) competing interests, and (4) uncertainty caused by missing information [2]. The following section will provide a brief overview of the game and how it is played. For more information on the development and research behind the game, please refer to Mayer et al. [45], and visit www.mspchallenge.info.

MSP Challenge 2050 transports players to the Sea of Colours, a fictional sea based on the North Sea, bordered by six countries: Green, Indigo, Orange, Purple, Red, and Yellow [2]. Anywhere from 18 to 40 people can play the game at once and a game can last from four to 40 hours depending on the number of people per team and the availability of the players. The game requires at least two facilitators: A technical facilitator that helps with the computer set up and a lead facilitator that serves as a game master. The lead facilitator divides players into groups (countries) of three to eight players; teams are all in the same room so as to facilitate inter-team discussions [2]. Each team member selects a specific role within their country (such as an environmental planner, military, oil and gas, etc.) based on cue-cards given to each team before the game starts [2]. Each team is given a minimum of one computer and is provided with written instructions, as well as goals specific to their country and roles (such as developing the oil and gas industry, focusing on fisheries' protection, etc.). Examples of the games materials can be found in Figure 1. Given this information, teams start the game by planning objectives according to their country's vision for the year, 2050 [45]. The goal of the game is for each team to develop a national maritime spatial plan and for the teams to work together on an international level to coordinate their objectives and plans. The planning phase ends once the teams present their objectives and preliminary national plan to the game overall director (G.O.D.), who provides them with feedback and adds country specific challenges. Players from different teams with similar roles must then interact with each other and work together to coordinate their plans and objectives on an international level for the Sea of Colours [45]. The game ideally ends when the clock reaches the year, 2050, although, due to time constraints, the lead facilitator may choose to end the game early. There are no clear winners or losers of the game, rather the focus is on the overall MSP process.

The geospatial data used in the game is based on real data from the North Sea [2,46]. The information is displayed on 55 different layers with information on commercial fishing areas, marine protected areas, pipelines, etc. Teams can choose to display the information on layers or hide these layers when the information is superfluous. When a team is building a plan, for example, to expand a wind farm, they draw the new items on a proposition layer. The teams' computers are connected by a local area network (LAN), and a graphical user interface (GUI) in the game allows the teams to publish their proposition layers for other teams to comment on and either approve or reject their plans and related projects [45]. Once the plans are implemented, the game computes the cumulative effects of the plans and the GUI shows their evolution on screen. This helps players understand the consequences of their decisions, not only on their own country, but on other countries as well [2,45]. The game does not keep a record of decisions made by the teams at each turn. An added challenge during the game is that in-game time passes slowly at first, but accelerates as the game progresses [45]. Given the time constraints, the teams cannot always develop a comprehensive national plan, but they can still start the process of international coordination. By the end of the game, each team can visualize the cumulative effects of their decisions on the GUI.

Figure 1. Game materials (adapted from [46]).

2.2. Game Events

A workshop using MSP Challenge 2050 was held in Newfoundland in late spring 2017 at Memorial University in St-John's. The event was organized as a collaboration between McGill University, Memorial University, and the MSP Challenge 2050 creators. The event was to act as a conclusion to the Memorial University MSP master's program, with invitations being extended to a group of McGill students with expertise in integrated water resource management and SG. A group of local stakeholders working in MSP in Newfoundland was also invited to participate. A total of 18 participants played at the event: Nine students from Memorial University, five local stakeholders, two postdoctoral researchers, and two graduate students from McGill University.

The event took place over two days and consisted of playing an MSP board game that served as an icebreaker during the morning of the first day followed by a day and a half of playing the MSP Challenge 2050 simulation. The moderators for the Newfoundland event consisted of the two facilitators (lead and programmer) and an expert in MSP playing the G.O.D. character. At the start of the simulation game, the lead facilitator gave players a brief explanation on how to navigate the game. The teams were then given an hour to get to know each other and to familiarize themselves with the software by clicking through the various layers and menus in the game with their teams while facilitators circulated the room to answer any questions that would arise. On day two, the game began. The simulation ended with a short lecture and a debriefing session. Teams with three players were provided with one computer, while teams with four players were given two computers. The breakdown of the teams can be seen in Table 1 below:

Table 1. Team composition and coding.

Teams	Team Code	Team Members	Member Codes
Indigo	IND	3	IND1, IND2, IND3
Purple	PUR	4	PUR1, PUR2, PUR3, PUR4
Orange	OR	4	OR1, OR2, OR3, OR4
Red	RED	3	RED1, RED2, RED3
Yellow	YEL	4	YEL1, YEL2, YEL3, YEL4

Events run by a different research team took place in Copenhagen and Venice. Supplementary survey data from these events were used to corroborate any trends discovered in the data from the Newfoundland event. The Venice event was organized in 2017 for students from the EU Erasmus Mundus Master's program in MSP at the Universitá luav di Venezia. The event was attended by 15 students with a variety of professional backgrounds from across Europe, most having less than two years of professional experience working in MSP. The Copenhagen event was organized as a kick-off event for the NorthSEE and BalticLines partnership meeting. A total of 34 professionals were in attendance from across the North Sea and Baltic Sea countries. Twenty-six of the participants recorded working in the field of MSP for at least one year; nine of which reported practicing MSP for over five years. From all three events, the Copenhagen event involved the most experienced MSP professionals. Both these additional events were one day long and had a facilitator team consisting of one lead facilitator, one programmer for technical support, and one facilitator playing the role of G.O.D. The Copenhagen cohort was the largest and therefore was divided into six teams, unlike the Newfoundland and Venice events where participants were divided into five teams. For the Venice event, teams ranged from three to four players, while the Copenhagen event had teams of three to five players.

2.3. Data Collection and Analysis Methods

Participants were asked to complete three short surveys by ranking their experiences on a Likert scale and by answering short answer questions. The Likert scale involves an incremental scale that was customized to the specific questions in the survey. A pre-game survey was given to participants from all three game events to gather personal and professional information. This survey gauged their knowledge of MSP and measured their willingness to try a novel tool, such as SG. During the Newfoundland event, a mid-game survey was completed by each team at the end of the planning phase. This survey gathered information on the players' experience using the game and their relationship with other teams. Finally, the post-game survey was given to all participants from the three game events and focused on their impression of the game, the power dynamics within their team, and the usefulness of SG as a learning and policy tool. All surveys were adapted from an earlier study by Zhou [47]. A Mann-Whitney U test was performed to determine the statistical difference between scores from different events and between participants with different levels of expertise in MSP. The Mann-Whitney U test was chosen because the responses from the surveys were not normally distributed [48].

During the Newfoundland event, exclusively, teams were audiovisually recorded to later perform an interaction analysis. The interaction analysis was performed based on methodology developed by Jordan and Henderson [49]. Both verbal and non-verbal interactions between players were studied and analyzed [49]. Also, the participants' interactions with the game artefacts were analyzed. For example, Jordan and Henderson [49] found that in a group setting, passing the mouse between participants is a collaborative problem-solving technique. The interaction analysis was performed on the planning phase of the game, which corresponded to the time before teams met with G.O.D. Once all teams had met with G.O.D., the implementation phase began. It should be noted that interaction analysis was not performed after the planning stage due to the increased movement of players often out of bounds of the audiovisual equipment. Following the planning phase, the teams started to disperse and talk amongst themselves. Although a previous study by Mayer, Zhou, Lo, Abspoel, Keijser, Olsen, Nixon, and Kannen [2] used survey data to research the quality of interactions and cooperation between players of MSP Challenge 2050, this study is the first to use audiovisual recordings to conduct an interaction analysis to count and qualitatively analyze the interactions between players throughout the game. Interactions and cooperation are important because they relate to knowledge co-creation through socialization and externalization. By putting the players in close proximity in a low-stakes environment, the likelihood of positive interactions and cooperation increases [50]. These interactions also simulate the socio-part of the socio-technical complexity of MSP [49].

During the planning phase, teams reflected on their goals and developed a strategy to achieve them, therefore, the interaction analysis of this phase provides insight into the potential of MSP Challenge 2050 as a PSS by providing information on how the teams analyzed the data and used the models in the game to create their plans. Subsequently, an interaction analysis was conducted on the debriefing session at the end of the game, providing insight into lessons learnt by participants. Interaction analysis was conducted by two researchers from the team to remove any biases and gain a deeper understanding of interactions. This approach provides deeper insight into player interactions facilitated through the SG event than the surveys and helps corroborate the answers given by the participants. The answers from the surveys and the interaction analysis can be found in the Supplementary Material provided with this article.

The first research question will be answered using data from the interaction analysis of the Newfoundland event, the post-game survey from all three events, and the survey given to the participants during the game at the Newfoundland event. To answer the second research question and to assess if the game indeed promotes cooperation and interactions between players, the number of interactions between team members at the Newfoundland event were counted. The interaction analysis of the debriefing session from the Newfoundland event contained important information and insights that helped answer this question. To quantify some of the findings of the interaction analysis of the planning phase, researchers performed an analysis of the quality of participant interactions. Survey answers from the mid-game survey at the Newfoundland event and the post-game survey from all three events were also used to answer this question. To answer the third question, the quality of the plans presented to G.O.D. by each team will be discussed. Their quality will be correlated to the number and quality of interactions presented in question 2. Together all three questions help determine if MSP Challenge 2050 provides added value as a PSS.

3. Results

3.1. Research Question 1: Does MSP Challenge 2050 Offer a Platform for Participants to Learn about MSP by Helping Them Understand Information Derived from Data, Analyses, and Models? (i.e., Individual Added Value)

A small sample of the different interactions between participants during the game can be found in Table 2 below. It identifies how these interactions show that the game helped participants' understanding of data, analysis, and models, and how they relate to the knowledge co-creation cycle. There were three main types of interactions recorded that totaled 102 interactions among the teams. The first involved players not understanding a certain word or concept presented in the game and asking their teammates for additional information (see, for example, 0:03:32, 0:18:00, and 0:21:40). It represented close to a quarter of interactions recorded (24.5%). This is externalization of knowledge that helps the participants understand the data and the model in the game. Another type of interaction observed involved players helping each other understand features of the game (see 0:09:30, 0:11:09, 1:24:40, and 1:27:40). These types of interactions represented the bulk (51%) of the interactions between players. These interactions helped the players understand the model in the game and corresponds to socialization, because they would not have had access to this knowledge had they not been experiencing the game together. The last major type of interaction recorded happened when teams extrapolated beyond the scope of the game to reflect on how their plans should be implemented in a real-world setting (see 0:07:30, 0:39:40, 0:45:00, and 1:00:10). These interactions are examples of analysis within the teams and the combination of knowledge provided by the game to apply it to more complex and realistic situations. This last type of interaction accounted for close to a quarter of interactions (24.5%) and mostly occurred when the teams discussed how to present their plan to the G.O.D.

Table 2. Sample of interactions between participants and how these interactions help them understand data, analysis, and models and the corresponding knowledge co-creation stage.

Time	Description of Interaction	Understanding of Data, Analysis or Model	Knowledge Co-Creation stage
0:03:32	IND2 asks their team "What does that mean, EEZ?" after seeing the term on a layer in the game, and teammates explain its meaning.	Data, model	Externalization
0:07:30	The team discusses what they will tell G.O.D. regarding stakeholder engagement, which is beyond the scope of the game.	Analysis	Combination
0:09:30	YEL1 explains to the team what information the layers concerning aquaculture can give them to plan for the next five years.	Data, model	Socialization
0:11:09	YEL2 asks how the team should go about combining wind farms and aquaculture farms: which one should be built first? Although this is not accounted for by the game, YEL1 explains that usually you build wind farms first and then use them as anchorage for aquaculture farms.	Analysis	Externalization, Combination
0:18:00	IND1 explains to IND2 what anchorages are and where they are located on the screen.	Data and model	Externalization, Socialization
0:21:40	RED3 explains to RED2 what a carbon sink is and what it corresponds to in the game	Data, model	Externalization, socialization
0:39:40	Team discusses what they plan to tell G.O.D. They go one step above what the game requires of them and discuss the type of institutions that would need to be put in place to reach some of the goals they have set for their country.	Analysis	Combination
0:45:00	When looking at placing offshore wind turbines, the team goes beyond the game and discusses the aesthetic implications of placing offshore wind farms next to certain touristic areas	Analysis	Combination
1:00:10	YEL1 says that they should bring up their plan to create more employment in the country during their meeting with G.O.D. The game does not track employment, so this strategy is beyond the scope of the game.	Analysis	Combination
1:24:24	OR1 needs to use the computer for the first time and OR2 shows them how to navigate the game and extract the information needed	Data, model	Socialization
1:27:40	IND2 Takes control of the mouse and asks IND3 how to use some features. IND3 shows IND2	Model	Socialization

The post-game survey questions and answers relating to learning outcomes from the three game events are presented in Table 3 below. The possible answers that participants chose from range from 1 (strongly disagree) to 5 (strongly agree), where a 3 is neutral, meaning the player neither agreed nor disagreed with the statement. The answers have been grouped based on the players' experiences working in MSP. The data is divided by experience because statistical analysis (i.e., Mann-Whitney U test) revealed that prior experience in MSP most influenced the players' answers. The participants had for choice: Less than a year, from one to two years, two to three years, three to five years, five to 10 years, and 10 or more years of experience working in MSP. When unsure, the participants were told to round down their level of experience. One column presents the average answer for players with less than two years of experience (less than two years), and the other for players with two years or more (two years and more). The first three questions ask the players how much they have learned about MSP from the game artefact, which corresponds to externalization of knowledge. A Mann-Whitney U test on the first three questions indicated that the players with less experience of MSP learnt significantly more about MSP from playing the game than the players with more experience (U = 116, p = 0.05). That being said, both cohorts chose, on average, an answer above "neutral", meaning that they report some learning about MSP from playing the game. Questions 3 to 7 pertain to the social aspect of playing the game, or the socialization part of the knowledge co-creation cycle. The questions relate to the interactions between the players, which acted as practitioners of MSP. There was no statistical difference between the answers of the two cohorts for these questions (U = 116, p = 0.05), with averages above 3 ("neutral"). Furthermore, the players agreed that the game helped them understand the different barriers to the development of a good MSP process (question 5), which can be linked to the combination of knowledge about different systems involved in MSP and how they can conflict with each other. Finally, survey question 8 asks the players about internal reflection throughout the game, which corresponds to an internalization of knowledge and information provided by the game. Although there is no statistical difference between the two cohorts for this answer, it is important to note that participants from the Copenhagen event scored significantly lower (mean = 3.05, SD = 0.94) on this question than players from Newfoundland (mean = 3.67, SD = 0.50, U = 48, p = 0.05) and Venice (mean = 4.20, SD = 0.77, U = 90, p = 0.05). The post-game survey written responses of the Copenhagen event indicate that the players wanted the game to be based on realistic events and conflicts. Two players suggested introducing different planning scenarios that would be specific to a region. Another player suggested using a real case study in the game. Finally, three players thought that the game was too focused on national planning and that there was not enough time for transnational planning. Overall, the interaction analysis and the post-game survey answers showed that playing MSP Challenge 2050 helped players with little work experience in MSP better understand the challenges of MSP and the game helped facilitate the MSP process within each team.

Table 3. Average question score with standard deviation in parenthesis for post-game survey questions that look at learning outcomes.

Post-Game Survey Question	Less than 2 Years of Experience (n = 29)	2–10 Year of Experience (n = 13)	Statistical Significance (U = 116, p = 0.05)
(1) I gained more insight into what the important factors in MSP are and how they (can) influence each other	4.31 (0.85)	3.38 (0.87)	Yes
(2) I gained more insight into conflicts and cooperation between different sectors (e.g. fisheries, energy, environment)	3.93 (1.07)	3.31 (0.75)	Yes
(3) I have a clearer picture on how MSP can be turned into an integrated process	3.90 (0.77)	3.23 (0.93)	Yes
(4) I gained more insight into how MSP decisions in different countries (can) influence each other	3.79 (1.05)	3.46 (0.88)	No
(5) I gained more insights in the problems and barriers of cooperation among countries in MSP	3.72 (0.96)	3.46 (0.88)	No
(6) I gained more insights in the various ways countries can cooperate in MSP	3.79 (0.94)	3.46 (0.88)	No
(7) I gained more insight into how decisions on different planning scales (local, regional, national, international) (can) influence each other	3.39 (1.13)	3.08 (0.95)	No
(8) As players, we did enough internal reflection and adjustment	3.59 (0.98)	3.23 (0.93)	No

The answers from the mid-game survey of the Newfoundland game event indicate that the teams were struggling to define their goals because of the overwhelming amount of data the game provided them. Team Indigo admitted that they did not know "How to geographically spread out some of the new facilities". To help work through this cognitive loading, the teams identified different strategies, such as "Us[ing] a more integrated approach" and "leaving on [the] most important layers (things that can't be moved) to plan our work" (team Indigo), "Taking on specific roles" (team Orange), and "Dividing tasks" (team Purple). These findings demonstrate that although the teams were overwhelmed by the amount of data at the end of the planning phase, they developed strategies to redress the situation, such as: Integrated planning, limiting the amount of information on the screen to only show what they deemed most important, and dividing the specific roles and tasks between the players within a team to limit information overload. The game was therefore a good learning tool that helped the players develop skills to adjust and adapt to working with large data sets required for MSP.

3.2. Research Question 2: Does playing MSP Challenge 2050 Promote Quality Interactions and Cooperation Between Participants Facilitating the Knowledge Co-creation Cycle? (i.e., Group Added Value)

First, from the videos of the Newfoundland event, a quantitative analysis of the interactions in each team was performed for the planning phase of the game (defined as the first 90 min of gameplay where the time in the game did not advance). To graphically represent the data, this 90-min timespan was divided into nine sub-phases that show how interactions changed over the course of the planning phase. Due to increased movement of players between teams after the planning phase, interactions past this point could not be counted. Table 4 below shows the quantitative evolution of interactions for each team during the initial planning phase of the MSP Challenge 2050 simulation. There are three points to keep in mind when looking at Table 4 below. The first is that the numbers represent the total number of interactions, although it should be noted that some teams had less players than others. Having one less player in a team surprisingly did not affect the number of interactions. Interaction analysis reveals that this could be because the teams with three players only used one computer instead of two and therefore had to communicate with each other more than a team that used two computers. The player on the computer tended to speak less than the others and instead take instructions from other teammates. Second, the bolded numbers represent moments where the team experienced technical difficulty and a facilitator had to come and reboot their computer. A technical difficulty is always accompanied by a dip in the number of interactions (see teams, OR, IND, YEL). A third thing to note is that each team met with G.O.D. individually towards the end of the first 90 min of gameplay to go over their country plans. These meetings took between 10–15 min, represented in the table as G.O.D.

From the interaction analysis, it was noted that after meeting with G.O.D., team members were more stressed and retreated into their chosen roles, which appeared to isolate them from other players within their team. Therefore, the number of interactions dropped as players focused their attention on changing their national plans to match G.O.D.'s demands before discussing with players with similar roles in other teams how to coordinate on an international scale. With that in mind, the table shows that interactions within teams are constantly changing, but that interactions stay relatively high throughout the simulation gaming event.

Table 4. Quantitative interaction analysis.

Teams	Phase 1	Phase 2	Phase 3	Phase 4	Phase 5	Phase 6	Phase 7	Phase 8	Phase 9
Indigo	76	80	89	80	G.O.D	78	89	**54**	65
Purple	91	86	88	74	94	74	G.O.D	G.O.D	75
Orange	68	86	100	97	**71**	G.O.D	*	*	92
Red	42	83	74	87	80	78	76	G.O.D	*
Yellow	65	59	80	75	75	78	70	**30**	G.O.D

* Teams were out of frame of the audio-visual recording devices during these phases.

The most important information coming from the quantitative interaction analysis is that MSP Challenge 2050 fosters numerous interactions between teammates during the planning phases. These interactions are necessary for the socialization stage of the knowledge co-creation cycle. Furthermore, it shows that the main hindrance to interactions throughout the game are technical difficulties. Therefore, the utmost care and caution must be exercised when planning these events to reduce instances of technical difficulties to promote the greatest amount of interactions. However, looking at quantity is not enough to come to definitive conclusions, instead, the nature and quality of these interactions must be evaluated. To do this, five qualitative indicators were developed for this research to describe the quality of the interactions. These indicators are described in Table 5 and the results are tallied in Figure 2 below.

Table 5. Quality of interaction indicators and descriptions.

Indicator	Description
Moments of Consensus	Moments when teammates come to an agreement on how to proceed.
Moments of Reflection	Moments when teams pause and reflect on their actions and consequences.
Shared Laughter	Moments when teammates laughed together.
Anecdotal and Opinion Exchange	Moments when one teammate shares an anecdote or opinion with another teammate or the team as a whole.
Explicit Knowledge Transfer	Moments wherein one teammate shares expertise or knowledge with another teammate or the team. Generally, takes the form of an answer to a question.

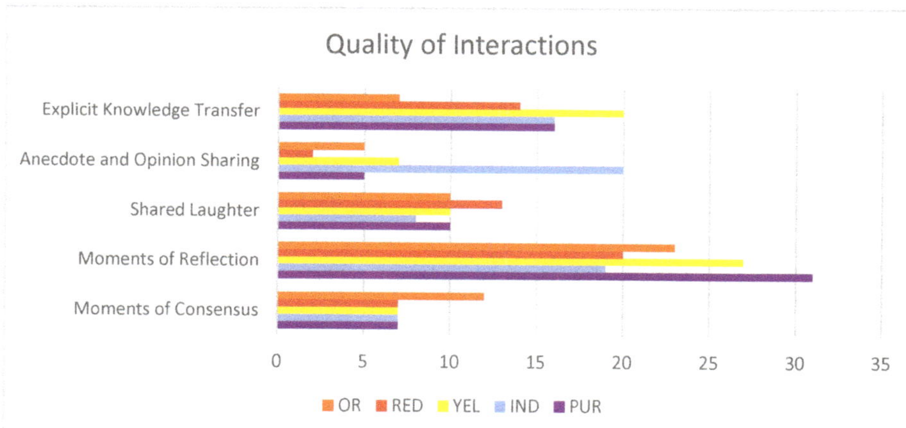

Figure 2. Summary of quality of interactions by team.

Once again, it is important to note that team Indigo and team Red had three players while the other teams had four players. Furthermore, interactions between teams during the planning phase were recorded though minimal in number (less than 10% of total interactions). The results for each indicator are similar for each team, with one outlier that outperformed the other teams in each category. Moments of explicit knowledge transfer and of reflection were the most common type of interaction between teams making up 22% and 37% of the quality interactions, respectively. This supports the hypothesis that playing MSP Challenge 2050 can lead to knowledge co-creation through explicit knowledge transfer and combination. Going back to the first research question, it also supports the idea that this SG could be used as a PSS providing added value while it helps participants learn to interpret the data, models, and analysis through interactions that lead to explicit knowledge transfer and reflection. These findings show that MSP Challenge 2050 can serve as a learning tool and as an innovative tool to support planning.

As can be noted by looking at Table 5 above, a great deal of interactions occurred between players and teammates; however, when looking at Figure 2 above, one can see a great deal less in terms of quality interactions. This is not to say that the game is ineffective at fostering quality interactions, rather, a large number of interactions are required to support these quality interactions. Previous research has shown similar results using a different SG (see [36]). Certain quality interactions are made of multiple smaller interactions, which can also explain the discrepancy in occurrence. Survey results from the mid- and post-game questionnaires were also analyzed to determine how the players felt the game helped them collaborate. The answers from the mid-game survey of the Newfoundland game event show that most teams identified team cohesion and collaboration as being helpful to develop and achieve MSP. Teams reported that much of their strategies relied on "lots of communication" (team Red), and "delegation, cooperation" (team Purple). When asked which strategies they were using to develop their MSP, team Yellow responded by saying: "Focus on communication between teams and between different ministers", while team Orange reported that: "[They] are still having fun working together easily and have chosen some key roles".

The answers to the mid-game survey indicate that teams valued cooperation during the game. The answers from the post-game survey in Table 6 below reinforce the findings of the mid-game survey and the interaction analysis, i.e. that cooperation took place between participants during the game event. Table 6 summarizes answers from the post-game survey from all three events to questions about the level of collaboration within teams. As before, the possible answers range from 0 (strongly disagree) to 5 (strongly agree), where a 3 is neutral. The Newfoundland and Venice groups ranked the questions similarly ($U = 34$, $p = 0.05$). However, we see a statistically significant dip in the values for the Copenhagen group for the second question relating to how well players worked together ($U_V = 90$, $U_N = 48$, $p = 0.05$). This may be attributed to the fact that for both the Newfoundland and the Venice event, the players were mostly students who knew each other well prior to the game, while the Copenhagen event was attended mostly by MSP professionals with more distant or non-existent relationships. The standard deviation for these questions remains under 1, meaning that most participants shared a similar experience. The high number of interactions and the teams' focus on collaboration indicates a potential for MSP Challenge 2050 to promote knowledge co-creation through socialization, externalization, and combination. In fact, as noted above, players identified communication as an important strategy for the next steps of the game. Overall, the game promoted ongoing collaboration within teams, despite each player choosing a distinct role for themselves.

Table 6. Average question score with standard deviation in parenthesis for post-game survey questions that look at collaboration.

Post-Game Survey Question	Newfoundland (N = 9)	Venice (N = 15)	Copenhagen (N = 19)
(1) The discussions between the players were good	4.22(0.44)	4.53 (0.52)	3.7 (0.73)
(2) As players, we worked together well during the game	4.22 (0.67)	4.4 (0.51)	3.4 (0.82)
(3) In general, other players (team members) played their roles well.	4 (0.50)	4.27 (0.80)	3.75 (0.79)
(4) I really put myself into my role.	4 (0.50)	4.53 (0.52)	3.65 (0.81)

The key insights from the debriefing session revolved around economic considerations, collaboration between different teams, and the feasibility of the task given to them. The players reflected on the complexity of MSP, which was made easier by removing economic components during the game, something the players were thankful for, but that removed from the realism of the task given to them. Even without economic considerations, the teams acknowledged that they had to make tradeoffs between different tasks and prioritize certain goals. In response to the complexity of MSP and learning to make decisions in a complex system, player RED2 noted that "It's not that people are useless—interactions between good and moral people make weird outcomes". The facilitator reflected on this statement by explaining that political and research cycles have different lengths, therefore, planners often find themselves trying to make decisions that fit within the short span of a political cycle without necessarily having as much scientific research to help them as they would like. This is a reality that most planners deal with, and the game illustrated that reality. From the interaction analysis of the debriefing session, it seems that the complexity of the subject mixed with the fast-paced game allowed some participants to reflect critically on what was happening, but that the game process could be improved to allow for more space for critical thinking. An interesting development occurred during the Newfoundland event, which can be explained by looking at the interaction analysis of the game event and the debrief. A feature of the game requires teams to make their plans visible to other teams to receive feedback and approval. As the game currently stands, it is suggested that teams seek approval from other teams before implementing plans in shared waters (based on the Kyiv (SEA) protocol that states that Member States must notify and consult each other on all major projects under construction that may have adverse environmental impacts across borders). However, the teams realized, one by one, that they did not need to wait for approval to go ahead with implementation of their plans (see Table 7 below).

Table 7. Interactions regarding consultation for project approval.

Time	Description of Interaction
0:031:16	RED2 asks team Indigo if they can see the plan they just proposed. Both teams have just discovered that they can share information. Indigo says that they don't need to approve it, that Team Red can just go ahead and do their plan without their approval.
0:37:17	IND 2 says that once everyone starts putting plans up for approval, the teams will be overwhelmed and approve everything and that it's probably like that in the real world
0:56:29	IND 2 asks if it is really necessary to ask for consultation for a plan that's very close to their coast. At 57:45 they just implement it.
0:02:40 (debrief)	YEL2: "It was interesting to see how when we thought we needed everyone to agree, we were all ready to do so but as soon as we realized we didn't have to come to a collective agreement it was just like every man for himself a little bit."

The role and importance of this feature of the game will be further explored in the Discussion Section. Overall the interaction analysis showed that players helped each other understand the terms used by the game, the game itself, as well as reflected on how their plans could be implemented in the real world.

3.3. *Research Question 3: What are the Characteristics of the Plans Developed While Playing the Game and How Do the Plans Differ From Team-to-team? (i.e., Outcome Added Value)*

To assess the outcome of added value of MSP Challenge 2050, the national plans developed by each team during the planning phase provide valuable information. Table 8 below provides an overview of the national plans developed by each team in the first 90 min of game-play and presented to G.O.D. The plans below are divided into five sectors: Ecology, fishing and recreation, oil and gas, renewable energy, shipping, and an additional miscellaneous category.

When looking at the above plans, some teams' plans are more comprehensive than others. For example, the Orange team planned for all five sectors as well as having miscellaneous goals. Similarly, the Purple team planned for all sectors, save shipping, although they managed to include timelines for most of their plans. Team Indigo also created a rather comprehensive plan. On the other end of the spectrum, we see that the Yellow and Red teams' plans are rather lacking, with the Yellow team failing to plan for two of the sectors while the Red team planned for all sectors, though vaguely.

Comparing the quantitative and quality interactions obtained in question 2 shows how these interactions may have supported the development of their plans. Table 9 below shows the total quality and quantitative interactions for all teams.

Table 8. Results of the planning phase.

Sector	Ecology	Fishing and Recreation	Oil and Gas	Renewable Energy	Shipping	Miscellaneous
Indigo	6% of marine area to become Nature 2000; 3% to become marine protected areas	Combine fishing with marine protected areas; Add 10 algae farms Add 5 fish farms	-	6000 MW Wind Energy	Add 2 anchorages and extend current anchorages by 2018; Add 3 dredging disposal sites	Focus on international discussion
Purple	30% of Marine areas to be marine protected areas by 2020	12 new algae farms; 6 new fish farms	Support existing fields and assure health and safety; three empty fields by 2020 (carbon capture)	Increase wind energy and explore new sites; 10 new tidal and wave energy locations by 2030	-	Keep the peace between fisherman dealing with new tidal and wave energy locations.
Orange	30% of sea to Nature 2000 Area (+5%)	Sufficient space for fishing and recreation; 3 new algae farms; 2 new fish farms; assure coast remains attractive and unobstructed; Preserve archeological values	Maintain oil and gas stack; Use at least two empty fields for carbon capture; Assure safety of platforms and transport	Increase wind energy production to 4500 MW; Create three test sites for tidal energy; Assure sufficient grid capacity on land; Establish international Sea of Colours Energy Grid	Extend 1 port; Add 2 more anchorages; 2 more dredge disposal areas	Sand and gravel extraction areas; 7% designated for military activities
Red	Increase marine protected areas	Add 7 algae farms; Add 4 fish farms	Implement CO_2 capture	Add Tidal Energy	Extend Anchorages	-
Yellow	Increase Nature 2000 areas by 2.5%; marine protected areas 10-year plan (increase area by 10%) (3 years)	-	-	Expand existing wind farms (increase capacity by 5000 MW); Asses sites for multi-use aquaculture areas (14 years)	Extend Seaports Add Anchorages every year; Increase research and consultation (5 years)	-

Table 9. Total quantitative and quality interactions.

Team	Quantitative Interactions	Quality Interactions
Indigo *	611	74 (12% of total interactions)
Orange *	514	57 (11% of total interactions)
Purple *	582	69 (12% of total interactions
Red	520	56 (10% of total interactions)
Yellow	532	71 (13% of total interactions

* Denotes teams with the most developed, comprehensive plans.

Table 9 above shows that the teams with more comprehensive plans interacted more throughout the game event. These teams generally had more quality interactions than other teams. The exception here being the Orange team that had less interactions (both quantitative and quality). However, it should be noted that for two phases, team Orange was out of view of the recording devices, and the team interacted more than the numbers show. In fact, in their mid-game survey, they reported "we looked at everything as a team, looked at what is missing, what needs more information, to make sure to talk with key partners to avoid conflict." The Yellow team is also an exception, with the second highest quality interactions, yet they still had an underdeveloped plan. Despite this, the Yellow team noted in their mid-game survey that one of the challenges they faced was "confusion over how to prioritize things", which may have led to confusion over how to develop their national plan. The Yellow team did, however, have the most instances of knowledge sharing, meaning that although their MSP plans may not have been as comprehensive as most, there was the highest exchange of knowledge occurring.

The outcome added value does not end at the planning phase, it extends into the implementation phase. Although not recorded, teams discussed the implementation phase during the debriefing period, which was recorded as well. The teams discussed an example of international cooperation that occurred during the implementation phase, namely the international Sea of Colours Energy Grid that was a joint creation by all teams. Proposed by the Orange team, a so-called "international summit" was held where representatives from each country were brought together. As a result of this summit, all five countries successfully linked their energy grids in a hub in the center of the Sea of Colours. The Orange team was composed of one student, two local stakeholders, and one postdoctoral fellow; it was the team with the most combined experience. The other teams benefitted from their expertise and were able to learn from their experience and create new knowledge between themselves, resulting in a successfully implemented Energy Grid for the Sea of Colours.

3.4. Summary of Research Results

Though the results and findings for each individual research question are essential to achieve the research objective, it is also important to look at the inter-connections between these questions. An important aspect of this study was to investigate the added values of MSP Challenge 2050 as defined by Pelzer, Geertman, van der Heijden, and Rouwette [14]. Namely, our research questions sought to investigate the individual, group, and outcome added values. By isolating these added values, we add credence to the idea that SG (specifically MSP Challenge 2050) can be used as PSS.

The findings from the first research question allowed for a deeper investigation into the individual added value that PSS can bring out in users. Results and findings relating to this first question indicated that by playing MSP Challenge 2050, participants are given the opportunity to acquire knowledge and information from the game artefact and game experience itself. The results have shown that this opportunity is greater for participants with less experience and formal training in MSP, though, generally, the game experience proved advantageous to all participants. The second research question allowed the researchers to investigate the added group benefits as defined by Pelzer, Geertman, van der Heijden, and Rouwette [14]. A quantitative analysis shows that players are offered a venue in which interactions are common and stable (unless teams experienced technical difficulty). Furthermore, an analysis of quality interactions has shown that knowledge between participants is exchanged either through explicit knowledge transfer or anecdotal exchange. The game experience also fosters moments of reflection, allowing for players to come together and reflect on the information being transmitted to them via the game artefact that was determined in research question 1. This knowledge transfer and reflection therefore allows for the different stages of knowledge co-creation to take place. Table 10 shows the stages of the knowledge conversion cycle and summarizes if and how these stages occurred during the MSP Challenge 2050 game event.

Table 10. Evidence of knowledge co-creation occurring during MSP Challenge 2050 gameplay.

Knowledge Co-creation Stage	Requirements	Achieved by MSP Challenge 2050?	If so, How?
Socialization	Physical proximity/direct interactions	Yes	MSP Challenge 2050 offers a space for Socialization to occur based on how the game is set-up and how players are grouped into small teams. **58** examples of socialization were noted in the planning phase of the Newfoundland event
Externalization	Peer-to-peer dialogue where individuals and groups engage in the creation of shared knowledge	Yes	The game asks the players to develop a national and international MSP process using complex information in the game. The players are required to take on different roles and explain to each other their logic and justify their opinions. **150** examples of externalization were noted in the planning phase of the Newfoundland event
Combination	(a) capturing and integration of new explicit knowledge, (b) dissemination of explicit knowledge among groups and networks, and (c) editing or processing of explicit knowledge to make it more user-friendly	Yes	While planning, the teams discuss how their plans could be implemented in real life, and which criteria are omitted in the game but would need to be considered in the real world. During the debriefing sessions, the tacit information of the experience is transformed into more explicit usable knowledge. **117** examples of combination were noted in the planning phase of the Newfoundland event
Internalization	(a) actualizing explicit knowledge in practice, (b) embodying explicit knowledge through simulations or experiments to trigger learning-by-doing, (c) active participation of all players.	Unknown	In order to determine this, more follow-up with participants needs to be performed to inquire if lessons learned in the game were used in the real world. No examples noted, follow-up required as it involves bringing new knowledge into one's everyday life.

Finally, research question 3 investigated the added benefits of the outcome level by analyzing the plans that were developed by teams throughout the planning process. These plans were informed by both the individual and group levels as they are a synthesis of all the information and knowledge developed by players throughout the planning stage of the game. Findings from this question indicate that teams that interacted more both in quantity and quality generally created more thorough MSP plans for their respective countries. Overall, it can be concluded that the MSP Challenge 2050 SG displays the added values required for a PSS on an individual, group, and outcome level. It offers opportunities for interaction and discussion within and between teams. This interaction and discussion increase the chances of creating new tacit or explicit knowledge on an individual and group level. The results of this study are preliminary, although this research provides insight into further research opportunities on how SG can effectively be used as a PSS or how PSS can be gamified to allow for learning outcomes that may not be present in traditional PSS.

4. Discussion

4.1. Improving MSP Challenge 2050 to Become a PSS

Although preliminary research results show that MSP Challenge 2050 offers many of the same added benefits as a PSS, there are still opportunities and areas for further improvements to increase the added value. Firstly, several players reported some difficulties in understanding the game early on, noting a learning curve. As mentioned in Section 3.1, one of the types of interactions most frequent during the game involved players helping each other navigate the game interface. Teams therefore spent a large amount of time learning how to use the game artifact instead of learning more about the MSP process. Since the individual added value of a PSS focuses on helping participants use and understand data, it is important to reduce this steep learning curve as much as possible [51]. MSP Challenge 2050, therefore, suffers from the same criticism as other PSS, which can often be characterized as user unfriendly [52]. This could be remedied by adding an instruction manual or by having more facilitators to help teams with the technical aspects of the game, which would allow participants to focus more on the MSP planning process. However, part of the game intends to overstimulate participants to help them learn to compartmentalize information and deal with uncertainty in decision-making [45]. In fact, Van Bilsen, Bekebrede, and Mayer [42] explain that SG are an important tool to teach decision-makers to make choices that affect complex systems. Therefore, the game should not be too simple or the participants will not learn these hard to teach lessons. However, the level of difficulty should always correspond to the types of users that participate in game events.

Although the technical complexity of the game detracted the participants from focusing on the planning process, several participants commented on the absence of financial considerations in the game, which made decisions easier to reach. These participants emphasized that the inclusion of economic analysis parameters in the game will increase the added value of the MSP Challenge 2050 SG as a PSS on the group and outcome level. On the group level, had the cost of decisions been included in the game, the teams would have had to discuss trade-offs in more depth, which could have led to more conflict or consensus building about setting national and international priorities. On the outcome level, the preliminary plans developed by the teams would have better reflected the realities of the MSP planning process. As it stood, the priorities of each nation were provided to teams at the beginning of the game event through cue cards and subsequently changed by G.O.D. during their one-on-one meeting with him. Without economic considerations, the MSP Challenge 2050 SG is unable to respond to the real needs of MSP stakeholders or decision-makers that play the game, because in reality, they require economic considerations to make decisions.

Biermann [52] explains that a common pitfall of PSS is their complexity, which makes it difficult to tailor their use for a specific purpose, therefore, they are often too general and do not respond to the very specific needs of its users. MSP Challenge 2050 experiences this common pitfall and if improved would increase added value on the outcome level. The game's focus is on teaching the MSP process, which makes it less effective to address more specific planning issues. This feedback was received from three players following the Copenhagen event that wished they could have used the game to test scenarios or learn about a specific case study that was more relevant to their work. Given that the Copenhagen players had the most MSP experience, their comments very likely reflect those of potential users of MSP Challenge 2050 as a PSS. Therefore, future versions of MSP Challenge 2050 should be tailored to meet more specific user needs of participants, whether it be by customizing the region in which they are playing or by setting specific initial boundary conditions that mimic a scenario the players are interested in exploring. This brings up an important concern when working with either SG, PSS, or hybrids of the two: Knowing your specific audience is essential. More experienced users may desire working on a tangible issue while those with less of an understanding of the MSP process benefit more from a game with a broad and more educational scope.

MSP Challenge 2050 relies on a simplified model that abstracts certain information [53]. Being aware of the strengths and weaknesses of a model can help its users make more informed decisions based on the information the model returns [54]. A recent review by Steenbeek, J. (2015) [46] concluded, however, that the model used in MSP Challenge 2050, the Ecopath food chain model, was insufficiently realistic to be used for decision-making purposes. This removes from the validity of the plans created using MSP Challenge 2050, and therefore reduces the added value on an outcome level. The next version of the game, called the MSP Challenge Platform, will be using more complex models with the aim to more closely reflect reality by giving players the option to choose between three real-world locations: The North Sea, The Baltic Sea, and the Clyde Marine region in Scotland. The new version of the game will also enhance roleplay options to allow for players to more fully immerse themselves in the game environment. Updated shipping and energy simulators will also add to the realism. Also, allowing participants to change certain parameters and see the projected outcome will allow them to develop and test scenarios, something that was identified as important by players. Overall, there are several changes that could be made to MSP Challenge 2050; though many of these changes will likely be made in the upcoming version of the game, MSP Challenge Platform.

4.2. Limitations of the Study and Avenues for Future Research

The current study has certain limitations and offers opportunities for further research involving the MSP Challenge 2050 SG and its newer platform version. Firstly, the information gathered in the post-game surveys provided insight into to how the participants felt immediately after playing the game. It should be noted that the research team was unable to determine whether there was a clear increase in knowledge in players after playing the game, which relates to added value on the individual level. To remedy this, future studies should include a survey gauging players' knowledge before and after the game to definitively show whether or not players increased their understanding of the MSP process through the game and game experience. This will also help researchers isolate which type of stakeholder (i.e., level of understanding of MSP processes) would benefit the most from the game. For example, if players score perfect on the pre-game questionnaire then it follows that the game will be of little use to them, at least in terms of increasing explicit knowledge related to MSP or individual level added value.

A second set of limitations involve improvements needed to assess the SG's group added value. First, the interaction analysis was performed only on the national planning phase of the game. An interaction analysis of the whole game would have gained more insight into how the game can promote collaboration between teams, and how international conflicts were dealt with during game-play. For future game events, cameras and microphones should be set up to record movement and discussions of participants throughout the room and further analyze interactions between participants within and between teams. Due to this limitation, only the intra-team interactions were analyzed. Had further stages of the game been recorded, researchers could have looked at inter-team interactions as well. Inter-team interactions more closely resemble the reality of decision-making and the MSP process and would allow for researchers to look at other indicators, such as moments of inter- and intra-team conflicts, capacity building, knowledge exchange, etc. Researchers could then look at the frequency of these indicators in both inter-team and intra-team contexts, allowing a greater understanding of the types of interactions that occur both within and between teams that may have conflicting objectives and goals.

Within the context of a simulation gaming event, moments of conflict, for example, can provide insight into the differing perspectives and viewpoints of players on other teams. Facilitating communication of diverging viewpoints directly refers to the group added value. Therefore, during the debrief, facilitators could explore the idea of conflict to give players a greater perspective on why and how these conflicts occur. Another important aspect that must be managed for future events is the fact that teams can publish their plans without the approval of all other countries. Once players became aware of this fact, a distinct decrease in inter-team collaboration was noted. In future versions of the game, it would be essential to evaluate what happens when teams are obligated to receive approval from all other affected teams prior to implementing their country plans. This would add realism to the game given that this feature is based on the Kyiv (SEA) protocol. By-passing approval removes the international dimension from the game; MSP, however, involves international, inter-disciplinary, and cross-actor collaborations that are better reflected by interactions that resemble a network of teams rather than the interactions occurring within each team separately [6]. This will also provide a game experience for participants that more closely resembles the reality and complexity of MSP decision-making processes.

Although we have shown that the added values of a PSS are present in MSP Challenge 2050, more research must be conducted to corroborate these results. Most notably the added value at the outcome level would greatly benefit from being further studied. We have shown that group and individual added values lead to better quality plans, but for the outcome level to be fully realized, these improvements must be transferable to real world settings. Therefore, follow-ups with players of the game will have to be completed in the months following gameplay to determine whether the game experience facilitated planning processes in the real-world. It will be interesting to look at a platform that allows for more in-depth planning where players can interact with data and each other both in person and virtually over the course of a longer time period. This platform may also be integrated into real world planning environments, thus allowing for a more in-depth assessment of the outcome level added value of a game, like MSP Challenge 2050.

After having investigated the benefits of MSP Challenge 2050 as a SG-PSS hybrid, it will also be beneficial to compare it to other non-gamified PSS tools. By organizing two parallel events, one using a gamified PSS and one using a more traditional PSS, it may be possible to further study the benefits that arise solely from the gamification and those that are common in both gamified and standard PSS.

5. Conclusions

Overall, this research outlines the benefits, opportunities, and further research required to study the added value of gamified elements to traditional PSS. On an individual level, the most important added value of gamification, such as that provided in the MSP Challenge 2050, appears to be that of learning [14]. The learning that occurs through gamification of PSS, such as with MSP Challenge 2050, takes several forms. Players of MSP Challenge 2050 indicate to have developed a greater understanding of the myriad of challenges and complexities that exist when developing a marine spatial plan. Game-based learning is increasingly recognized as an essential element for sustainable water and environmental management that seeks to include not only more, but also a broader variety of stakeholders into decision making processes [55]. By using SG, managers can facilitate stakeholder meetings that enhance the socialization of new stakeholders while assuring that all stakeholders have access to the same level of baseline information. This process also allows stakeholders to gain a deeper understanding of the views and perspectives of their peers as well as the context they are working within.

On a group level, it is important that the added value of a gamification tool fosters communication, collaboration, and consensus building. It has been shown in this study and others [36] that SG possess immense potential to foster the quantity as well as the quality of interactions between participants. Specifically, it has been shown that SG can create a space where participants can laugh together, share thoughts, ideas and stories, and reflect on their experiences. This allows for the participants to develop and enhance their collaboration skills while activating the knowledge co-creation cycle and developing new tacit and explicit knowledge through these interactions. Gamifying the collaborative decision-making process removes the risk of real-world consequences, allowing for players to explore, test, and discuss scenarios they may not have been able to otherwise. The outcome level refers to the resultant plans stemming from a PSS. It has been shown that teams that collaborate more and shared more quality interactions developed more comprehensive plans. The outcome in these cases can be said to be based on a deeper consideration of the information provided [14]. The human resource is an important one and allowing players the time to interact and discuss with their teammates allows for an externalization and combination of knowledge that leads teams to create more comprehensive marine spatial plans. Although further research is required to fully understand the extent of the outcome level added value of gamification, a preliminary argument can be made that SG have the ability to improve planning outcomes by fostering spaces wherein the individual and group added values are able to manifest.

Overall, the MSP Challenge 2050 has indicated to foster all three levels of added values; as such an argument can be made that the potential of SG should be further studied for their utility as PSS. SG should be used in the early phases of the MSP process to assure all involved are given the same baseline information (individual level), to allow all involved a chance to socialize and work together on a specific problem (group level), and to aid in the creation of more comprehensive and widely agreed upon plans (outcome level). The results of this study indicate that SG have the potential to be used as tools for stakeholder development and planning. Specifically, MSP Challenge 2050 is a promising tool for planners and future versions of the game may provide more effective tools to enhance MSP processes.

Supplementary Materials: The following are available online at http://www.mdpi.com/2073-4441/10/12/1786/s1.

Author Contributions: I.M. and X.K. have moderated the MSP Challenge game sessions in which data collection for this paper has taken place. S.J., L.G., A.I. and X.K. conducted the research activities; W.M., J.A. and I.M. designed and supervised this research project. S.J. and L.G. wrote the paper; W.M. and X.K. provided ongoing support and reviews to the paper drafts.

Funding: This study was funded by a Social Sciences and Humanities Research Council of Canada (SSHRC) Partnership Development Grant held by Adamowski and contributed to by Nguyen from the McGill University Brace Centre for Water Resources Management. The research is also part of the PhD thesis by author X.K. on the use of Serious Gaming in (transboundary) Maritime Spatial Planning at Wageningen University, The Netherlands, with support of Rijkswaterstaat.

Acknowledgments: The original idea of the MSP Challenge board game is from Lodewijk Abspoel, of the Netherlands Ministry of Infrastructure and Water Management. The concept of the board game and digital game has been further developed by a team, including the authors I.M. and X.K. The authors acknowledge the contributions by Linda van Veen and Bas van Nuland in the design and development of the game materials and software. The authors furthermore wish to thank Francesco Musco, Federica Appiotti and the Erasmus Mundus MSP students at IUAV (Italy), Geoff Coughlan and students at Memorial University (Canada), and the EU Interreg NorthSEE and BalticLINes project partners.

Conflicts of Interest: The authors declare no conflict of interest.

References

1. Kannen, A. Marine spatial planning in the context of multiple sea uses, policy arenas and actors. 2014. Available online: https://www.climateservicecentre.de/imperia/md/images/gkss/institut_fuer_kuestenforschung/kso2/kso_kannen.pdf (accessed on 3 December 2018).
2. Mayer, I.; Zhou, Q.; Lo, J.; Abspoel, L.; Keijser, X.; Olsen, E.; Nixon, E.; Kannen, A. Integrated, ecosystem-based marine spatial planning: Design and results of a game-based, quasi-experiment. *Ocean Coastal Manag.* **2013**, *82*, 7–26. [CrossRef]
3. Arbo, P.; Thủy, P.T.T. Use conflicts in marine ecosystem-based management—The case of oil versus fisheries. *Ocean Coastal Manag.* **2016**, *122*, 77–86. [CrossRef]
4. Douvere, F.; Ehler, C.N. New perspectives on sea use management: Initial findings from European experience with marine spatial planning. *J. Environ. Manag.* **2009**, *90*, 77–88. [CrossRef] [PubMed]
5. Ehler, C.N. Chapter 1: Marine spatial planning: An idea whose time has come. In *Offshore Energy and Marine Spatial Planning*; Katherine, L., Yates, K.L., Bradshaw, C.J.A., Eds.; CRC Press: New York, NY, USA, 2018.
6. Abramic, A.; Bigagli, E.; Barale, V.; Assouline, M.; Lorenzo-Alonso, A.; Norton, C. Maritime spatial planning supported by infrastructure for spatial information in Europe (inspire). *Ocean Coast. Manag.* **2018**, *152*, 23–36. [CrossRef]
7. Pomeroy, R.; Douvere, F. The engagement of stakeholders in the marine spatial planning process. *Mar. Policy* **2008**, *32*, 816–822. [CrossRef]

8. Gazzola, P.; Onyango, V. Shared values for the marine environment–developing a culture of practice for marine spatial planning. *J. Environ. Plan. Policy Manag.* **2018**, *20*, 1–14. [CrossRef]
9. Deal, B.; Gu, Y. Resilience thinking meets social-ecological systems (SESs): A general framework for resilient planning support systems (PSSs). *J. Dig. Landsc. Archit* **2018**, 200–207. [CrossRef]
10. Harris, B. Beyond geographic information systems. *J. Am. Plan. Assoc.* **1989**, *55*, 85–90. [CrossRef]
11. Kuller, M.; Bach, P.M.; Ramirez-Lovering, D.; Deletic, A. Framing water sensitive urban design as part of the urban form: A critical review of tools for best planning practice. *Environ. Model. Softw.* **2017**, *96*, 265–282. [CrossRef]
12. Geertman, S.; Stillwell, J. Planning support systems: An inventory of current practice. *Comput. Environ. Urban Syst.* **2004**, *28*, 291–310. [CrossRef]
13. Zulkafli, Z.; Perez, K.; Vitolo, C.; Buytaert, W.; Karpouzoglou, T.; Dewulf, A.; De Bievre, B.; Clark, J.; Hannah, D.M.; Shaheed, S. User-driven design of decision support systems for polycentric environmental resources management. *Environ. Model. Softw.* **2017**, *88*, 58–73. [CrossRef]
14. Pelzer, P.; Geertman, S.; van der Heijden, R.; Rouwette, E. The added value of planning support systems: A practitioner's perspective. *Comput. Environ. Urban Syst.* **2014**, *48*, 16–27. [CrossRef]
15. Brömmelstroet, M.T.; Schrijnen, P.M. From planning support systems to mediated planning support: A structured dialogue to overcome the implementation gap. *Environ. Plan. B Plan. Des.* **2010**, *37*, 3–20. [CrossRef]
16. Amara, N.; Ouimet, M.; Landry, R. New evidence on instrumental, conceptual, and symbolic utilization of university research in government agencies. *Sci. Commun.* **2004**, *26*, 75–106. [CrossRef]
17. Innes, J.E. Planning theory's emerging paradigm: Communicative action and interactive practice. *J. Plan. Educ. Res.* **1995**, *14*, 183–189. [CrossRef]
18. Pelzer, P.; Geertman, S.; van der Heijden, R. Knowledge in communicative planning practice: A different perspective for planning support systems. *Environ. Plan. B Plan. Des.* **2015**, *42*, 638–651. [CrossRef]
19. Chew, C.; Zabel, A.; Lloyd, G.J.; Gunawardana, I.; Monninkhoff, B. *A Serious Gaming Approach for Serious Stakeholder Participation*; The City University of New York: New York, NY, USA, 2014.
20. Mayer, I.; Bekebrede, G.; Warmelink, H.; Zhou, Q. A brief methodology for researching and evaluating serious games and game-based learning. In *Psychology, Pedagogy, and Assessment in Serious Games*; IGI Global: Hershey, PA, USA, 2014; pp. 357–393.
21. RATAN, R.A.; Ritterfeld, U. Classifying serious games. In *Serious Games*; Routledge: London, UK, 2009; pp. 32–46.
22. Mayer, I.S. The gaming of policy and the politics of gaming: A review. *Simul. Gaming* **2009**, *40*, 825–862. [CrossRef]
23. Barreteau, O.; Le Page, C.; Perez, P. *Contribution of Simulation and Gaming to Natural Resource Management Issues: An Introduction*; Sage Publications: Thousand Oaks, CA, USA, 2007.
24. Aubert, A.H.; Lienert, J. Gamified online survey to elicit citizens' preferences and enhance learning for environmental decisions. *Environ. Model. Softw.* **2019**, *111*, 1–12. [CrossRef]
25. Merriam, S.B.; Caffarella, R.S.; Baumgartner, L.M. *Learning in Adulthood: A Comprehensive Guide*; John Wiley & Sons: Hoboken, NJ, USA, 2012.
26. Polanyi, M. Tacit knowing: Its bearing on some problems of philosophy. *Rev. Mod. Phys.* **1962**, *34*, 601. [CrossRef]
27. Roux, D.J.; Rogers, K.H.; Biggs, H.C.; Ashton, P.J.; Sergeant, A. Bridging the science–management divide: Moving from unidirectional knowledge transfer to knowledge interfacing and sharing. *Ecol. Soc.* **2006**, *11*, 1–20. [CrossRef]
28. Gröhn, P.; Kasu, D.; Swiac, M.; Zafar, A. Organizing the organization: Recommendation of development for explicit and tacit knowledge sharing at a university library in north america. 2017. Available online: http://www.diva-portal.org/smash/record.jsf?pid=diva2%3A1085189&dswid=-7231 (accessed on 17 August 2017).
29. Polanyi, M. The logic of tacit inference. *Philosophy* **1966**, *41*, 1–18. [CrossRef]
30. Sudhindra, S.; Ganesh, L.; Arshinder, K. Knowledge transfer: An information theory perspective. *Knowl. Manag. Res. Pract.* **2017**, *15*, 400–412. [CrossRef]
31. Nonaka, I.; Von Krogh, G. Perspective—tacit knowledge and knowledge conversion: Controversy and advancement in organizational knowledge creation theory. *Organiz. Sci.* **2009**, *20*, 635–652. [CrossRef]

32. Nonaka, I.; Takeuchi, H.; Umemoto, K. A theory of organizational knowledge creation. *Int. J. Technol. Manag.* **1996**, *11*, 833–845.
33. Nonaka, I.; Byosiere, P.; Borucki, C.C.; Konno, N. Organizational knowledge creation theory: A first comprehensive test. *Int. Bus. Rev.* **1994**, *3*, 337–351. [CrossRef]
34. Alavi, M.; Leidner, D.E. Knowledge management and knowledge management systems: Conceptual foundations and research issues. *MIS Q.* **2001**, *25*, 107–136. [CrossRef]
35. Nonaka, I.; Takeuchi, H. *The Knowledge-Creating Company: How Japanese Companies Create the Dynamics of Innovation*; Oxford University Press: Oxford, UK, 1995.
36. Jean, S.; Medema, W.; Adamowski, J.; Chew, C.; Delaney, P.; Wals, A. Serious games as a catalyst for boundary crossing, collaboration and knowledge co-creation in a watershed governance context. *J. Environ. Manag.* **2018**, *223*, 1010–1022. [CrossRef] [PubMed]
37. Medema, W.; Wals, A.; Adamowski, J. Multi-loop social learning for sustainable land and water governance: Towards a research agenda on the potential of virtual learning platforms. *NJAS-Wageningen J. Life Sci.* **2014**, *69*, 23–38. [CrossRef]
38. Medema, W.; Adamowski, J.; Orr, C.J.; Wals, A.; Milot, N. Towards sustainable water governance: Examining water governance issues in québec through the lens of multi-loop social learning. *Can. Water Resour. J.* **2015**, *40*, 373–391. [CrossRef]
39. Keijser, X.; Ripken, M.; Mayer, I.; Warmelink, H.; Abspoel, L.; Fairgrieve, R.; Paris, C. Stakeholder engagement in maritime spatial planning: The efficacy of a serious game approach. *Water* **2018**, *10*, 724. [CrossRef]
40. Kaufman, S.; Smith, J. Framing and reframing in land use change conflicts. *J. Architect. Plan. Res.* **1999**, *16*, 164–180.
41. Akkerman, S.; Bruining, T. Multilevel boundary crossing in a professional development school partnership. *J. Learn. Sci.* **2016**, *25*, 240–284. [CrossRef]
42. Van Bilsen, A.; Bekebrede, G.; Mayer, I. Understanding complex adaptive systems by playing games. *Inf. Educ.* **2010**, *9*, 1–8.
43. Crookall, D. Serious games, debriefing, and simulation/gaming as a discipline. *Simul. Gaming* **2010**, *41*, 898–920. [CrossRef]
44. Guillén-Nieto, V.; Aleson-Carbonell, M. Serious games and learning effectiveness: The case of it'sa deal! *Comput. Educ.* **2012**, *58*, 435–448. [CrossRef]
45. Mayer, I.; Zhou, Q.; Keijser, X.; Abspoel, L. Gaming the Future of the Ocean: The Marine Spatial Planning Challenge 2050. 2014. Available online: http://www.mspchallenge.info/uploads/3/1/4/5/31454677/msp_game_2014_sgda.pdf (accessed on 16 June 2018).
46. Steenbeek, J. *Integrating ecopath with Ecosim into the MSP Software—Conceptual Design*; Figshare: London, UK, 2015.
47. Zhou, Q. The Princess in the Castle: Challenging Serious Game Play for Integrated Policy Analysis and Planning. 2014. Available online: http://citeseerx.ist.psu.edu/viewdoc/download?doi=10.1.1.869.6212&rep=rep1&type=pdf (accessed on 30 September 2017).
48. MacFarland, T.W.; Yates, J.M. Mann–whitney u test. In *Introduction to Nonparametric Statistics for the Biological Sciences Using R*; Springer: New York, NY, USA, 2016; pp. 103–132.
49. Jordan, B.; Henderson, A. Interaction analysis: Foundations and practice. *J. Learn. Sci.* **1995**, *4*, 39–103. [CrossRef]
50. Levesque, V.R.; Calhoun, A.J.; Bell, K.P.; Johnson, T.R. Turning contention into collaboration: Engaging power, trust, and learning in collaborative networks. *Soc. Nat. Resour.* **2017**, *30*, 245–260. [CrossRef]
51. Rodela, R.; Bregt, A.K.; Ligtenberg, A.; Pérez-Soba, M.; Verweij, P. The social side of spatial decision support systems: Investigating knowledge integration and learning. *Environ. Sci. Policy* **2017**, *76*, 17–184. [CrossRef]
52. Biermann, S. Planning support systems in a multi-dualistic spatial planning context. *J. Urban Technol.* **2011**, *18*, 5–37. [CrossRef]
53. Nordhaus, W.D. *The Challenge of Global Warming: Economic Models and Environmental Policy*; Yale University Press: New Haven, CT, USA, 2007.
54. Christensen, V.; Walters, C.J. Ecopath with ecosim: Methods, capabilities and limitations. *Ecol. Model.* **2004**, *172*, 109–139. [CrossRef]

55. Pahl-Wostl, C.; Craps, M.; Dewulf, A.; Mostert, E.; Tabara, D.; Taillieu, T. Social learning and water resources management. *Ecol. Soc.* **2007**, *12*, 1–19. [CrossRef]

water

MDPI

Article

Capacity Building for Water Management in Peri-Urban Communities, Bangladesh: A Simulation-Gaming Approach

Sharlene L. Gomes [1,*], Leon M. Hermans [1,3], Kazi Faisal Islam [2], Sheikh Nazmul Huda [2], ATM Zakir Hossain [2] and Wil A. H. Thissen [1]

1 Delft University of Technology, Faculty of Technology, Policy, and Management, Building 31, Jaffalaan 5, 2628 BX Delft, The Netherlands; L.M.Hermans@tudelft.nl (L.M.H.); W.A.H.Thissen@tudelft.nl (W.A.H.T.)
2 Jagrata Juba Shangha, 35/8, TB Cross Road, Khulna-9100, Bangladesh; faisaljjs@gmail.com (K.F.I.); nazmuljjs@gmail.com (S.N.H.); atmzakir@gmail.com (A.Z.H.)
3 IHE Delft, Integrated Water Systems and Governance Department, Westvest 7, 2611 AX Delft, The Netherlands
* Correspondence: S.L.Gomes@tudelft.nl; Tel.: +31-152-784-149

Received: 29 August 2018; Accepted: 17 November 2018; Published: 21 November 2018

Abstract: Peri-urban areas in the global south are experiencing over-exploitation and contamination of water resources as a result of rapid urbanisation. These problems relate to the ineffectiveness of the underlying institutions in this dynamic, multi-actor context. Institutions need to be considered during problem solving; however, peri-urban communities have limited insight into their institutional context. This research examines the extent to which problem solving capacity can be improved through gaming-simulation methods. A game-based approach is tested in a capacity building workshop with peri-urban communities in Khulna (Bangladesh). A role-playing game designed from game theory models is used to examine local drinking water problems through an institutional lens. Workshop evaluation shows that through role-play, participants learned about strategies in drinking water supply (in both the current and future scenarios) and about the potential to address water quality issues through cooperative groundwater monitoring. Results also show improved problem understanding with regards to institutions, actor strategies, and problem-solving constraints. Participants valued the interactive medium for comparing and evaluating strategies. This paper highlights limitations in game design and its implementation, and offers ways to address this in future applications.

Keywords: drinking water management; peri-urban; institutions; gaming-simulation; groundwater; capacity building

---

1. Introduction

Water resource management during urbanisation is becoming increasingly problematic in the Global South given the rate at which development is occurring. In the coming decades, Asia and Africa will experience rapid urbanisation compared to other regions, and 66% of the global population is expected to be urban by 2050 [1]. Therefore, attention must be paid to water governance, particularly in peri-urban contexts, where the transition from rural to urban landscapes occurs. "Peri-urban" is defined, according to Narain [2], as a transition zone in close proximity to urban centers experiencing a two way flow of goods, services, and population. Here, urbanisation processes are visible in the changing land use, economic activities, and population [2,3]. Achieving sustainable and equitable water management in peri-urban areas is important, given that they are the cities of tomorrow.

The institutional context is relevant for understanding and addressing peri-urban water management problems. Here, institutions are defined as formal and informal 'rules' that structure interactions and behaviour in society [4]. Formal rules include laws, policies, and regulations, while informal rules refer to customs, codes of conduct, and taboos [5]. With regards to water management, institutions exist to facilitate coordination over its use and allocation. For example, service providers can use water tariffs, a type of rule intended to regulate water usage. In this way, institutions serve a function in society by offering guidance during the problem solving process.

In peri-urban areas, however, these institutions are often ineffective. Typically, institutions are arranged along rural and urban (administrative) boundaries, with peri-urban governance defined by rural institutions [6]. These traditional rural institutions are unable to support the needs of communities in such a dynamic context. In some cases, peri-urban institutions are fragmented or overlapping. As a result of this, identifying which rules apply, or who has what roles and responsibilities in a given situation, is challenging. Moreover, peri-urban areas are much more socially-heterogeneous than rural villages [3]. Often, multiple actors with varying interests are competing for the same resources. Balancing actor needs is an important part of peri-urban governance, but requires supportive institutional arrangements. Without it, peri-urban communities can become marginalised from essential public services.

Examples of ineffective institutional arrangements can be found in peri-urban Khulna (Bangladesh). Here, urbanisation is rapid and largely unregulated. Despite being located in the water-rich Ganges delta, peri-urban communities face drinking water insecurity. Access to safe drinking water supply is limited in many communities. Conflicts have also emerged between peri-urban and urban water users over access to groundwater, an important source of drinking water. For example, in nearby Phultala (north of Khulna city), peri-urban residents organised protests and filed a legal case against a large-scale groundwater abstraction project by Khulna city [7]. Declining groundwater levels and contamination from iron, salinity, arsenic, industrial, and wastewater contamination is also reported in this region [8,9]. Together, this provides a clear signal that existing institutions are unable to support sustainable or equitable water management outcomes.

Even though peri-urban communities are adversely affected by the outcomes of institutional design, they have limited insight into their institutional context. This is the result of isolation from formal decision-making arenas that limits access to information about institutions and the actors operating within decision-making arenas who apply these institutions. As a result, peri-urban communities have limited ability to formally intervene. This has been observed in peri-urban Khulna, where residents have previously resorted to informal strategies given their isolation from formal policy arenas [7]. Research from other peri-urban contexts also signals a lack of participatory interventions [10]. Thus, there is a need to support the problem solving efforts of peri-urban communities by closing this knowledge gap and building capacity to navigate policy arenas.

This paper explores the use of gaming-simulation methods to support problem solving through capacity building at the local community level. A game-based capacity building workshop is piloted with residents from peri-urban Khulna. During this workshop, participants used role-play to explore strategies to address the drinking water problems that are currently affecting them. Evaluation findings from this workshop are used to answer the following research question:

To what extent does the game-based workshop improve communities' problem understanding and capacity to intervene in peri-urban problems?

The paper is structured as follows. The theories in Section 2 explain the potential of using gaming-simulations to support peri-urban communities. Section 3 describes the materials and methods used to design the role-playing game and the workshop, as well as the evaluation framework and protocols used in this study. A brief introduction to the case study follows in Section 4, specifically, the drinking water problem in peri-urban Khulna. Results in Section 5 describe the role-play activities by participants during the workshop and evaluation of the game, workshop, and learning outcomes. The discussion in Section 6 reflects upon the pilot study, its limitations, and potential future research. Conclusions on the use of gaming-simulation methods with peri-urban communities are presented in Section 7.

2. Theory and Concepts

Gaming-simulation methods have appeared in the policy analyst toolkit since the 1960s in response to the need for human-centered approaches that incorporate the socio-political complexity of public policy issues [11]. Since then, different kinds of games have emerged in fields relevant to the area of application in this research: resource management [12–14], urban planning [15], and peri-urban conflicts [16].

Game design handbooks by Duke [17] and Greenblat [18] explain that design starts by defining the game's purpose. The purpose helps select the most appropriate gaming-simulation method. Bots and van Daalen [19] describe different functional uses of games to support policy processes. Here, gaming-simulations can serve as a laboratory for research and analysis, a medium for design and recommendation, a practice ring for strategic advice, a negotiation table for mediation, a consultative forum for democratisation, and finally, a parliament to clarify values and arguments. It is common for there to be more than one purpose, in which case, the designer must prioritise one, as the process of game design involves making trade-offs such as this [18].

In situations where stakeholders are marginalised from policy processes, gaming-simulation can be used for learning to achieve integrative negotiation within policy processes [16]. Here, learning also extends to the interactions occurring within policy arenas. Role-playing games offer a useful medium to achieve learning in this regard. Players can examine a problem by experiencing how different actors behave (within a set of institutions). One function of games is to serve as a 'virtual practice ring' for stakeholders to experiment with different strategies before formally entering the policy arena [9]. Here, the game can simulate decision-making arenas, where players use role-play to build negotiation skills before entering real-world negotiations. In this way, role-play can offer marginalised actors strategic support during the problem solving process.

Designing a role-playing game that simulates a real-world problem requires abstracting relevant details about the problem. These details help select the appropriate boundaries and inputs in the game. In this way, real-world phenomena can be mimicked in the roles, rules, and incentives in the game (Meijer and Hofstede, 2003 cited in [20]). One way of abstracting from the real world is with the help of game theory models. Game theory modelling is a method used to structure and analyse strategic behavior. The logic and analytical rigor of game theory is useful for interpreting real-world decision making phenomena in multi-actor situations [21,22]. They help illustrate how outcomes result from strategic interactions between actors based on their preferences, values, and objectives [23].

The literature highlights several similarities between the design schemas of game theory models and gaming-simulations. The same inputs used for constructing game theory models are also needed for game design. Models formalise strategic behaviour in the form of a 'game' consisting of players, actions, and payoffs on resulting outcomes [24]. These same inputs may be used by game designers to map the formal structure of a game [25]. Moreover, the focus on decisions and outcomes in game theory provides the building blocks for meaningful play [25].

Game theory models implicitly specify the 'rules' of the game. This offers additional design details for game designers with regards to the order of play, information in the game, and resources available to players. In the real-world, these rules are given by formal and informal institutions. In this way, the institutional context of the problem can be incorporated into the game's design. There are two ways of specifying rule conditions in the model. Non-cooperative game theory is used to structure conflicts within a fixed set of rules, where actors adopt a self-optimising attitude to meet their own objectives [23,26]. In cooperative game theory, the game assumes a willingness to communicate, coordinate actions, and pool resources by actors in the game [26,27]. Thus, different types of game theory models may be used to design role-playing situations where players can experiment with different strategies under the existing set of institutions, or experiment with collective action by changing or creating new institutions.

This study explores the use of gaming-simulation methods to support peri-urban communities in on-going problem solving efforts. In peri-urban contexts, communities have limited insight into formal policy arenas and negotiation experience. Simulating peri-urban problems as a role-playing game offers a medium for communities to explore the problem through its multi-actor environment, and experiment with problem solving strategies in a safe, virtual environment before engaging in real-world negotiations. Therefore, a role-play game is well-suited to the capacity-building needs of the beneficiaries in this context.

Designing a role-playing game for this purpose requires integrating unique features about the problem and the peri-urban context. The authors have previously developed a structured and participatory approach to help in this regard [28]. It begins by identifying the community's most pressing problem. Then, this problem is framed through an institutional lens, and relevant formal and informal institutions are mapped using the Institutional Analysis and Development framework [29]. Next, the focus shifts to the decision making arenas. Here, game theory models are used to examine how institutions are operationalised through actor interactions and the outcomes of these interactions. This research experiments with translating game theory models into a role-playing game to improve problem-solving capacity within peri-urban communities.

As peri-urban problems are multi-actor in nature, game theory models serves as the starting point in game design. Inputs from the models are used to define roles, rules, and incentives in the game, and identify potential scenarios for players to explore through role-play. The dynamic nature of peri-urban contexts is also incorporated by designing a game where players start with the existing peri-urban problem (starting conditions), after which new elements are introduced during the game [17]. In this way, players can identify appropriate strategies to peri-urban problems as they evolve.

3. Materials and Methods

3.1. Data Collection and Model Development

The village that participated in this study is situated approximately 7 km outside Khulna city. To protect the privacy of the study area, the village's name is kept confidential. This research is conducted as part of the Shifting Grounds project, that aims to support institutional change for pro-poor, sustainable, and equitable peri-urban groundwater management in the Ganges delta [30]. It is through this project that the authors first connected with the village in October 2014. Since then, extensive research and capacity building initiatives have been conducted. One of the main concerns in this village is access to safe drinking water supply. This problem was highlighted during early discussions with the community as part of the Shifting Grounds project. This game-based intervention was designed to offer insight into this drinking water problem.

Primary data about this problem was collected during two field visits by the primary author in 2015 and 2017. The 2015 field visit comprised 13 key informant interviews with government agencies and academic institutions, and a focus group discussion with 27 village residents. During this time, discussions covered the institutions, actors, interactions, and outcomes associated with drinking water supply. The second field visit in 2017 was used to identify actor values and preferences in resolving both water supply and water quality aspects of this problem, now and in the future. It consisted of 19 key interviews with residents (from peri-urban and urban areas), government agencies, a bottled water company manager, and a focus group discussion with 10 residents from the village. All meetings lasted between 30 and 90 min, were semi-structured, and were organised using snowball sampling methods. Co-authors from the local NGO partner Jagrata Juba Shangha (JJS) served as translators. To supplement this primary data, secondary sources such as government websites and Shifting Grounds reports were also used as inputs.

Using this data, game theory models were constructed. Detailed steps followed in early versions of these models were based on the initial 2015 field visit, and are available in Gomes, Hermans, and Thissen [28]. Since then, models have undergone several revisions as new details emerged about the problem. Over time, participatory research revealed the existence of several smaller but inter-connected problems, such as the lack of drinking water infrastructure, poor drinking water quality, and future changes in water services as a result of urban expansion from Khulna city. These sub-problems were modelled as individual games, each defined by their own institutional arrangements and multi-actor interactions. This study uses inputs from three game theory models. Model 1 is on the existing (peri-urban) drinking water supply situation, and model 2 is on the future (urban) drinking water supply situation. Both are modelled as non-cooperative games using Gambit [31]. Model 3 focuses on water quality aspects of the problem, more specifically, groundwater quality, and explores cooperative strategies to improve monitoring. Solutions in this cooperative game theory model were analyzed using R [32].

3.2. Game Design

Three short role-playing games were designed using inputs from the corresponding game theory model. This included the players, their roles, actions, resources, and potential outcomes in the game. Game 1 is about the existing drinking water supply situation, game 2 on the future drinking water supply situation, and game 3 on groundwater monitoring.

Initial game design proposals were discussed with local experts and project partners for context-specific feedback. Thereafter, two test sessions were conducted at Delft University of Technology (Netherlands). The final game was also played with researchers from Khulna University of Engineering Technology in March 2018. As this game was designed for peri-urban residents, it was translated into the local language (a dialect of Bangla) by the co-authors from JJS. Two half-day training sessions were held with facilitators at JJS offices in Khulna to familiarise them with the game and other provided supporting materials, such as the facilitator script.

In each game, participants are assigned a role. Each game has several strategic players. These players consciously made decisions in the game based on their objectives, using the actions and resources available to them. In games 1 and 2, there were also non-strategic (or chance players). Their actions during the game were selected by rolling a dice. Each player was given various cards that described their role, actions, and resources (assets) that a player can use during the game. All players had at least two actions to choose from. Action cards also specified the conditions for each action. For example, some actions could only be used in combination with a resource card. Examples of the aforementioned game materials can be found in the supplementary materials.

Role-play in each game was based on the game's objective. The objective in games 1 and 2, was for residents to obtain drinking water supply. Therefore, residents moved first in these games. Their actions determined who would move next in the game, and so on. The combination of actions by different players produces an outcome. The outcome could be a type of drinking water supply or no water supply at all. From game theory models, we had identified the set of all possible outcomes in each game. After each round, players had to rate their satisfaction (happy, indifferent, or sad) with the outcome using a scorecard (see supplementary materials for an example). The scoring was based on that player's values. Multiple rounds allowed players to draw comparisons between the strategic outcomes.

The objective in game 3 was slightly different. The first round of this game focused on non-cooperative strategies. It represents the status-quo of groundwater monitoring in peri-urban Khulna. Each player in this round had to identify the locations where groundwater data was individually collected. A groundwater crisis situation was then introduced in round 2 of the game. It prompted players to explore more collaborative groundwater monitoring strategies. Participants were given 15 min to negotiate agreements. Thereafter, they had to describe the locations, terms, and conditions for cooperative monitoring that was agreed to. All possible non-cooperative and cooperative outcomes in this game were identified from the model. Participants used their scorecards to rate their level of satisfaction with both non-cooperative vs cooperative groundwater monitoring strategy.

All three games were facilitated with the help of a game board (Figure 1). It described the physical geography of a fictional peri-urban area outside a city boundary (demarcated by a red line). Its design was closely based on the geography of peri-urban Khulna. The game board was used to indicate the location of players (orange circles) and their actions during the game (white rectangles). Outcomes were also visualised on the game board using drinking water icons, which described the type of drinking water supply that resulted in each round (Figure 1 top row). In game 3, groundwater monitoring icons (Figure 1 bottom row) were used to identify the locations of non-cooperative (different coloured icons) and cooperative groundwater monitoring (blue icons). The game board was customised for each game as needed. In session 2, urban expansion was visualised by extending the city boundaries with colored tape and adding city icons (Figure 1 middle row) in the expanded areas to indicate urban development. During game 3, players who had agreed to cooperate were connected on the board with coloured tape.

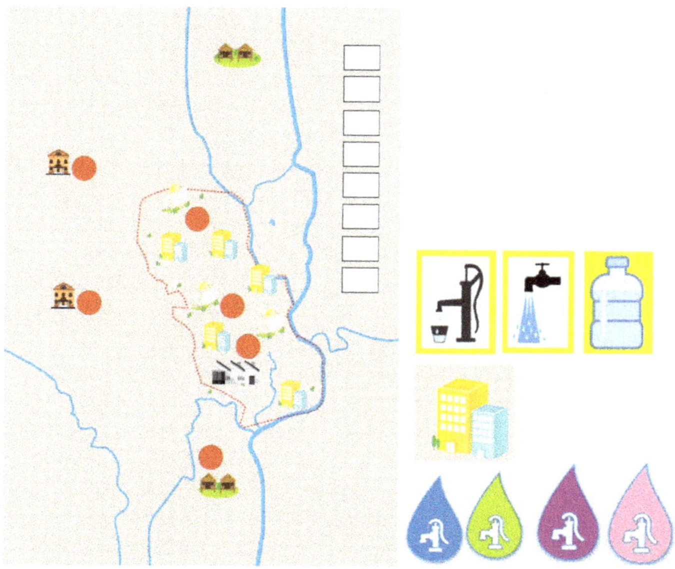

Figure 1. Game board (**left**); Game board pieces: (**top row**) drinking water icons, (**middle row**) city icons, (**bottom row**) groundwater monitoring icons.

3.3. Design and Evaluation of Gaming-Simulation Workshop

All three games were played as part of a gaming-simulation workshop. The one-day workshop was held on 27 March 2018 at a venue in Khulna city. Four facilitators from JJS were present, including a main facilitator who conducted the workshop, one observer who recorded workshop proceedings, and two co-facilitators who organised game materials, assisted participants, and recorded discussions during the debriefings. The game designer (primary author) was also present to answer questions related to the game or workshop activities.

The workshop was attended by 9 participants from the peri-urban village that were formally invited by JJS. They included 6 female and 3 male participants. Participants represented key stakeholder groups from the community. Six participants were from the negotiation group that regularly participated in community meetings organised by JJS as part of the Shifting Grounds project. The remaining 3 represented recent migrants and low-income households from the village. All participants signed a consent form and were remunerated for travel costs to the venue.

The overall workshop design, including its structure, specific activities, and time management plan, is shown below in Figure 2. In the introductory session, facilitators presented the workshop's objectives, introduced participants to the role-playing format, and conducted the pre-workshop evaluation. This was followed by the three role-playing games. The sequencing of games was important. We assumed that the existing situation is most familiar to peri-urban residents, and hence, is a suitable starting point. The (future) urban water supply game was played next, given that it closely resembled game 1, extending it with a long-term perspective to the problem. This was followed by game 3, where participants focused on water quality aspects of the problem, and explored non-cooperative and cooperative groundwater monitoring strategies.

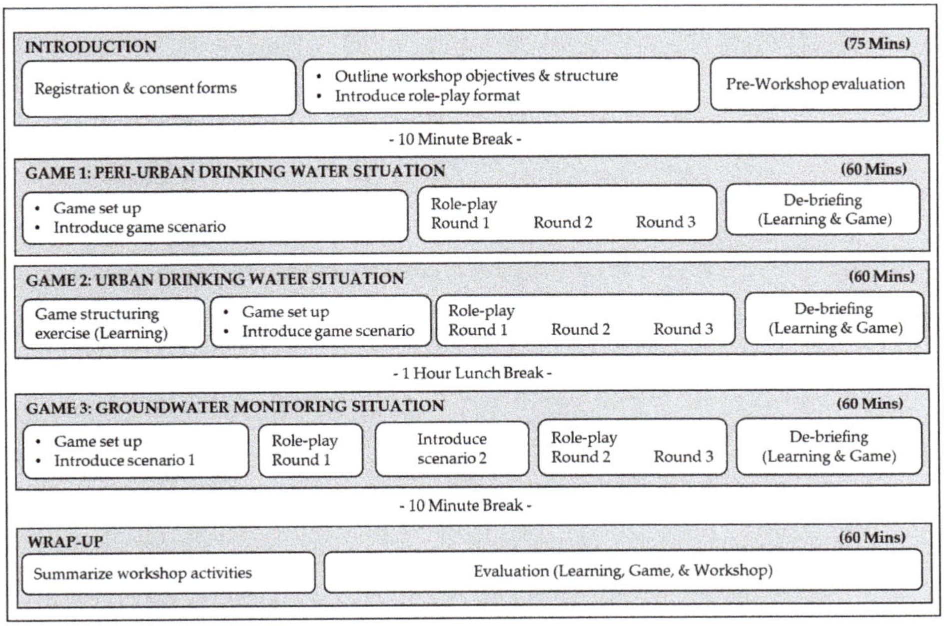

Figure 2. Overview of workshop structure.

Facilitators began each game with an introduction to the drinking water situation to be examined and the objective of the game. Next, game materials were distributed and roles were assigned. Depending on the number of roles available, participants were individually assigned a role, or worked in groups of 2–3 persons. Multiple rounds were played in each game, and each outcome was evaluated by the players. In game 1, a trial round was played at the beginning to familiarise players with the game materials. Each game concluded with a de-briefing exercise, led by the facilitator.

A wrap-up session concluded the workshop with a summary of activities and final evaluations of the game, workshop, participant experience, and learning.

Assessing the impact of this gaming-simulation workshop required a suitable evaluation framework. Different frameworks offer suitable starting points for this [33–35]. Evaluation in this study, shown below in Figure 3, draws from Thissen & Twaalfhoven [34]. It is designed around three components: the input, activity, and result. Specific evaluation criteria and sub-criteria are specified for each component, taking cues from Midgley et al. [35].

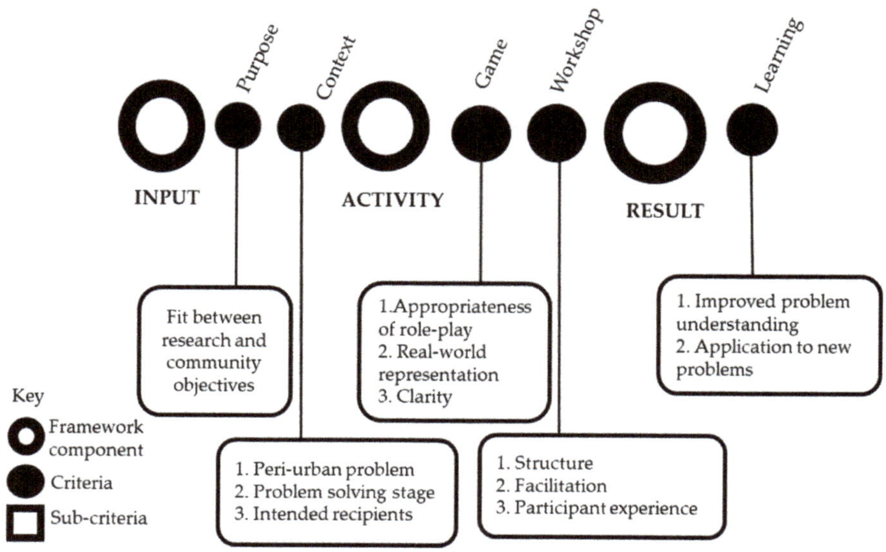

Figure 3. Evaluation design (adapted from [33–35]).

The inputs refers to the purpose of the game. Examining the fit between designer and community objectives is a principle of good game design and offers a basis for reasoning the successes or failures experienced. Context refers to inputs beyond the control of the designer, thus representing important design boundaries. Contextual attributes included appropriateness to the type of problem being addressed, stage of problem solving, and intended recipients. The activity is evaluated with respect to the game, workshop, and overall participant experience. Game criteria refers to the appropriateness of role-play, real world representation, and clarity of the game. Workshop criteria includes its structure and facilitation.

Finally, results refer to participant learning. Learning was evaluated in two ways. First, we examined the extent to which participants' problems with understanding were improved. This was done by comparing problems of understanding at the start of the workshop (as a pre-workshop discussion) and after each game during the workshop (de-briefing). Second, we identified higher-order learning outcomes. For this, a game-structuring exercise was conducted after the first game. Participants were presented with a new problem situation and asked to name the actors, actions, resources, and values. This exercise was used to assess whether knowledge and insights from the previous game could be applied to similarly structure other types of peri-urban problems.

The evaluation medium was facilitated group discussions using flip charts. Different evaluations were structured throughout the workshop (Figure 2). Participants also filled out a post-workshop questionnaire in the wrap-up session. It was used to assess their overall workshop experience. During the workshop, observations were recorded by one of the facilitators on a process record form. This included the questions asked by participants during the games and outcomes of role-play activities. Evaluation results were later translated into English, coded in Excel, and analysed qualitatively.

4. Case Study: Drinking Water Problems in Peri-Urban Khulna

In this peri-urban village, groundwater is the primary source of drinking water. The village is officially part of the rural jurisdiction. Currently, residents have two strategic options to access groundwater. One is to apply for a public tube well from the Water and Sanitation (WATSAN) committee. The other option is to invest in a private tube well. Compared to public tube wells that are subsidised by the government, private tube wells are expensive. Therefore, most households prefer public over private tube-wells, as not everyone has the financial means to invest in the latter.

Approval from the WATSAN committee is needed before a public tube-well can be installed. This committee operates at the sub-district (upazilla) level, as per the formal rules of the rural administration. They can either reject an application or approve it by issuing a tube-well licence. There are two reasons why the committee might not approve an application. First, given that most members of the WATSAN committee are elected, there is a need to appease their local constituency. This can lead to an inequitable distribution of tube-well licenses, with preference given to villages with stronger ties to WATSAN members. Second, license quotas (or tube-well funding) are limited. In this village, applications are rarely approved.

If a license is approved, the Department of Public Health Engineering (DPHE) manages the installation process. This begins by selecting a site to install the public tube-well. This can be done based on an assessment of local groundwater conditions, or based on the preference of the applicant. Discussions highlight a preference for the latter, as local aquifer data is limited. Thereafter, DPHE hires a mechanic to install the tube-well. Mechanics are hesitant to accept these installation contracts, as the terms state that payment depends on the success of tube-well installations. In other words, mechanics are not paid for tube-wells that supply poor quality or limited groundwater volume. If sites are selected without assessing local groundwater conditions first, then installation contracts can be risky for mechanics.

As mentioned, the village can also invest in a private tube well. Officially, this also requires a licence; however, it is currently unregulated, as the government is aware of the gap in rural drinking water supply. Here also, mechanics are hired to install a private well, except they are paid irrespective of the viability of the tube-well site. Therefore, there is no risk for mechanics to accept private tube-well installation contracts. The risk instead is borne by the village, as a non-viable tube-well would result in financial losses for the investor.

In the future, drinking water supply in this village is likely to change from rural to urban service providers as proposals to expand city boundaries are currently under review. Thereafter, peri-urban villages will become part of the urban jurisdiction. In this urban scenario, residents have four strategies to obtain drinking water supply. First, they can apply for piped water supply at the city's Water and Sanitation Authority (WASA). Currently, Khulna WASA supplies treated surface water via pipelines. To reduce the pressure on groundwater, they are investing in new supply projects to increase the capacity and coverage of the piped supply network. However, these on-going projects are designed to meet the needs of the current population. Piped supply in these future urban areas depends on the availability of additional water supply projects. Therefore, in future, this option may not be available to residents.

A second option is to apply for a public tube-well from the city's WASA. Here, there is the risk that public-tube-wells might be unuseable if the installation site is non-viable. However, the costs for using a public-tube well are less than those of piped supply. It is also less convienient compared to a household connection. Depending on their financial capacity, residents might prefer public tube-wells. A third option is to invest in private tube-wells. Similar to the peri-urban scenario, mechanics are hired to install private tube-wells, and there is a risk for residents that the site will be non-viable. Moreover, the costs of a private tube-well are significantly more than those of a public-tube well, making it a less affordable option.

Finally, the fourth option is to purchase drinking water from (private) bottled water companies. In Khulna, there are several informal suppliers. They typically sell large jars of packaged groundwater. This option is only accessible to households who can afford it. These companies are expected to test drinking water quality; however, it is likely that this is not always done, especially by those operating illegally. Similarly, in the future, bottle water companies can only supply if they are able to meet the demand and recover the higher distribution costs (from supplying to customers in more remote areas of the city).

The village's drinking water problem also relates to groundwater quality. Residents have the impression that good quality water is only accessible from deeper aquifers. Rural providers are unable to ensure safe drinking water quality due to limited knowledge of groundwater conditions. This points to a gap in groundwater monitoring. This monitoring gap is also a concern for other government actors. Several actors highlighted a need for coordinated groundwater monitoring. In Bangladesh, groundwater is monitored by several government agencies as part of their mandate. Residents also informally monitor groundwater to meet daily household needs. They mentally record the quantity and quality available from different tube-wells in the village. This is done by observing seasonal changes in the volume available from tube-wells, and using sensory signals like taste, odor, and color of groundwater. Health issues from drinking contaminated water is also an indicator of poor water quality.

A game theoretic representation of the existing and future drinking water supply and groundwater monitoring problems, can be found in the supplementary materials. The role-playing games in this study are designed to help communities explore this problem.

5. Results

5.1. Strategies Explored through Role-Play in the Capacity Building Workshop

Participants played multiple rounds of each role-playing game during the workshops. The results of each round in games 1, 2 and 3 are summarised below in Table 1.

Table 1. Overview of strategies explored during the three games.

Game	Strategy	Outcome
	1. Peri-urban drinking water supply	
Round 1	License for public tube-well application approved but aquifer assessment pre-installation finds site non-viable, so installation is halted.	No drinking water supply in peri-urban village available
Round 2	Investment in private tube-well is successful, as installation site is discovered to be viable for groundwater.	Peri-urban village receives drinking water supply via private tube-well
Round 3	Application for public tube-well not approved by WATSAN committee as all licences are issued elsewhere.	No drinking water supply available for peri-urban village
	2. Urban drinking water supply	
Round 1	Application for piped water supply is approved by urban drinking water provider.	Residents get drinking water supply via piped network supplying treated surface water
Round 2	Bottled water companies refuse to sell water to residents of former peri-urban area.	No drinking water supply available for newly urban residents
Round 3	Failed investment in private tube-well as installation site is discovered as non-viable after installation.	Residents must either use the poor quality, unreliable groundwater from private tube-well or look for other alternatives
	3. Groundwater monitoring	
Round 1	Individual (non-cooperative) monitoring by residents, DPHE, DOE	Fragmented groundwater data from different areas is insufficient for decision-making purposes
Round 2	Joint (cooperative) monitoring by residents and DPHE, while DOE monitors groundwater separately.	Sharing of groundwater data and better knowledge of aquifer conditions in peri-urban areas only

5.1.1. Game 1: Peri-urban Drinking Water Supply

In this game, participants explored the existing (peri-urban) drinking water supply situation. The players in this game included peri-urban (residents), WATSAN committee (chairman), DPHE (engineer), a mechanic, and the two chance players—nature and other unions. Nature determined the viability of a tube-well installation site, whereas the chance player 'Other unions' decided the total number of applications received by the WATSAN committee.

In round 1, residents applied for a public tube-well, the cost for which was paid using monetary resources. The WATSAN committee approved a tube-well license for the village. Thereafter, DPHE opted to assess aquifer conditions beforehand, as groundwater data was available (as a resource card). However, nature revealed the site to be non-viable, so installation was halted. Therefore, the outcome of round 1 was a return to the status quo or no drinking water supply for the village. Figure 4 shows how game play was depicted on the game board in this round. Although this strategy is the village's preferred solution to the drinking water problem in reality, scorecards in the game showed mixed feelings towards this outcome. Residents and the mechanic were unhappy, whereas the WATSAN committee and DPHE were satisfied overall, despite the outcome.

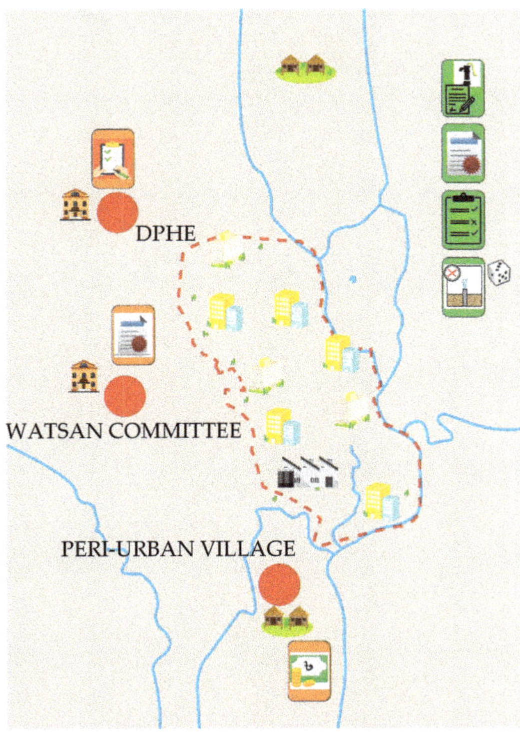

Figure 4. Game play in round 1 (Game 1).

In round 2, residents explored a different strategy. Residents used their monetary resources to invest in a private tube-well by hiring a mechanic. As groundwater data was unavailable (as a resource card) to both residents and the mechanic, no aquifer assessment was conducted prior to installation. Nature discovered the site as viable, resulting in reliable, good quality, although more expensive drinking water supply. Everyone except DPHE was satisfied with this outcome. Therefore, private tube wells are a more favorable strategy for peri-urban residents, despite the higher costs and associated risks. It is possible that residents are willing to pay for reliable, safe drinking water services. However, their decision to re-visit a more affordable option in round 3 also suggests otherwise.

In round 3, residents decided to re-visit the public drinking water option. This time, other unions also submitted tube-well applications to the committee. In this case, the committee decided to allocate all licenses to the other unions. This meant no drinking water supply for residents. As expected, they were unhappy with this outcome, while the committee and DPHE had mixed feelings, and the mechanic was satisfied.

5.1.2. Game 2: Future Drinking Water Supply

In game 2, participants explored the (future) urban drinking water supply situation. The players included former peri-urban village (residents), WASA (engineer), bottled water company (manager), the mechanic, and nature (chance player).

In round 1, residents applied for a piped water supply connection from WASA (paid using resource cards). The availability of piped water supply projects (a resource for this player) made it possible for WASA to extend good quality, reliable piped water supply (Figure 5). In round 2, residents opted to purchase bottled water; however, they were refused by the bottled water company due to low production volumes (as stated on their resource card). This meant a return to the status quo for residents. Thereafter, they reverted to a familiar strategy in round 3 and paid a mechanic to install a private tube-well. However, following installation, the site was discovered to be non-viable by nature. Thus, the outcome of this round is a failed tube-well installation, financial losses, and no water supply.

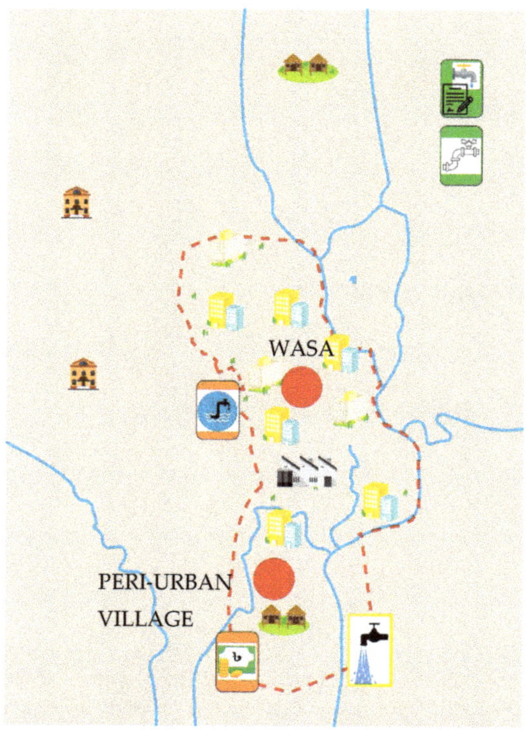

Figure 5. Game play in round 1 (Game 2).

The scorecards in game 2 shows that residents were successful only in the piped water supply strategy (round 1). This outcome was considered satisfactory by both WASA and residents, although it scored poorly with respect to affordability by residents. The mechanic was indifferent, and the bottled water company was unhappy with this outcome. By comparison, round 2 was considered unsatisfactory by almost all players except for the mechanic, who was indifferent to the result. In round 3, the outcome was poorly scored by residents as well as the bottled water company, while WASA was indifferent and the mechanic was happy, as they were paid for installation services despite the outcome.

5.1.3. Game 3: Groundwater Monitoring

There were three players in game 3: DPHE (engineer), village (residents), and the (water quality analyst) from the Department of Environment (DOE). In round 1, players identified the locations in which they were each collecting groundwater data as part of their monitoring efforts. This outcome is shown in Figure 6. The action cards indicated that DPHE collects groundwater data only in non-urban areas from public tube-wells at limited times of the year. Residents monitor both private and public tube-wells used for drinking purposes within their village. The DOE's monitoring locations meanwhile are only within city areas, as per the instructions from higher levels of the department. Scores in this round varied. Residents were indifferent to this outcome, as informal methods are, to some extent, helpful in managing daily water needs. DOE was satisfied overall, while DPHE was only partially satisfied, presumably as the data collected is insufficient and unreliable for decision-making purposes.

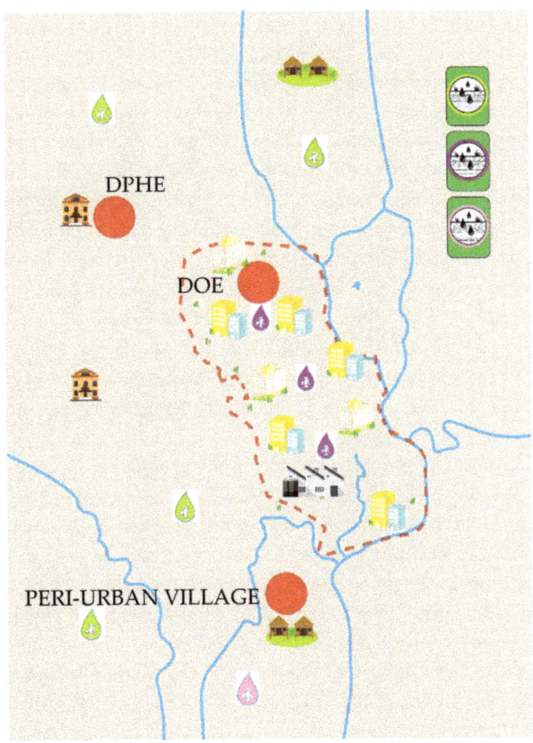

Figure 6. Game play in round 1 (Game 3).

In round 2, players explored cooperative groundwater monitoring strategies in response to a groundwater crisis. Following negotiations, participants explained that a full consensus between all 3 players could not be reached, as they were unable to partner with the DOE. Only residents and DPHE agreed to share knowledge and information about groundwater conditions from their respective monitoring areas. Periodic meetings were arranged for this. Compared to round 1, all 3 players were satisfied with the outcome of round 2. Only groundwater availability was scored poorly by residents. A likely explanation for this is that collaborations with DPHE might improve groundwater quality given the shared interest, more so than groundwater level monitoring.

5.2. Evaluation of Role-Playing Game

The appropriateness of role-play was evaluated during the workshop. In the introductory session, participants expressed their familiarity with role-play through local dramas (e.g., theatre) and cultural events. Moreover, local experts had highlighted that in this context, 'games' are typically associated with informality and playfulness. Therefore, we purposefully referred to the games only as role-play activities during the workshop. Despite participants familiarity, significant time was spent initially to ensure understanding of the game materials (30 min) and to complete role-play activities (50 min) in game 1. Games 2 and 3 by comparison were completed much more quickly (e.g., 10 min for role-play in game 2), with less assistance and greater confidence observed among the participants.

The wrap-up discussion revealed that participants enjoyed the role-play very much, as it provided information about the problem and the institutions. Participants expressed an interest in disseminating knowledge gained from this workshop with others in their village. They requested support from JJS to provide game materials for this. This indicates that knowledge sharing is one of the wider impacts of this workshop. With regards to problem solving, the workshop highlighted the need for collaboration and showed how residents can conduct this with authorities. One participant also suggested that the game-based approach could be applied to examine other problems and used to facilitate multi-stakeholder workshops between the community and authorities.

Local residents from this village varied in their level of education, literacy, and socio-economic background. Therefore, significant effort was made to ensure materials were easy to understand by translating them using a previously-developed glossary of local terms. Game testing and facilitator training sessions helped identify potential questions that might be raised, or points of confusion. During the debriefings, we evaluated the clarity of all three games. Participants felt that role of description cards could be simplified further, and in some cases, be more specific. On the other hand, they were positive about the game board, particularly the visualisation of rural-urban boundaries (Game 2) and negotiation outcomes (Game 3). Similarly, scorecards were well received, and were considered useful for comparing strategies.

The game's representation of the real-world had mixed reviews. Some participants highlighted missing details such as, for example, politics at the WATSAN committee (due to pressure from higher level authorities and local communities). Similarly, the practice of groundwater monitoring was, in reality, much more limited compared to what was observed in session 3. On the other hand, preferential allocation of tube-wells by the committee to other unions over the village was considered accurate. On the positive side, participants experienced the complexity of problem solving through this game. Both nature and player's resources (for example, financial constraints) were found to strongly influence the outcome, especially in game 2. In game 3, participants compared negotiations in the game to multi-stakeholder meetings held by JJS as part of the Shifting Grounds project. In other words, participants recognised that real-world negotiations to address peri-urban problems often requires third-party involvement and support.

5.3. Workshop Evaluation

The three games were explained well according to the participants, despite problems understanding the materials in game 1. They noted that some participants needed more help than others, as not everyone had the same level of problem understanding. Those who had participated in project meetings in the past felt more comfortable discussing local issues, whereas additional background information about actors was required by first-time participants. The time allocated for each game was considered appropriate, aside from the negotiation round in game 3 and in game 1, where more time was needed. It was suggested that the game materials could also be shared prior to the workshops to give residents more time to understand them.

Facilitation was overall positive, although it was highlighted that facilitators needed further training on the order of play. This is also visible in the video footage of the workshop. Despite the use of local terminology, participants misinterpreted certain elements of the game. Efforts to involve facilitators—who have knowledge about the local context and culture—earlier in the design process might prevent this. Significant time and effort was spent preparing facilitator scripts and conducting training sessions. This was found to be very valuable during the workshop. Although facilitators had no prior experience with gaming-simulation or game theory modelling, their expertise and relations with the community were valuable during the workshop. Participants felt comfortable, and did not hesitate to ask questions or communicate their feedback as a result.

A post-workshop questionnaire was used to evaluate overall participant experience. Respondents expressed that the workshop made them feel happy (75%), sad (18%), and inspired (9%). The workshop was convenient to attend for peri-urban residents, at an appropriate venue, and with sufficient refreshments and breaks. Finally, 60% would have preferred the workshop to be longer in duration, while the remaining considered it appropriate.

5.4. Evaluation of Learning Outcomes

The pre-workshop assessment in the introductory session (Figure 2) represents the initial problems with understanding among the participants. Community members explained that their drinking water problem relates to several aspects: poor groundwater quality (due to salinity and iron contamination), inconvenience of water collection from distant sources, groundwater depletion (seasonal scarcity), and wasted investments in tube-wells (as residents do not know the exact layer to abstract groundwater from). When asked which actors were involved in this problem, participants only listed women. It is true that women are most affected given their responsibility for household water collection. As 6 of the 9 participants in this workshop were female, this response is not entirely surprising. In reality, however, the problem involves several actors. It is possible that this question was misinterpreted by participants as the actors 'affected' by the problem, or it represents a very narrow perception of the problem's scope or boundary.

Participants described two potential solutions to this problem. First, they see a need to identify the appropriate groundwater layer. This strategy requires support from the sub-district chairman and local Members of Parliament. Until now, the village has been unable to achieve this. Another solution would be to install a water treatment plant nearby, although it was not mentioned who would be responsible for providing this solution. The community recognises that they have a role to play in the problem solving process. Participants explained that they need to present a united front while requesting solutions from the authorities. They added that this type of collective action is not new to the village, but was practiced regularly, i.e., every time they apply for a public tube-well.

The de-briefings conducted after each game highlighted the learning that resulted from role-playing activities. Participants' understanding of the tube-well licensing process was improved through game 1 (peri-urban situation). They learned that access to financial resources does not guarantee access to groundwater, given the important role of nature in the success of tube-well installations. Tube-well quotas was also a previously unknown factor in the decision making process of the committee prior to this workshop. When asked if they discovered new actors through this game, several were mentioned, such as the DPHE, WATSAN chairman, nature, and villages. Participants likely misunderstood this de-briefing question, as these actors were previously mentioned during field interviews and in the pre-workshop evaluation. Although game 1 did not result in a new solution strategy, the strategy in round 2 was considered most satisfactory, and is reflected in the scorecards. Furthermore, participants learned that not all strategies lead to satisfactory outcomes (especially for peri-urban residents), and that different strategies require interaction with different actors.

Game 2 (future urban situation) provided insight into the application process for piped water supply. It highlighted the fact that the option to purchase bottled water depends on the availability of financial resources. This was a newly-discovered problem for participants. Moreover, in the event that future drinking water options (piped water supply or bottled water) are unavailable, residents need to fall back on existing strategies (private or public tube-wells) to access drinking water. Participants learned that not all actors are satisfied with the same outcome. For example, bottled water companies were unhappy in all 3 rounds, although generally, round 1 (piped supply) was considered satisfactory by most players. In this way, participants discovered a new solution to the drinking water problem from game 2, namely, piped water supply. The session also led to the discovery of two new actors: bottled water companies and WASA.

The main learning from game 3 relates to coordination with regards to groundwater monitoring. Participants noted that, in reality, both DPHE and residents are unwilling to monitor groundwater. But residents now feel they have a role to play, by informing the authority about the (local) groundwater situation. They see a benefit from collaborating with the other players to identify a suitable groundwater layer. In round 2, residents could collaborate with DPHE but not the DOE. Participants did not see a need to collaborate with the DOE, as DOE operates primarily in city areas. It is likely that the DOE's role was not fully understood during the game. Players found resources helpful while negotiating agreements. However, it is possible that player resources cards (assets) were mistaken for physical water resources by some participants and/or facilitators. The response to this question suggests this.

Evaluating the transferability of problem structuring to other local problems was based on a game structuring exercise at the start of session 2. Participants were asked to list the basic game elements (actors, their actions, values, roles, and resources) in the future drinking water scenario. Only two actors were listed during this exercise: Khulna WASA and KCC, together with some details about their role, values, actions, and resources. This exercise shows that learning from game 1 could be easily applied to structure other peri-urban problems, although only to a limited extent. This is understandable, given the availability of information about this future scenario.

6. Discussion

Workshop results show clear improvements in problem understanding, as reported by participants. At the same time, this gaming-simulation workshop was part of a larger project, and the impacts of both the larger project activities and this specific intervention are hard to disentangle. The understanding of participants prior to the workshop was already partly fed by earlier interactions with the project's researchers and trainers about the drinking water problems and about negotiation plans. Even then, it is clear that the workshop was successful in communicating some of the key insights about institutions and strategic interactions to the local community members.

We are aware that this pilot study provides a limited basis for more general conclusions about the uses of gaming-simulation methods to support peri-urban communities. The authors have conducted this workshop with government actors from peri-urban Khulna on 28 March 2018. A follow-up workshop was also conducted by JJS in July 2018 with the same village. We are also currently exploring opportunities to use game-based capacity building to strengthen polder management elsewhere in Bangladesh. These case studies lie outside the current scope of this paper, however, they provide a basis for broader reflection in the future.

Although this pilot study shows the benefits of role-playing methods, this intervention had its limitations, given the constraints of the methods and boundaries selected for its design. Game theory models were the starting point in game design. These models are limited by the number of players that are used to structure the game; this necessitated simplifying the game's design and the level of detail about the problem. Therefore, it was not surprising that participants highlighted missing details. We argue that the problem boundaries used to design this game were appropriate at this stage of the problem solving process. At the same time, it is clear that a game theoretic analysis, and using a game to communicate its results, likely produces a different experience for participants compared to a game designed more for play.

The role-playing game focused problem examination onto the institutional context. The peri-urban system is far more complex, with biophysical changes (land use, groundwater use) and socio-economic changes (livelihoods, population, economic conditions) occurring simultaneously. In other words, the 'wicked' nature of this problem is only partially captured in the game. The physical and socio-economic context was still being researched by project partners, and therefore, could not be used as inputs in the game's design. On the other hand, had these results been available, their full inclusion might also have led to a more complex game that ran the risk of not being playable or easily understandable by the community. Striking the right balance here, therefore, remains an issue for further deliberation.

There is scope to improve the game design process as well. Through field visits, the community contributed inputs for the game used in this study, but was not directly involved in the game's design. A model building workshop could be used to more directly co-create the game theory models with community representatives, which would thereafter translated into a game by researchers and played by the community in a follow up workshop. Another possibility would involve creating a semi-structured game, where participants have the ability to add new game materials (actors, roles, actions, or resources) during the workshop. Follow up research can explore this and other participatory game design approaches.

We do not want to claim that this single gaming workshop could have been done in isolation of other community capacity building activities. We believe that this workshop has made a valuable contribution though, as part of a stream of work and activities in a larger project. However, even this larger stream of work is limited in its expected results—and these limitations thus also apply to our game-based workshop. Community empowerment and capacity building is not something to be achieved in a single workshop, or even by one project that runs for a few years. It requires much more than this. In this case, for instance, some of the very powerful and influential actors have not been willing to truly engage with the project, such as private sector companies and land development actors. This is not unique for this case, and also, does not mean that communities cannot be supported so that they can have a stronger voice and role in future processes.

In this case, government agencies have been more responsive and willing to participate. Here, the game-based workshop also offers a vehicle for direct interaction, playing it jointly with local community and government representatives. Similarly, now that this medium is available and known, it can be used within the community to extend the dialogue and understanding to community groups beyond the workshop participants and to other neighboring communities. Both these ideas for future use were put forward by the community participants, suggesting that they do see the value of the gaming workshop.

7. Conclusions

The pilot workshop in peri-urban Khulna indicates that gaming-simulation (in this case, a role-playing game) offers a practical and visual medium for peri-urban communities to examine local problems through an institutional lens. This method was well-suited to the context and the workshop participants. Role-play fostered interactions and active discussions on a complex multi-actor problem. It also helped communicate insights about institutions, often an abstract concept, in a way that is easy to understand and apply to strategic problem solving. Moreover, it revealed the dynamic nature of peri-urban problems as a result of urbanisation. Participants' understanding of problems improved with regards to the multi-actor interactions that exist in peri-urban problems, evaluation and comparison of strategies to address local problems, and the benefits and challenges of cooperation.

The use of gaming-simulation methods to support peri-urban communities requires a simple game design by selecting the appropriate level of detail. In this study, game theory models were used to design games of this nature. Key elements of game theory models such as players, actions, outcomes, and payoffs can be easily translated into a role-play game. However, while simple role-play is recommended to start with, we also recognise that, over time, capacity building activities should address more complex aspects of the problem.

This research provides a basis for improving the game design process and replicating this approach elsewhere. Future uses of gaming-simulation methods for strategic exploration can experiment with even more participatory game design processes with the local community. Overall, this pilot study shows that gaming-simulation methods can be a suitable and useful medium to support problem solving by peri-urban communities in the global south.

Supplementary Materials: The following are available online at http://www.mdpi.com/2073-4441/10/11/1704/s1, Figure S1: Non-cooperative game theory model of peri-urban drinking water supply, Figure S2: Non-cooperative game theory model of future (urban) drinking water supply, Figure S3: Cooperative game theory model of groundwater monitoring, Figure S4: Role description card for peri-urban residents, Figure S5: Action card for residents in session 2 (**a**) Front; (**b**) Back, Figure S6: Resource card for residents (**a**) Front; (**b**) Back, Figure S7: Scorecard for residents in session 1, Table S1: Description of outcomes in the model of peri-urban drinking water supply, Table S2: Description of outcomes in the non-cooperative model of urban drinking water supply, Table S3: Description of outcomes in the cooperative model of groundwater monitoring.

Author Contributions: Conceptualisation, Methodology, formal analysis (including workshop facilitation, evaluation, and analysis of results), data curation, and writing (original draft and revisions) contributed by S.L.G.; Conceptualisation, Methodology (game testing), Writing (review & editing), project administration, supervision, and funding acquisition contributed by L.M.H. and W.A.H.T.; Formal analysis (workshop facilitation, planning, and evaluation), resources (logistical support, and translation) contributed by K.F.I., S.N.H. and A.Z.H.

Funding: The research reported here was part of the Shifting Grounds project, which was funded by the Urbanising Deltas of the World Programme of the Netherlands Organisation for Scientific Research (NWO), under Grant No. W 07.69.104.

Acknowledgments: We extend our sincere thanks to S.R. (former employee of JJS) for assisting in this workshop; TU Delft colleagues for inputs and participation in the test sessions; to S.G. for contributing to the design of the game board, game materials, and Figure 4, Figure 5, Figure 6 in this paper; and to the Shifting Grounds project team for support at various stages of the project. We wish to thank the workshop participants of the 2018 International Workshop on Public Policy for their feedback on earlier drafts of this paper. Finally, we are indebted to the anonymous reviewers for their valuable comments and suggestions.

Conflicts of Interest: The authors declare no conflict of interest.

References

1. UN DESA. *World Urbanization Prospects: The 2014 Revision, Highlights*; United Nations: New York, NY, USA, 2014; Available online: https://esa.un.org/unpd/wup/publications/files/wup2014-report.pdf (accessed on 15 December 2015).
2. Narain, V. *Peri-urban water security in a context of urbanization and climate change: A review of concepts and relationships*; Peri Urban Project; SaciWATERS: Secunderabad, India, 2010; pp. 1–16. Available online: http://saciwaters.org/periurban/newsletter/pdfs/discussion_papper_1.pdf (accessed on 22 October 2014).
3. Allen, A. Environmental planning and management of the peri-urban interface: Perspectives on an emerging field. *Environ. Urban.* **2003**, *15*, 135–148. [CrossRef]
4. North, D.C. *Institutions, Institutional Change and Economic Performance*; Cambridge University Press: Cambridge, UK, 1990.
5. Williamson, O.E. Transaction Cost Economics: How it works; where it is headed. *De Economist* **1998**, *146*, 23–58. [CrossRef]
6. Gomes, S.L.; Hermans, L.M. Institutional approaches to peri-urban research: Added value and conceptual approach. *S. Asian Water Stud. J.* **2016**, *5*, 103–108.

7. Gomes, S.L.; Hermans, L.M. Institutional function and urbanization in Bangladesh: How peri-urban communities respond to changing environments. *Land Use Policy* **2017**. [CrossRef]
8. Kumar, U.; Khan, M.S.A.; Rahman, R.; Mondal, M.S.; Huq, H. *Water Security in Peri-Urban Khulna: Adapting to Climate Change and Urbanization*; Peri Urban Water Security Discussion Paper Series; SACI WaterS: Telangana, India, 2011; 20p, Available online: http://saciwaters.org/periurban/2%20idrc%20periurban%20report.pdf (accessed on 29 December 2014).
9. Thissen, W.A. Diagnosing Policy Problem Solutions. In *Public Policy Analysis: New Developments*; Springer: New York, NY, USA, 2013; Volume 179, ISBN 978-1-4614-4601-9.
10. Ducrot, R.; van Paassen, A.; Barban, V.; Daré, W.; Gramaglia, C. Learning integrative negotiation to manage complex environmental issues: Example of a gaming approach in the peri-urban catchment of São Paulo, Brazil. *Reg. Environ. Change* **2015**, *15*, 67–78. [CrossRef]
11. Mayer, I.S. The gaming of policy and the politics of gaming: A review. *Simul. Gaming* **2009**, *40*, 825–862. [CrossRef]
12. Meinzen-Dick, R.; Chaturvedi, R.; Domenech, L.; Ghate, R.; Janssen, M.A.; Rollins, N.; Sandeep, K. Games for groundwater governance: Field experiments in Andhra Pradesh, India. *Ecol. Soc.* **2016**, *21*, 1–53. [CrossRef]
13. Onencan, A.; Van de Walle, B.; Enserink, B.; Chelang'a, J.; Kulei, F. WeShareIt game: Strategic foresight for climate-change induced disaster risk reduction. *Procedia Eng.* **2016**, *159*, 307–315. [CrossRef]
14. Zhou, Q.; Mayer, I.; Bekebrede, G.; Warmerdam, J.; Knepflé, M. Integrating water management and spatial planning through serious gaming: The water game in the Netherlands. *Next Generation Infrastructures 2010* **2010**. Available online: https://pdfs.semanticscholar.org/666b/1bad67af2cfc5436445babe60c275ebab26a.pdf (accessed on 5 June 2018).
15. Mayer, I.S.; Carton, L.; de Jong, M.; Leijten, M.; Dammers, E. Gaming the future of an urban network. *Futures* **2004**, *36*, 311–333. [CrossRef]
16. Ducrot, R. Gaming across scale in peri-urban water management: Contribution from two experiences in Bolivia and Brazil. *Int. J. Sustain. Dev. World Ecol.* **2009**, *16*, 240–252. [CrossRef]
17. Duke, R. A paradigm for game design. *Simul. Games* **1980**, *11*, 364–377. [CrossRef]
18. Greenblat, C.S. Setting Objectives and Parameters. In *Designing Games and Simulations: An Illustrated Handbook*; Sage Publications: Newbury Park, CA, USA, 1988; pp. 35–39.
19. Bots, P.; van Daalen, E. Functional design of games to support natural resource management policy development. *Simul. Gaming* **2007**, *38*, 512–532. [CrossRef]
20. Meijer, S. *The Organization of Transactions: Studying Supply Networks Using Gaming Simulation*; International chains and network series; Wageningen Academic Publishers: Wageningen, The Netherlands, 2009; Volume 6, ISBN 978-90-8686-102-6.
21. Hermans, L.; Cunningham, S.; Slinger, J. The usefulness of game theory as a method for policy evaluation. *Evaluation* **2014**, *20*, 10–25. [CrossRef]
22. Osborne, M.J.; Rubinstein, A. Introduction. In *A Course in Game Theory*; The MIT Press: Cambridge, MA, USA, 1994; pp. 1–8.
23. Madani, K. Game theory and water resources. *J. Hydrol.* **2010**, *381*, 225–238. [CrossRef]
24. Rasmusen, E. The Rules of the Game. In *Games and Information: An Introduction to Game Theory*; Blackwell Publishing: Malden, MA, USA, 2007; pp. 11–33.
25. Salen, K.; Zimmerman, E. Games as Game Theory Systems. In *Rules of Play: Game Design Fundamentals*; The MIT Press: Cambridge, MA, USA, 2003.
26. Slinger, J.H.; Cunningham, S.W.; Hermans, L.M.; Linnane, S.M.; Palmer, C.G. A game-structuring approach applied to estuary management in South Africa. *EURO J. Decis. Processes* **2014**, *2*, 341–363. [CrossRef]
27. Hermans, L.M.; Cunningham, S.W. *Actor and Strategy Models: Practical Applications and Step-wise Approaches*; Wiley & Sons: Hoboken, NJ, USA, 2018; ISBN 978-1-119-28473-4.
28. Gomes, S.L.; Hermans, L.M.; Thissen, W.A.H. Extending community operational research to address institutional aspects of societal problems: Experiences from peri-urban Bangladesh. *Eur. J. Operational Res.* **2018**, *268*, 904–917. [CrossRef]
29. Ostrom, E. *Understanding Institutional Diversity*; Princeton University Press: Princeton, NJ, USA, 2005.

30. Thissen, W.A.; Hermans, L.M.; Prakash, A.; Banerjee, P.; Khan, M.S.; Salehin, M.; Narain, V.; Kempers, R.; Hossain, A.Z.; Banerjee, P. *Shifting Grounds: Institutional Transformation, Enhancing Knowledge and Capacity to Manage Groundwater Security in Peri-Urban Ganges Delta Systems;* Urbanising Deltas of the World—Integrated Project—2013 Full Application; Delft University of Technology: Delft, the Netherland, 2013.
31. McKelvey, R.D.; McLennan, A.M.; Turocy, T.L. *Gambit: Software Tools for Game Theory*. 2014. Available online: http://citeseerx.ist.psu.edu/viewdoc/download;jsessionid=37F12E46D2D0E3D0AAD54B7DE58C6409?doi=10.1.1.468.4103&rep=rep1&type=pdf (accessed on 7 November 2018).
32. R Development Core Team. *R: A Language and Environment for Statistical Computing;* R Foundation for Statistical Computing: Vienna, Austria, 2008.
33. Mayne, J. Contribution analysis: Coming of age? *Evaluation* **2012**, *18*, 270–280. [CrossRef]
34. Thissen, W.A.; Twaalfhoven, P.G. Towards a conceptual structure for evaluating policy analytic activities. *Eur. J. Operational Res.* **2001**, *129*, 627–649. [CrossRef]
35. Midgley, G.; Cavana, R.Y.; Brocklesby, J.; Foote, J.L.; Wood, D.R.R.; Ahuriri-Driscoll, A. Towards a new framework for evaluating systemic problem structuring methods. *Eur. J. Operational Res.* **2013**, *229*, 143–154. [CrossRef]

water

MDPI

Article

Outstanding Videogames on Water: A Quality Assessment Review Based on Evidence of Narrative, Gameplay and Educational Criteria

Laura Galván-Pérez [1], Tania Ouariachi [2], M.ª Teresa Pozo-Llorente [1] and José Gutiérrez-Pérez [1,*]

[1] Department of Educational Methodology Research, University of Granada, 18071 Granada, Spain; lauragp@correo.ugr.es (L.G.-P.); mtpozo@ugr.es (M.T.P.-L.)

[2] Professorship Communication, Behavior and the Sustainable Society, Hanze University of Applied Sciences, 30030 Groningen, The Netherlands; t.ouariachi.peralta@pl.hanze.nl

* Correspondence: jguti@ugr.es; Tel.: +34-958-243757

Received: 11 August 2018; Accepted: 26 September 2018; Published: 9 October 2018

Abstract: Videogames have become educational, communicative and social tools among the young, favouring the acquisition of skills, abilities and values, encompassing an endless number of themes, and helping them to experience and to face, in the first person, a great diversity of environmental situations and ecology problems. Thus, the present article aims: (a) to evaluate a sample of 20 educational videogames about water, making use of some empirical criteria of quality; and (b) to design, validate and apply an integrated quality indicator of educational videogames on water, based on the aspects of narrative, gameplay and education, which allows us to obtain a ranking. The findings reflect a ranking of games allowing us to suggest that the nature of the game (simulation, adventures, platforms or questions) does not determine the quality of the game, although generally simulations and adventure games are placed in a range of medium- or high-quality, as well as those games that pursue objectives related to the design and management of a territory in a sustainable way. The paper provides teachers with quality criteria based on narrative and gameplay that complement and enrich the pedagogical dimension.

Keywords: assessment; educational videogames; online games; water; ecology education

---

1. Introduction

At present, we are immersed in the Anthropocene, the new geological era of the planet, characterized by the influence of the human species on Earth [1–4]. This influence is exerting great pressure on water and aquatic ecosystems, resulting in water pollution, loss of aquatic biodiversity, overexploitation of aquifers, and ultimately putting the survival of the species and human welfare at risk. The recent Intergovernmental Panel on Climate Change (IPCC) reports [5] point out that the most serious consequences of climate change (CC) are problems related to water resources. For this reason, the United Nations has just proclaimed the International Decade for Action "Water for Sustainable Development" [6]. This plan seeks to promote international cooperation partnerships that allow an efficient use of water, reduce tensions and disputes between the territories as a result of water deficits, and at the same time contribute to the achievement of the objectives of the 2030 Agenda for Sustainable Development, among which water is a priority goal. Agenda 2030 includes several objectives related to the sustainability of water and aquatic ecosystems, as well as recognizing youth as a key actor in achieving these objectives. In this regard, it specifies that youth have the ability to participate widely in raising awareness and promoting the Sustainable Development Goals (SDGs) both in their communities and through social networks and other digital communication tools, as they are capable of providing innovative approaches and solutions for their implementation [7].

To face this bleak panorama, citizens have united around the movement "Get wet for the water, Get involved for your community", whose objective is to mobilize society to act in defense of water and aquatic ecosystems through actions that favor the maintenance of ecosystem services, human welfare and the recognition of water as a human right, with the slogan: "Do you dare to participate in this social movement in your community? We are waiting for you. Act!"

Accordingly, it is well known that young people spend a great part of their leisure and free time using social networks and videogames, besides online games having a lot of acceptance among this public [8]. Videogame electronic games which use computer technology and allows a player to interact with the machine in real time promote motivation for learning, as well as cognitive potentialities [9–11]. In addition, videogames favour personal autonomy, social and cultural commitment [12]. In McGonigal's words [13]: "Videogames can make us better people and help us to change the world". Some other studies on the effects of video games on young people conclude: "The usefulness of video games to complete some of the skills acquired in the formal education system, favour the formation of identities and knowledge of social rules, as well as supporting the development of organizational skills or the development of some critical capacities" [14–18].

New generations of "digital natives" [19] and "digital wisdom" [20] need educational tools that cover learning needs, new to previous generations, and videogames as a powerful educational resource [21]. In terms of the formal educational context, the use of videogames (called "online serious games" when they are played on the Internet) by teachers in the classroom has experienced significant growth, since its use as a curricular element or resource has allowed students to achieve greater motivation towards knowledge, making the classroom a dynamic, participatory and attractive space [22]. "Serious games, gamification and virtual reality can be seen as a response of policy analysis to a growing social need to get 'engaged and entertained', also in public policy making" [10] and the sustainability of water.

Recent research on games and learning points to the need to carry out shared studies that determine which games are better or worse when contributing to student learning [23]. This research shows a classification of a video game on water according to its narrative, gameplay and educational potential. This classification can help teachers to choose those excellent games and differentiate them from others of lower quality. The possibility of integrating these didactic resources as an ordinary element of the curriculum and analysing their pedagogical impact opens interesting ways to educate future generations of young people about sustainability.

2. Background

The need to promote more attention to sustainability issues, games and youth people learning and demands to develop more empirical research in the field of social sciences and especially focusing on the education for sustainability has become an explicit agenda of the academic literature [24,25]. In this agenda, the field of videogames in their different genres [26,27], is an explicitly recognized hot topic of research [28–30]. In this sense, the IPCC Working Group III has more recently tried to deal with socio-environmental research demands and to move beyond the topics of hard sciences; to this end, it proposes focusing social research on reducing epistemic uncertainty surrounding the perceptions and social responses to environmental topics such as water and climate change. The report identifies more than 20 future research topics in this line, including those located more specifically in the fields of behavioural sciences, education and communication [5]. Among these lines, the field of videogames stands out as one of the 20 lines of the agenda, being a topic with its own entity with the need to promote evaluative research in this regard: "Evaluate the effectiveness of experiential methods such as simulations, games and films to improve the public's understanding and perception of processes of CC" [5]. The view of social scientists on these debates on education for sustainability is essential, since they must provide answers based on evidence of relevant empirical research that meets the new evaluation standards of IPCC initiated in 2015 and that will extend to 2022 within those debates [24].

Playing computer games is linked to a variety of impacts and perceptual, cognitive, behavioral, affective and motivational results; with the most frequent results and impacts being the acquisition of knowledge and understanding of the content and the affective and motivational results [23]. Videogames that are intended to convey ideas, values and influence the thoughts and actions of players in real-life contexts have been called serious games [31]. These games, also called "games of change" [32,33] or "social impact games" [17], have experienced a rapid rise in the last decade, due to the popularity of videogames as audiovisual technology for the media, non-governmental organizations (NGOs), politicians, activists, teachers, professionals and the art sector, because they can encompass multiple learning objectives, encompass diverse areas and target different age groups [34]. Attending to their gameplay characteristics, serious games exist in a variety of genders and formats such as simulations, simulating aspects of a real or fictional reality; quizzes, answering questions with the goal of obtaining points- platforms-navigating one's environment while avoiding obstacles; or adventures, solving puzzles by interacting with people or the environment in a non-confrontational way [23,27–30]. The choice of the type of game influences the success of digital educational games, where success is defined as a significant gain of knowledge in combination with the positive experience of the player. In this regard, a study made a game genre map, where five types of games were defined: mini-game, action, adventure, resources and role play. This research concluded that those role-plays that combine the elements of the adventure, action and resource genres are the most appropriate for educational learning. However, it must focus on what attributes of the different types of games genre are most appropriate depending on the type of learning to be achieved [30].

Our research evaluates electronic games on water sustainability that use computer technology and allow a player to interact with the machine in real time to promote the motivation for learning as well as cognitive potentialities.

Videogames have great potential to favor scientific education among young people, awakening the motivation to learn concepts and scientific processes through new challenges adapted to the particular needs and interests of each one. They can allow learners to visualize, explore, and formulate scientific explanations for scientific phenomena that would otherwise be impossible to observe and manipulate. Increasing learning time, focusing instruction toward individual learning needs and opportunities, and providing ongoing formative feedback have been shown to support learning generally and science learning specifically [35].

While it seems that games do enhance student motivation, are engaging and can be associated with behavioral change, more active design studies are needed to ensure that the best interests of the learner are met in different contexts [28]. As with other educational tools, it is important to consider how games are integrated into the student's learning experience to influence outcomes related to cognition, effect, and behavior [30].

Meta-analysis research on sustainability videogames like that conducted by Soekarjo and Van Oostendorp [36] suggest that limited empirical evidence is currently available to prove the effectiveness of games in attitude change: first of all, evidence of successful change in attitude after playing a persuasive game can only be found in five of 60 papers reviewed. Secondly, using a pre-test–post-test design, they tested whether change in attitude was different for people playing the persuasive game EnerCities compared to a control condition where participants read a document with highly similar information, finding no significant differences.

Reeves et al. [37] built a social game about energy use in a virtual home and, in a field test, smart meter data showed a significant decrease in electricity usage when comparing 30-day periods before and after playing. A comparative study examines the influence of 2020 Energy among north American and Spanish teenagers 12 age old (n = 108); using a pre-test–post-test research design in which participants are divided into two conditions: the experimental condition (playing) and the control condition (not playing). Results show that there have not been statistically significant differences after playing the game, although when looking closer at each factor, some positive consequences can be found. the game seems to have had a bigger impact on the American experimental group than on the Spanish group regarding positive attitude towards energy-saving behaviors, and that moving by bicycle is the activity that increased the most after playing the game, both among Spanish and North American students [38].Learning based on videogames has great advantages in the training processes by allowing students to be active and direct their process; in this type of learning we obtain immediate feedback, we learn through problem solving, and the students focus only on their learning [39]. "A well-designed serious game environment provides a feedback mechanism that allows the player to reflect on his or her actions and adopt different approaches or strategies. The internalisation of actions and reactions stimulates learning, often resulting of an increase in self-learning and knowledge retention" [40]. Therefore, serious games can play a fundamental role in the promotion and awareness of the sustainability of water and aquatic ecosystems among youth.

If videogames aspire to be validated as useful and constructive tools to foster learning, social change, or anticipated understanding of social issues, the quality of their design must be evaluated. In this sense, some research has been carried out with the aim of establishing evaluation tools that allow the identification of criteria for serious games in various dimensions [41–44], as well as the study of the characteristics of good commercial videogames [45,46] and assessment about educational aspects of commercial videogames in the teaching-learning processes [47–52]. "Research on how videogames actually contribute to, or even influence, policy making and management is scarce, perhaps because it requires an evaluation type of research that is quite difficult to set up. It would need to build on a comparative analysis of a rich and varied set of cases with such innovative approaches" [10]. Just a few of these studies have been concerned, therefore, with evaluating the quality of videogames from perspectives beyond the pedagogical dimension, integrating quality criteria based on evidence of other dimensions such as narrative or gameplay. Other research, concludes that it is important to focus on the development of interdisciplinary research that can address the different variables involved in the processes of design, development and evaluation of educational videogames [28,53].

Many teachers demand from researchers and game designers the criteria that allow them to differentiate between good and bad games from different characteristics and quality dimensions. Research on the subject is an emerging field that requires an approach that is not exclusively pedagogical, as the gameplay aspects and the narrative potential to build stories must also be considered as influential variables in the characterization and selection of quality games. In relation to the theme of water and aquatic ecosystems, there are scarcely references to the evaluation of videogames in general, and the evaluation of videogames produced in languages other than English.

Research has revealed the need for providing a greater discursive complexity to the theme of water in the different initiatives, policy, practices and educational resources that from environmental education are carried out with the aim of favouring the conservation of water resources and aquatic ecosystems [54–60]. Therefore, it seems extremely important to know how the issue of water is being treated from the window of virtual games, and to investigate if, as in other initiatives, it is necessary to take a step in the construction of a discursive approach towards a complexity of the same kind, besides identifying which are the characteristics that gather the online games that are approaching water [10]. Some authors have done different studies on: (1) classifications of serious games according to different criteria [27,61–63]; (2) use of serious games as a tool for teaching and evaluation of generic competences in higher education, by designing levels, indicators and descriptors [64,65] and (3) how online games can influence individual lifestyles of the players [60]. However, no study has yet been carried out that addresses the need to establish indicators to assess the quality of educational videogames in order to establish different quality scenarios in which to classify such games [66] taking into account integrated dimensions in a multimodal way.

Research findings suggest that the games used for teaching sustainable development have generally increased players' understanding of issues around sustainability and have enhanced their knowledge of sustainable development strategies. Use of videogames for teaching sustainable development is an interesting way to acquire the managerial skills required to effect change and to develop and increase attitudes, knowledge and awareness in a sustainable perspective [67].

Games present great opportunities as tools of edutainment (educational entertainment) for teaching and training, with positive effects on learning outcomes [68].

With this objective, on the one hand, we considered indispensable to evaluate educational videogames on water produced today, from a holistic and multidisciplinary perspective which encompasses both aspects of narrative (discursive construction around water under the paradigm of complexity); gameplay (set of properties that describe the player's experience with a given game system); and pedagogy or education (referred to competencies, skills and learning). On the other hand, it also seemed of paramount importance, to carry out the design and application of a quality indicator that allows users to know which games are better in relation to three fundamental aspects such as narrative, gameplay, and education. In this way, it will be possible to obtain a ranking in games, as well as to detect strengths and weaknesses in each of the aspects to be evaluated.

3. Materials and Methods

3.1. Mixed Method Research

The evaluation of the sample of selected videogames was carried out using a mixed methodology research, qualitative and quantitative [69,70], based on the evaluation of pairs and the classification of the games according to a series of scores assigned to its educational, narrative, and gameplay quality. The dimensions emphasize factors external to the technological and architectural quality of the game design, although they integrate key elements of the video game's internal character such as the story play and narrative structure. Based on a Delphi consultation that consists in a systematic, iteractive and group process aimed at obtaining expert opinions and analyzing consensus after several rounds of consultation. A matrix of quality criteria for the evaluation of videogames has been validated after the round of consultations. This matrix was applied to each of the games by three independent evaluators who discussed and agreed on the discrepancies. Afterwards, a hierarchical classification was carried out based on which it was possible to establish a ranking of excellence by quartiles. This methodology is similar to the methodology used in other procedures of honorable mention and recognition of merit, considered as modalities of subjective peer evaluation [71].

A very popular example of this type of model is applied by the Hollywood filmaker industry to assign the annual awards in their Oscar Awards to the best films, depending on how different aspects of the film quality are valued such as costumes, music, photography, revelation actor, better director, etc. Although, for the film industry, the most objective index is the box office, in terms of collection and number of viewers who watch and pay for films, for video games one could think of similarity in the number of buyers, downloads or users who play and make use of it. As in the cinema, these quantitative indices that turn a film into a "box office" success are filtered and conditioned by other ex ante evaluation systems of prior reputation, subject to peer consensus evaluation. The publicity that the nominated films receive in the case of the Hollywood Oscar Award contributes to increasing its reputation based on the judgment of experts (before being even awarded) and is complemented by the ratings of the users in web and media communication, comments, criticisms and "likes" received in social networks. In this way, experts and users are those who assign merit to some relevant dimension of the artistic quality of that work. Advertising contemporary social media marketing also contributes to all this. Videogames assessment also imitate these evaluation procedures used in the cinematographic world and grants more or less fame and publicity to the games through awards and prizes such as those granted, for example, by the International Education Games Competitions (IEGC, 6th edition, 2018) [72], the Best of Swiss Web [73]; and the Continental Media Competition-Prix Europe [74].

3.2. Videogame Excellence and Metrics for Quality Evaluation

We understand the "qualty concept" as "a measure of excellence" or a "state of being free from defects, deficiencies and significant variations" [75]. The classic reference standard on quality, ISO/IE

9126, 2001 defines quality as "the totality of features and characteristics of a product, process or service that bears its ability to satisfy stated or implied needs" [76]. This excellence "is brought about by strict and consistent commitment to certain standards that achieve uniformity of a product (videogames) in order to satisfy specific customer or user requirements" [75].

The ISO standard has more recently developed specific proposals applied to the world of video games that use heuristics and metrics for the evaluation of quality; and have been applied to the production and use of video games in different ways. This diversity of metrics is justified from the need to objectify relevant aspects of video games according to the perspective of creators, sellers and users. The scopes of these metrics have focused initially on dimensions such as design, production and marketing and more recently on aspects such as education, gameplay and usability. These more recent dimensions put an emphasis on the assessment of more qualitative aspects such as satisfaction, learning, efficiency, reliability or efiencency.

The standard, ISO/IEC 25010-3/2009 [77], addresses both internal aspects, specific to software quality based on criteria of navigability, functionality, portability, flexibility, reliability etc., as well as other external criteria such as usability and related quality of use (Figure 1) with satisfaction, motivation, learning, immersion, attractiveness, effectiveness, efficiency, etc.

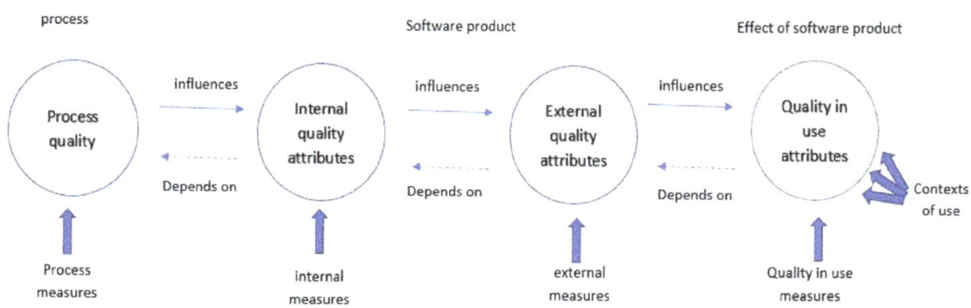

Figure 1. ISO/IEC 2501-3/2009: internal and external quality games aspects.

The classic ex-ante videogame evaluation metrics [78,79] focus on internal evaluation processes centered on the design (game mechanic, game engine) and on the architecture of the game (game interface, narrative game and and plot development). Other types of more recent external metrics have focused on aspects external to the design and architecture of the videogame [80–82].

Some recent trends suggest the use of metrics based on the analysis of educational facets and playability attributes integrated with usability and quality of use (Figure 2), based on integrated metrics or multicriteria [83,84].

Figure 2. Integrated attributes of games: usability and use.

3.3. Sample Study

Videogames that are available on the Internet are also called online games. In order to select the sample of our study, we focused on those educational videogames on water that were online. We used the following web search: "videogame" or "online game" and "water"; "water cycle"; "aquatic ecosystems". Different search engines were used in different languages, such us: Google, Yahoo, Ask. On the other hand, it was also taken into account that they fulfilled the following requirements: that they are hosted on a web platform and are free to access (thus favoring social inclusion and reaching a greater number of players); that they have communicative and educational objectives; that the focus of the storyline is water or related aspects such as cycle of water, water management, ecosystem aquatic, the human right to water; that they are aimed at an adolescent audience (+12); and that they are produced in several languages, thus covering different countries and recipients.

The sample of games varies in the possible strategies to carry out, the languages, as well as in the duration of the game, thus favoring the evaluation of different types of games and valuing their pedagogical contribution. A sampling of maximum variability in all these variables has been considered. The sample consists of 20 games (Table 1), which are detailed below: "Water alert" (Unicef); "Stop disaster" (United Nations); "Simbiocity" (Government of Sweden); "Floodsim" (Government of England); "Kingfisher" (Water Consortium of Guipúzcoa); "SAIH Ebro" (Hidrographyc Confedarcy of the Ebro river); "Plant it Green" (National Geography); "Darfur is dying" (International Crisis Group and Reebok Humanrights Foundation); "Fish game" (Cloud Institute); "Catchment" (ABC Catchment Australia SA); "Pipe dreams" (Government of Englad); "Riverbed" (Mary wharmby); "Citizen Science" (Game, Learning and Society GSL); "Fluvi" (Council of Zaragoza); "Climántica" (Regional Government of Galicia); "Where the River Meets the Sea" (National Oceanic and Atmospheric Administration); "Dive in the Guadiana" (Hidrographyc Confederacy of the Guadiana River); "Project Wet" (Water Education for Teacher Foundation); "Moving through the water" (Red Cross); Water cycle (Meteorology Statal Agency, Government of Spain, AEMET).

Table 1. Study simple of games.

Game	Creator Language Website	Description
	National Geography Languages: English	Simulation game with the objective of designing a green city, managing the resources in an ecological way, applying environmental criteria to urban planning and promoting the creation of ecological jobs. You decide where to place each building, factory and power plant. Deciding whether to build a park or a nuclear power plant is a hard choice, but now you are the mayor and it is all in your hands. As a mayor you have to decide between providing power and keeping a clean and sustainable city.
Plant it Green	https://www.nationalgeographic.org/game/plan-it-green-generation-station/	
	ABC Catchment Australia SA Languages: English	Simulation game related to the management of the River Basin and a sustainable city. It is an online game where you're in charge of the whole catchment. You get to decide what activities you undertake—whether to plant crops, log forests, build factories or set up national parks. The aim is to avoid environmental problems and provide food and wealth for the population.
Catchment	http://www.catchmentdetox.net.au/	
	United Kingdom Government Languages: English	Simulation game related to the productive management of a territory and its environmental quality. You are the territory manager, you decide where to grow, where to plant, and where to build, from the upper basin to the mouth. The objective of the game is to make a sustainable management of the territory.
Pipe Dreams	http://www.pipedreams-online.co.uk/	
	Water consortium of Guipúzcoa Languages: Spanish and Euskera	Game related to questions about the new water culture, personified by means of the bird: the Martin Fisherman. This kingfisher is the symbol of the new culture of water. There has been a great drought and now it is in danger. If you want to help Martintxo and his world, select the correct answer.
Kingfisher	http://www.gipuzkoakour.eus/Martintxo/juego.asp?idioma=E	
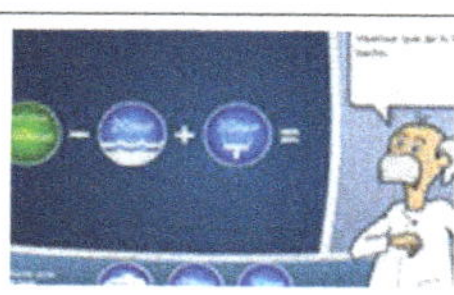	Hydrographic Confederacy of the Ebro River Languages: Spanish	Simulation-skill game related to the management of the Ebro basin. The reservoir's technical scientist teaches us how to manage the reservoir volume of the dam. The objective is to avoid a flood and properly manage the water reserves of the reservoir.
SAIH Ebro	http://www.chebro.es:81/educativo/juego/index.html	

Table 1. *Cont.*

Game	Creator Language Website	Description
 Darfur is Dying	International Crisis Group and Reebok Human rights Foundation Languages: English and Spanish http://www.gamesforchange.org/game/darfur-is-dying/	Simulation game based on the narrative of a displaced refugee, overcome the forces that threaten the survival of the camp, such as water management and sanitation. It is a web-based, viral video game that provides a window into the experience of the 2.5 million refugees in the Darfur region of Sudan. It is designed to raise awareness of the genocide taking place in Darfur and empower college students to help stop the crisis.
 Fish Game	Cloud Institute Languages: English https://cloudinstitute.org/fish-game/	Simulation game on sustainable fisheries management and ecosystem conservation. You have 10 days to catch as many fish as you can. The money you make from these fish will need to support your family for the next month. The object of the game is to have as many fish as possible at the end of the game.
 Citizen Science	Game, Learning and Society GSL Languages: English http://www.sciencegamecenter.org/games/citizen-science	Adventure game with the purpose of recovering eutrophic lake through different measures of ecological conservation and citizen awareness. The protagonists of the game warn about the eutrophication of the aquatic ecosystem to citizens, trying to find solutions for the restoration of it. The objective is to implement both individual and collective awareness actions for the conservation of aquatic ecosystems.
 Fluvi	Zaragoza Local Government Languages: Spanish http://www.expozaragoza2008.es/juegofluvi/	Platform game whose objective is to help Fluvi to save the rivers of the contamination, to take care of the ground waters and to watch the operation of the factories. My name is Fluvi and I am a small creature of water. My mission is to get water to everyone. The objective of the game is to pass each of the phases, avoiding the monsters and picking up trash in your path.
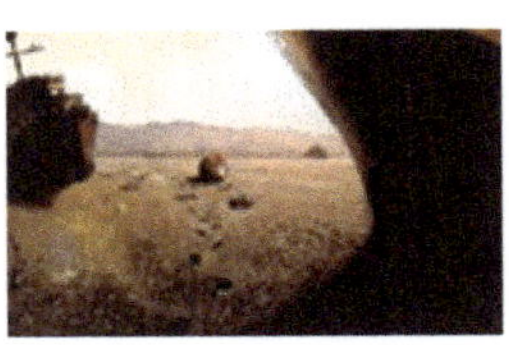 Riverbed	Mary wharmby Languages: English http://www.gamesforchange.org/game/the-riverbed-an-eco-noir-mystery/	Adventure game about the water crisis, focused on improving water management and promoting awareness with this problem. Water is the new oil. The Riverbed is a first-person interactive experience: part game, part story. The fictional murder-mystery set in a land devastated by water scarcity. The Riverbed is designed to raise awareness of this looming crisis and help players to better understand the dynamics at work in water scarcity situations. Issues like upstream/downstream, sustainability, conservation and the security dilemma are explored in a fun and engaging way.

Table 1. *Cont.*

Game	Creator Language Website	Description
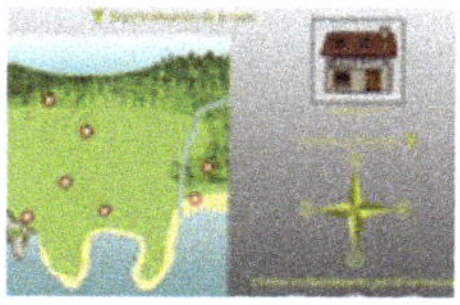	Government of Galicia	Simulation game to manage a territory in a sustainable way. Decisions must be made about how to manage a territory in areas such as water, energy, protected natural spaces, among other topics. The objective of the game is to manage a territory in a sustainable manner.
	Languages: English, Spanish and Galego	
Climantica	http://xogo.climantica.org/	
	The National Oceanic and Atmospheric Administration, Alabama	Adventure game whose objective is to raise awareness about the problem of pollution of aquatic ecosystems. An otter and a girl will try to raise awareness about the problems of water pollution. To do this, you must pass several tests and ensure that the water is clean for human consumption and your home: the river.
	Languages: English	
Where the River Meets the Sea	http://games.noaa.gov/oscar/game/welcome.html	
	United Nations	Simulation game whose objective is to plan and build a safe city in the face of natural disasters. You decide where and how you build a city, keeping in mind that the area is affected by floods and tsunamis. The objective of the game is to know how to build a safe city in the face of natural catastrophes.
	Languages: English, Chinese, French, Spanish and Russian	
Stop Disaster	http://www.stopdisastersgame.org/es/home.html	
	UNICEF	Adventure game whose aim is to ensure the survival of the inhabitants of a village by accessing drinking water. The protagonists of a village try to get drinking water and build infrastructure for an adequate sanitation of wastewater.
	Languages: English, French and Spanish	
Water Alert	http://www.enredate.org/eng/juegoseng/water_alert	
	Swedish Government	Simulation game whose objective is to manage a city in a sustainable way in the face of the various economic, social and environmental challenges and conflicts. Create an attractive city for citizens and business by improving health, comfort, safety and quality of life for you and future generations in harmony with nature, balancing economical, social and environmental effects of your decisions. The goal is to create your sustainable city.
	Languages: English	
Simbiocity	http://www.btslearning.com/app/eBS/symbiocity/index.asp	
	UK Government	Simulation game based on political decision-making and citizen awareness around flood management. The intention of the game is to know how to adequately manage the territory of a city in front of the floods. The objective will be to avoid floods in the city through direct actions of water management and aquatic ecosystems.
	Languages: English	
Floodsim	http://playgen.com/play/floodsim/	

Table 1. *Cont.*

Game	Creator Language Website	Description
	Hydrographic Confederacy of the Guadiana River	Adventure game about the cycle of water, aquatic ecosystems and the Guadian river. The main characterthe will show the use of water in the home and in the city, with actions for its improvement. The objective of the game is to raise awareness among the population about good water management practices in the home and in the city.
	Languages: Spanish	
Dive in the Guadiana	http://www.chguadiana.es/corps/chguadiana/url/swf/cdsumergete/index.html	
	Water Education for Teachers Foundation	Game of questions about the water cycle, aquatic ecosystems, water footprint and good habits at home. The game tries to discover the role that water plays in our lives through different tests and topics such as: the water cycle, the use of water in the home, aquatic ecosystems, among others.
	Languages: English	
Project Wet	http://www.discoverwater.org/	
	International Red Cross Association	Adventure game about the importance of water in our quality of life and conservation of the planet. The main character with the help of a drop of water discovers the important role that water plays in people's lives. It will also face episodes of drought and, therefore, make the public aware of the importance of efficient water management in the city.
	Languages: Spanish	
Moving through the Water	http://www.cruzroja.es/juego_del_agua/	
	Meteorology State Agency, Government of Spain-AEMET	Game of questions about the water cycle and efficient habits in its use. The intention of the game is to know more about the role that water plays on Earth through different tests.
	Languages: Spanish	
Water Cycle	http://www.aemet.es/documentos_d/conocermas/recursos_educativos/juegos/n1/agua/agua_espanol.html	

The sample of games reveals that most water games are located in a section of the website of the institution or author, specifically 65% of games versus 35%, that are located on an independent website, and none of them have access to the game through any mobile application. In relation to the language, 10 games are exclusively in English (50%), 7 of them in Spanish—along with other official languages like Galician (Climatic) and in Euskera (*"Kingfisher"*)—(35%) and 3 of them in several languages such as English, Spanish, Chinese, Arabic and French (15%). On the other hand, the study indicates a wide variety of types of producers ranging from national (20%), regional (10%) and provincial (10%) governments, as well as intergovernmental institutions, mainly the United Nations (Water Alert, Darfur is Dying, Stop Disaster); to educational institutions (10%), scientific (5%), university (10%), companies (10%), media (5%) and NGOs (5%). With regard to the typology of the game according to its nature, it was found that most of them correspond to games of simulation (35%), followed by games of questions (15%) and games of action and adventure (15%). There are also mini-games, which include skill games, questions and strategy (10%), platform games (Fluvi) and games of strategy (*"Fish Game"*). Finally, the majority communicative proposal is to provide general information on the subject (71%), followed by the objective of providing information on causes and consequences (60%), as well as favouring reflection, critical thinking and the development of ideas and solutions (50%), and lastly, encourage the change of attitudes and behaviours (35%).

3.4. Evaluation Instrument

In order to evaluate the sample of videogames about water and have a better understanding of their communicative and educational nature, we used the validated criteria identified by [85]. The study applies the Delphi method, a structured and interactive process to collect opinions that establish consensus based on the experiences and judgments of experts, to develop a set of indicators to analyse communicative and educative features of online games on climate change related issues (including water). A total of 13 experts were chosen for their theoretical and practical knowledge, motivation to participate in the study, and feasibility of contact; their areas of expertise are communication, education, games and climate change. The technique consisted of three consulting rounds: in the first round, an open question was sent to the experts to assess the relevance and usefulness of preliminary dimensions (categories of evaluation) and criteria (indicators of evaluation), identified through an extensive literature review and pre-selected for their potential to provide useful information from a narrative and ludological point of view, inspired by the "Social Discourse of Videogames Analysis Model" by Pérez-Latorre [86], which integrates both analysis perspectives. In this round, the five dimensions were identified: identification, narrative, contents, gameplay and educational. The second consulting round consisted of a questionnaire in which experts assessed in ordinal terms (high, medium, low) the relevance and usefulness of the evaluation criteria. The criteria that obtained a consensus of 90% (high + medium) were selected. In the third and last round, a final assessment of these criteria that did not pass the second phase took place, confirming whether or not they should remain outside of the final set of criteria (Table 2).

Table 2. Dimensions and evaluation criteria [adapted from reference].

Evaluation Criteria of IDENTIFICATION DIMENSION	
✓ **Game title.** ✓ **URL:** ***Link to the website; and*** **availability of mobile app.** ✓ **Language/s.**	✓ Type of creator: *author behind the creation of the game and type of institution.* ✓ Communicative purpose: *communicative intentions and objectives of the game.* ✓ Brief description: *summary according to the genre, objectives and back story.*

Evaluation Criteria of NARRATIVE DIMENSION	Evaluation Criteria of SCIENCES CONTENTS DIMENSION	Evaluation Criteria of GAMEPLAY DIMENSION	Evaluation Criteria of EDUCATIONAL DIMENSION
✓ Relevance of narrative: *narrative elements can acquire importance or be irrelevant.* ✓ Global storyline: *the story in its entirety, the logical or causal succession of the events.* ✓ Character depiction and role: *characteristics and qualities of the character/avatar.* ✓ Representation of the environment: *the world in which the character/player develops.* ✓ Dimension/space/scale: *general context and scale of the scenarios* ✓ Dimension/time: *period in time that the story spans.*	✓ Term used: *terminology used to describe the phenomenon being studied.* ✓ Existence of false concepts and misconceptions. ✓ Explicit use of scientific concepts: *definition of climate change terms.* ✓ Explicit use of information sources: *the sources of information and data are cited.* ✓ Convergence with other media or social networks: *links to social networks are included.* ✓ Message framework: *themes, causes/consequences/actions, style and tone.*	✓ Number of players and type of use: individual or multi-player. ✓ Player type: *players' profile depending on their interests.* ✓ Degree of interactivity: *user intervention in the content.* ✓ Length of playing: *time employed to play the game.* ✓ Game mission. ✓ Game dynamics and mechanics: *structure, rules and basic elements.* ✓ Feedback system: *message that the player receives in light of certain actions.* ✓ Reward system: *actions that incentivize and the rewards themselves.* ✓ Availability of game instructions and possibility of saving the game.	✓ Competences: *knowledge and attitudes that students can reach.* ✓ Abilities: *mental operations that students can reach.* ✓ Problem resolution conditions: *type of reasoning to solve problems.* ✓ Need for previous knowledge ✓ Learning curve: *level of learning difficulty.* ✓ Possibility of group work ✓ Accessibility: *availability of the game for students with functional diversity.* ✓ Interdisciplinarity: *combination of two or more academic disciplines.* ✓ Availability of educational guidelines: *document or link with educational information.*

3.5. Peer Review Evaluation

The qualitative evaluation procedure that we have developed has had the evaluations carried out by three specialists with different training and professional trajectory (teacher, gameplay expert and game research expert). These three qualified evaluators are experts in scientific communication (journalist with experience and motivation in the use of video games), environmental education (researcher specialized in the topic and with extensive research experience), and secondary school teacher (with experience in the integration of serious games in the science classroom). These evaluators have independently assessed the different facets of quality of each of the games and agreed upon the dimensions of quality, focusing on the analysis of the scientific-educational content, the educational potential, the playfulness and entertainment value, the plot of stories and implicit narratives that give structure and serve as a thread to the story that recreates the video game. Independently, each evaluator has assigned scores and assessed the quality of the dimensions and criteria on a scale of three values (high, medium, low). When consensus has not been reached among the evaluators, a debate has been held on the extreme ratings and adopted a consensus assessment.

The 20 games evaluated have been classified according to their excellence based on the calculation of scores in quartiles. The procedure used in this final phase to evaluate videogames imitates, in a certain way, the heuristic used to differentiate and order scientific publications from their impact index by quartiles [87]; although our methodology incorporates other previous elements of more qualitative assessment focused on the evaluation of pairs that judge different quality dimensions of videogames.

In this work, we propose as dimensions to evaluate in pairs a series of elements for assessing the quality of the game in terms of different dimensions such as a ludic experience experienced by the player, or other types of complementary perspectives of its usability and its formative impact on the player himself or his pedagogical potential; at the time of being used as an educational resource to address a certain curriculum content. The three dimensions in which we focus our attention are the gameplay, the narrative potential and the intrinsic educational value.

We classify the games based on a scalar score given to a set of criteria that integrate each of the dimensions of playability, narratological and educational potential (Table 3). The sum of these scores allows the assignment of values to the different dimensions and in turn an integrated global score was obtained resulting from the sum of the different dimensions. This procedure will allow indexing the games according to each one of the considered dimensions and an integrated multilevel classification that agglutinates them in a global index.

The indexation and its classification in quartiles is an operative way of ranking games based on this global index and its different dimensions depending on the quartile in which they are located with respect to the score obtained by the rest of the sample elements. Those games evaluated in their different dimensions by independent evaluators are qualified according to their excellence and from these scores a range of quartiles is established. The excellent games located in the Q1 are those that differ from the others because they have been highly valued by experts with high scores. The lowest quality games are placed in lower ranges of medium or low quality as they exceed a score threshold established from the normalized scores.

Table 3. Matrix with levels of specification for high, medium and low quality according to indicators and categories.

Level 1. Excellent Games of High Quality in Narrative, Gameplay and Education		
Narrative	**Gameplay**	**Education**
Indicator Standard	**Indicator Standard**	**Indicator Standard**
✓ Nº of elements: includes 5 or more elements in which water fulfills a specific role or function (hydraulic, hydrological, cultural, social, ecological, etc.) ✓ Spatial level: detects 3 levels (micro, meso, macro) ✓ Grade of organization: detects various complex networks ✓ Grade of evolution: evolution over time with changes and uncertainties ✓ Language: uses a broad language with different looks ✓ Story: presents a story with emotional impact and inspiring characters ✓ Objective: favors the change of attitudes and behavior ✓ Topic: focused on socio-ecosystem elements ✓ Causes: human causes ✓ Actions: promotes changes in attitudes and behaviors, both individual and collective	✓ Player profile: 2 or more profiles ✓ Level of demand: the player is pushed to the limit of his abilities to reach the goal, having to pass several levels in the game ✓ Feedback: the player receives feedback immediately after making a decision in the game, either positive or negative ✓ Dynamics: use more than two dynamics (eg: decision making, object collection, memory retention, aiming, etc.) ✓ Rewards: rewards and praise are awarded when making a correct decision.	✓ Competences: more than 2 competences ✓ Skills: the player is able to evaluate, plan and produce, getting to create something new (evaluate and create) ✓ Learning: is based on participatory learning

Table 3. *Cont.*

Level 2. Good Games of Medium-High Quality in Narrative, Gameplay and Education		
Narrative	**Gameplay**	**Education**
Indicator Standard	**Indicator Standard**	**Indicator Standard**
✓ Number of elements: includes 3 or 4 elements ✓ Spatial level: detects two levels ✓ Grade of organization: greater number of linear relationships and some complex relationships ✓ Grade of evolution: intermediate situation ✓ Language: language in transition ✓ Story: presents a story without emotional impact or inspiring characters ✓ Objective: a linear relationship of cause and consequences dazzles ✓ Topics: focuses on ecosystem elements ✓ Causes: human causas and natural causes ✓ Actions: changes in attitudes and behaviors only collective	✓ Player profile: 2 profiles ✓ Level of demand: the game does not demand enough effort and is limited to only one level ✓ Feedback: the player receives feedback only at the end of a game or mission ✓ Dynamics: uses two dynamics ✓ Rewards: praises is given but no rewards when making a correct decision	✓ Competences: 2 competences ✓ Skills: the player is able to apply learned knowledge, break it down into parts and think about how they relate to its global structure (apply and analyze) ✓ Learning: it is based on an interactive learning
Level 3. Games of Lower in Narrative, Gameplay and Educational Quality		
Narratology	**Gameplay**	**Education**
Indicator Standard	**Indicator Standard**	**Indicator Standard**
✓ Number of elements: includes 1 or 2 elements ✓ Spatial level: detects a level ✓ Grade of organization: some linear cause-effect relationships ✓ Grade of evolution: static look, without changes in time ✓ Language: language riddled with topics, myths, misconceptions, using an alarmist and sensationalist tone ✓ Story: does not present history ✓ Objective: general knowledge ✓ Topic: focuses on hydraulic and hydrological elements ✓ Causes: does not present ✓ Actions: changes in attitudes and individual behaviors	✓ Player profile: only one profile ✓ Level of demand: the level of difficulty does not correspond to the cognitive level of the player, becoming too easy or too difficult ✓ Feedback: the player does not receive feedback after making a decision in the game ✓ Dynamics: a single dynamic ✓ Rewards: they are not granted	✓ Competences: promotes only one ✓ Skills: the player brings to the memory relevant information and is able to interpret meanings (remember and understand) ✓ Learning: theoretical learning.

3.6. Quality Indicators

The indicators have been constructed based on a series of categories that refer to three fundamental aspects such as: narrative, gameplay and education. The category model consists of a series of thematic items (TI) associated to a scoring system (SP) of 1 to 3 points, which allows a ranking of quality about videogames on water to be obtained. In relation to the narrative aspects, it has mainly been developed incorporating the paradigm of complexity in the construction of knowledge around water, always under the look of the new water culture and ecosystem services, understanding that these currents of thought make possible the change of paradigm towards a complexiation of the content in this case that concerns us, water. Regarding the gameplay aspect, the following items have been selected: the profile of the player, the level of exigency, the feedback, and the rewards. Finally, in relation to the educational aspect, the basic competences defined in the compulsory educational systems have been taken into account; as well as the abilities that appear in the taxonomy of Anderson et al. [22] and the digital adaptation of Churches [88]. They propose a pyramid in ascending order: to get to the stage of "create", it must pass through "evaluate", "analyse", "apply", "understand" and "remember". Regarding the curricular integration of the games, according to the typology of teaching methodologies it is observed that the games that encourage active/interactive learning seem to get higher quality scores than those that have more traditional/passive methods. Finally, it also includes the type of learning that the game fosters (Table 3). The system of categories conformed by their indicators with a certain score would reflect different scenarios of quality of the games (low, medium, high), and would allow us to outline quality intervals in which a sample of games can be placed.

4. Results

Below are developed each of the aspects taken into account in order to characterize the sample of videogames on water, which have been arranged in a general ranking of three categories: excellent games (Q1), medium-quality games (Q2 and Q3) and low-quality games (Q4) (Table 4). The excellence is identified with their position in the corresponding quartile according to the assessment.

We obtain the integrated ranking of games dimensions (narrative, gameplay and education), sorted from highest to lowest, placing each of the games in different quality intervals. If we observe the results we have:

(a) Q1, green colour (excellent quality): games are placed in a low-quality scenario, whose score would range from 42 to 62 points;
(b) Q2, yellow colour (medium–high quality) and Q3, orange colour (medium–low quality): games are placed in a scenario of medium quality that would go between 63 and 83 points;
(c) Q4, red colour (low quality): games would be placed in a scenario of high quality, with a score above 84 points.

Table 4. Global integrated ranking of video games about water according to quantitative evaluation of experts: game classification and position by quartile.

Excellent games of high quality in narrative, gameplay and education
Position Q_1 in the global integrated ranking

Simbiocity	Citizen Science	Climántica	Catchment	Stop Disaster

Good games of medium-high quality in narrative, gameplay and education
Position Q_2 in the global integrated ranking

Plant it Green	Where the River Meets the Sea	Moving under Water	Fishgam	Riverbed	Floodsim

Good games of medium-low quality in narrative, gameplay and education
Position Q3 in the global integrated ranking

Darfur is Dying	Water Alert	Dive in the Guadiana	Project Wet

Low quality games in narrative, gameplay and education
Position Q4 in the Global Integrated Ranking

Pipe Dreams	SAIH Ebro	Fluvi	Kingfisher	Water Cycle

The cut-off point to include a game as excellent in the integrated ranking is given by the corresponding normalization in each dimension and in the overall score.

Table 5 presents a detailed ranking of the 20 games in terms of narrative, gameplay and education. The first three columns represent the sorting by categories and the last the general ranking from the summation score of the ratings in games of high, medium or low quality. If we look at column 4 (global ranking) we can observe that 5 of the games (in green) are rated as excellent, since they are positioned in quartile 1 (punctuated with ratings above 84 points); in quartile 2, the next 6 games are placed (in yellow) with ratings between 70 and 83 points; below 63 points are the games of low average quality (quartile 3, orange color); finally, with less than 62 points, are the games of lower narrative, educational and educational quality (quartile 4, red color). If we consider the three dimensions of evaluation in Table 5 (columns 1, narrative; column 2, gameplay; column 3, education) and their equivalence, in order to be classified in green as Q1, we observe that the ranking remains stable and that the order is consistent in relation to those games considered excellent (games 1 to 5), which are positioned in Q1 homogeneously for the three dimensions evaluated. We can observe the exception of game 5, which in the gameplay dimension occupies position Q2 (yellow) because it is considered of high average narratological quality by the narratives (38 points), and position Q3 (orange) because it has a low–mid range assessment (16 points); maintaining its position Q1 in the educational dimension (30 points) and Q1 in the global ranking (84 points). The stability in the arrangement of games according to the excellence in the criteria depends on the dimensions in relation to its normalized position: from game 6 onwards alternation is observed in the rankings according to the position occupied in each of the dimensions. For example, game 10 of the ranking can be considered globally of medium-high quality, although from the gameplay point of view it is a game of low quality, it is excellent narrative and of medium-high quality from the educational point of view.

Videogames that are located in a high quality scenario, in terms of narrative, gameplay and education, refer to simulation games ("Simbiocity", "Climatic", "Catchment"), as well as an adventure game ("Citizen Science"). In this sense, these games are primarily intended to create, design and manage (whether a territory, a basin or a city) in a sustainable manner including different ecological, social and economic aspects over time, whose main tool is the decision making between different variables and models (energy, agriculture, urban planning, waste management, conservation of natural spaces, etc.). Also, as a high-quality game, we encounter "Citizen Science", an adventure game focused on the fight against eutrophication of a lake, encompassing both ecological conservation measures and citizen awareness, being able to incorporate numerous and diverse elements from a narrative view, as well as different profiles (citizenship, environmentalists, scientists, users of the lake, among others) feedback and dynamics (understanding of the phenomenon, measurement and data collection, conversation with local agents, etc.) in relation to the gameplay. Finally, in relation to the educational aspect, these high-quality games promote different competences (interaction with the physical, social and civic environment, cultural and artistic, etc.), and different skills (understanding, analyzing, evaluating), favouring participatory learning.

Table 5. Ranking of videogames by dimensios and integrate indicator.

Dimensions	Narrative Quartile (Scores)	Gameplay Quartile (Scores)	Education Quartile (Scores)	Integrate Ranking Quartile (Scores)
Games				
1. Simbiocity 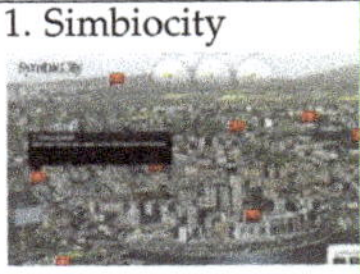	Q_1(50) Excellent narrative quality	Q_1(19) Excellent gameplay quality	Q_1(30) Excellent educational quality	Q_1(99) Excellent global quality
2. Citizen Science 	Q_1(50) Excellent narrative quality	Q_1(18) Excellent gameplay quality	Q_1(30) Excellent educational quality	Q_1(98) Excellent global quality
3. Climantica 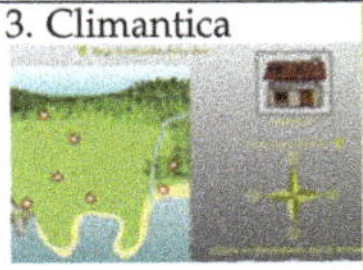	Q_1(42) Excellent narrative quality	Q_1(20) Excellent gameplay quality	Q_1(30) Excellent educational quality	Q_1(92) Excellent global quality
4. Catchment 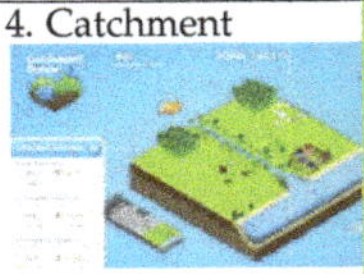	Q_1(43) Excellent narrative quality	Q_1(18) Excellent gameplay quality	Q_1(30) Excellent educational quality	Q_1(91) Excellent global quality

Table 5. *Cont.*

Dimensions	Narrative Quartile (Scores)	Gameplay Quartile (Scores)	Education Quartile (Scores)	Integrate Ranking Quartile (Scores)
Games				
5. Stop Disaster 	Q_2(38) Medium–high narrative quality	Q_3(16) Medium–low gameplay quality	Q_1(30) Excellent educational quality	Q_1(84) Excellent global quality
6. Plant it Green 	Q_2(37) Medium–high narrative quality	Q_1(20) Excellent gameplay quality	Q_2(26) Medium–high educational quality	Q_2(83) Medium–high global quality
7. Where the River Meets the Sea 	Q_2(41) Medium–high narrative quality	Q_3(16) Medium–low gameplay quality	Q_3(16) Medium–low educational quality	Q_2(73) Medium–high global quality
8. Moving through the Water 	Q_2(40) Medium–high narrative quality	Q_2(17) Medium–high gameplay quality	Q_3(16) Medium–low educational quality	Q_2(73) Medium–high global quality
9. Fish Game 	Q_3(36) Medium–low narrative gameplay quality	Q_4(11) Low gameplay quality	Q_2(26) Medium–high educational quality	Q_2(73) Medium–high global quality
10. Riverbed 	Q_1(42) Excellent narrative quality	Q_4(10) Low gameplay quality	Q_2(18) Medium–high educational quality	Q_2(70) Medium–high global quality
11. Floodsim 	Q_3(34) Medium–low narrative gameplay quality	Q_1(18) Excellent gameplay quality	Q_2(18) Medium–high educational quality	Q_2(70) Medium–high global quality
12. Darfur is Dying 	Q_3(36) Medium–low narrative gameplay quality	Q_3(14) Medium–low gameplay quality	Q_2(18) Medium–high educational quality	Q_3(68) Medium–low global quality

Table 5. *Cont.*

Dimensions	Narrative Quartile (Scores)	Gameplay Quartile (Scores)	Education Quartile (Scores)	Integrate Ranking Quartile (Scores)
Games				
13. Water Alert 	Q_3(34) Medium–low narrative gameplay quality	Q_1(20) Excellent gameplay quality	Q_3(14) Medium–low educational quality	Q_3(68) Medium–low global quality
14. Dive in the Guadiana 	Q_2(38) Medium–high narrative quality	Q_3(16) Medium–low gameplay quality	Q_3(14) Medium–low educational quality	Q_3(68) Medium–low global quality
15. Project Wet 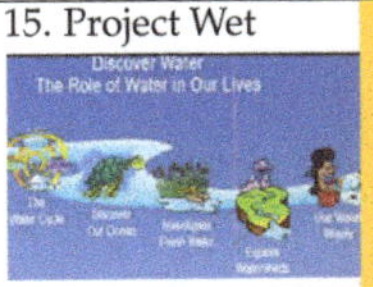	Q_3(34) Medium–low narrative gameplay quality	Q_1(18) Excellent gameplay quality	Q_4(12) Low educational quality	Q_3(64) Medium–low global quality
16. Pipe Dreams 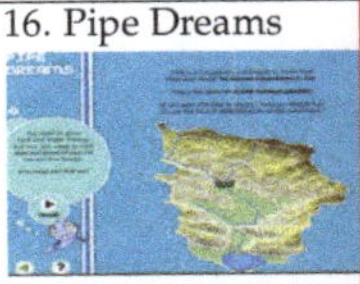	Q_4(31) Low narrative quality	Q_2(17) Medium–high gameplay quality	Q_3(14) Medium–low educational quality	Q_4(62) Low global quality
17. SAIH Ebro river 	Q_4(25) Low narrative quality	Q_3(14) Medium–low gameplay quality	Q_2(18) Medium–high educational quality	Q_4(57) Low global quality
18. Fluvi 	Q_4(24) Low narrative quality	Q_3(15) Medium–low gameplay quality	Q_3(16) Medium–low educational quality	Q_4(55) Low global quality
19. Kingfisher 	Q_4(28) Low narrative quality	Q_4(13) Low gameplay quality	Q_4(12) Low educational quality	Q_4(53) Low global quality
20. The Water Cycle 	Q_4(24) Low narrative quality	Q_4(13) Low gameplay quality	Q_3(14) Medium–low educational quality	Q_4(51) Low global quality

Narratology [$Q_1(X \geq 42)$, $Q_2(X \geq 37)$, Q_3 $(X \geq 32)$, Q_4 $(X \leq 31)$]; Gameplay [$Q_1(X \geq 18)$, Q_2 $(X \geq 17)$, Q_3 $(X \geq 14)$, $Q_4(X \leq 13)$]; Education [$Q_1(X \geq 29)$, $Q_2(X \geq 18)$, $Q_3(X \geq 14)$, $Q_4(X \leq 13)$]; Global Ranking [$Q_1(X \geq 84)$, $Q_2(X \geq 70)$, $Q_3(X \geq 63)$, $Q_4(X \leq 62)$].

A correlational analysis of the three dimensions evaluated (Table 6) shows that there is a statistically significant relationship (r_{xy} = 0.653) between the narrative and educational categories, finding a slight correlation (r_{xy} = 0.301) between the gameplay and educational categories and very similar (slightly higher than $r_{xy} = 0.364$) between gameplay and narrative. This shows that although the three dimensions considered contribute to conform the integrated concept of game quality, the object of evaluation, the narrative and educational dimensions have more weight in terms of correlated dimensions that give meaning to the possible learning processes triggered by these games. This aspect acquires special relevance in the case of serious games in general, and in our research whose thematic content is water, which are not intended for the sole purpose of entertainment and fun.

Table 6. Pearson correlations by dimensions.

		Narrative	Gameplay	Education
Narrative	Pearson correlation	1	0.364	**0.653** **
	Sig. (bilateral)		0.114	0.002
	N		20	20
Gameplay	Pearson correlation		1	0.301
	Sig. (bilateral)			0.197
	N			20
Education	Pearson correlation			1
	Sig. (bilateral)			
	N			

** (Level of significance $\alpha = 0.01$).

Aware of the importance of the gameplay dimension and its different integrating features for young people, as the main recipients of these games, it seems appropriate to note the interest of this result as a scenario for future research: can serious games not be fun? It will be necessary to analyse, in greater detail to what extent this lack of correlation reflects the slogan of seriousness with which video games designed with more pedagogical intent than playful have been classified, commonly called "serious games", is a relevant issue that deserves more attention. In any case, the gameplay dimension should not be subordinated to the exciting opportunity to transform learning processes into situations of amusing inquiry and construction and discovery of rigorous scientific knowledge. This justifies the fact that some scholars prefer the term games for training or learning [89], because they provide learners with an authentic learning experience where the entertainment and learning are seamlessly integrated [90,91], while others prefer to refer to these games as serious games [63,92].

The games located in a scenario of medium quality encompass an endless variety of typologies such as simulation games, adventure games and questions, with different objectives ranging from sustainable fisheries management, awareness of the importance of water and pollution as a problem, the fight against floods and other natural disasters, and the survival of refugees, among others. Depending on the game to be dealt with, some of them get a higher or lower score depending on the aspect being analyzed, although most are in the following medium range: narrative (32–41 points) being the maximum 50 points; gameplay (14–17 points) being the maximum 20 points; and didactic (14–28 points) being the maximum 30 points (Table 7).

Table 7. Standard scores by dimensions.

Quartil	Standard Score Narrative	Standard Score Gameplay	Standard Score Education	Standard Score Global Integrated Ranking
Q_1	41,75	18	29	83,75
Q_2	36,50	16,5	18	70
Q_3	31,75	14	14	62,5
Q_4	$X \leq 31$	$X \leq 13$	$X \leq 13$	$X \leq 62$

For example, "Project Wet" gets 34 points in narrative, 12 points in didactic, being the lowest scores in that range, but it is excellent in gameplay (18 points), because it contains different dynamics and profiles. By contrast, it can be observed that "Darfur is Dying" obtains 36 points in narrative and

18 points in education, which means a good score in that rank, and obtains 14 points in gameplay, which means quite a poor score in the dynamics that offers the game, the feedback or the possible profiles to develop.

In relation to low-quality games, you will find games such as "Fluvi", "Kingfisher", "Pipe Dreams", "SAIH Ebro" and "The Water Cycle". These games are of different nature as platform game and quiz, and include only a few narrative elements of poor content, mainly generalist, as well as a few dynamics, mechanics profiles, and feedback between the game and the player. Finally, if we analyse the educational aspect, we find that it promotes few competences, skills and learning is mainly theoretical.

The resulting dendogram of the cluster hierarchical analysis, according to the quadratic averages estimation model, shows two large clusters that represent the two main levels of polarization of the quality of the videogames on water analysed: the first group groups the games classified as excellent and good against the second group that includes games of medium and low quality (Figure 3).

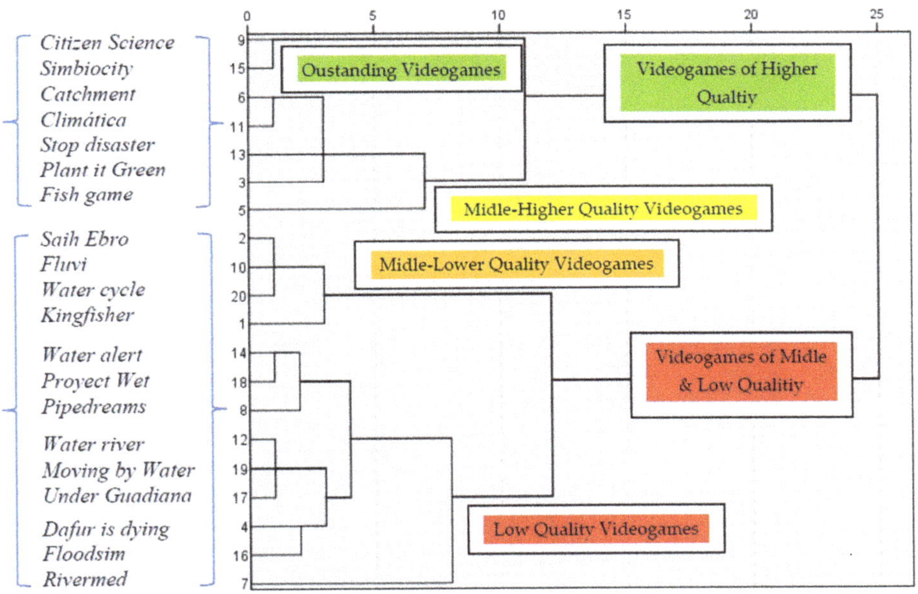

Figure 3. Cluster of Quality Videogames on Water.

5. Discussion

5.1. Narrative Dimension

The results of the study in relation to the narrative aspect indicate the existence of a narrative of medium relevance, being the majority with 40%, followed by a high relevance (35%) and characterized by a low relevance (25%) of the sample. Some games with high relevance are: "Simbiocity", "Water Alert", "Where the River Meets the Sea", "Citizen Science" and "Moving through the Water". The latter uses a narrator who accompanies the main character through different settings (home, school, city), showing the importance of water for daily life and different economic sectors (industry, agriculture, tourism), as well as showing different problems and proposing different measures (ecological agriculture, purification of the waste water, good practices of the use of the water in the home). In contrast, there are games of low relevance such as: "The Kingfisher", "Dive in the Guadiana", "Pipe Dreams" and "Fluvi", the latter the narration is almost non-existent because it is a platform game. On the other hand, the existence of a narrator appears in 55% of the gamers. In those games in which a narrator appears, we encounter the figure of a citizen (35%), followed by mayor

(15%) and manager of the urban water cycle (15%) and other diverse figures such as farmer, fisherman, animal and scientific. In this study, the space and time where the game was played were also analysed: results show that in 65% of the games the setting was fictitious, compared to 35% in a real setting (Africa, Sweden, Spain, England, etc.). In this sense, the story ran on a local scale in 70% of games (city, home, town) compared to 20% on a local and global scale and 10% on a regional hydrographic scale). In relation to the time scale, the stories took place in the present (50%) and in several scalespast, present and future at once (50%). For example, this is the case of "Climatica", which begins in 2015 and the game evolves until the year 2034. In this game as in others, connections between past, present and future are established in order to teach how to depend of the management model of a territory, problems can be solved or worsened.

5.2. Contents Dimension

The main generic contents of the games deal with water management (55%), aquatic ecosystems (40%), water cycle (35%) and ecosystem services (35%). In relation to the use of misconceptions, only the game "Moving through the Water" refers to water as scarce (we understand from the prism of the new water culture, that water is not scarce, but is given according to the climate in each territory); the use of scientific concepts appear in 60% of the games as in the case of "Citizen Science" with terms like eutrophication. On the other hand, only 35% of the games use information sources, such as UN reports, compared to 65% that do not indicate any source. More striking is that only one game, "Plant it Green", converges with social networks. If we look at the more specific contents, the following are listed in order of importance: water management (50%), urban planning and biodiversity (35%), aquatic ecosystems and water consumption (30%), citizen participation and water as economic resource (25%), water and water cycle properties (20%) and natural disasters, river basin management and water as a human right (10%). Regarding the problems, the games show the following: water pollution (45%, examples: "Where the River Meets the Sea"), loss of biodiversity (35%, ex.: "Fish Game" and "Citizen Science"), loss of ecosystem services (25%, ex.: "Simbiocity", "Catchment", "Climantica"), diseases and threats to "Darfur is Dying" (20%), natural disasters, rising temperatures and desertification (15% examples: "Floodsim" and "Stop Disaster"), change in species behaviour (10%, ex.: "Project Wet") and sea level rise, social and political crises (5%, "Water Alert"). In this sense, the study reveals that games, to a large extent, focus the origin of problems on human causes (75%), followed by both human and natural causes (15%), only natural causes (5%) and No cause (5%). In relation to action proposals developed by the games: 40% of the games do not show any example, 30% propose individual actions, 20% show individual and collective actions, and just 10% only show collective actions.

Some examples of individual actions would be those related to good practices in the use of water in the home ("Project Wet"), to recycle or to donate used clothes ("Plant it Green"); examples of collective actions would be: wastewater treatment, energy efficiency in the city, more sustainable transport models, re-vegetation of urban areas, among others ("Simbiocity"). Finally, the tone used in the games is characterized by being essentially informative (95%), proactive (55%), alarmist and ethical moral (30%), and also protest (15%).

5.3. Gameplay Dimension

The sample of games is characterized by a high degree of interactivity (which allows the player a great capacity for intervention) (40%), medium grade (35%) and low grade (25%). In relation to the dynamics of the game prevails progression, discovery, challenge, and skill; and with respect to the mechanics of the game, it is mainly the decision making in 15 of the 20 games, as well as the collection of resources, the pressure of time, and the retention of memory. In this sense, we can highlight games in which decisions are taken repeatedly in different aspects in order to sustainably manage a city or territory or solve an environmental problem through the analysis and evaluation of the information given in a particular setting and time. On the other hand, if the type of player is observed, the most popular profile is that of explorer (45%), followed by the creator (35%), and equal collaborator and competitor (10% each).

In relation to the feedback system, which evaluates the player's actions and their impact, the study reveals that 60% of the games have a mixed response system, sending both positive and negative messages; and with respect to the rewards system, 60% of the games do not have any system that rewards the actions of the player compared to 40% that values the actions through different mechanisms like extra points or new resources. Finally, the number of players is one in all games, mainly for individual use (14), with a duration of less than 30 minutes (50%), followed by games lasting more than 30 mt (35%) and one hour (15%).

5.4. Educational Dimension

The development of games mainly favours educational competences such as: knowledge and interaction with the physical world (70%), social and citizenship (55%) and linguistics (50%), among others. In relation to the promoted skills they emphasize understanding, analysis and application. On the other hand, the level of difficulty of the games is characterized by being mainly middle level (45%, e.g., "Moving through the Water"), low level (35%, e.g., "Water cycle") and high level (20%, e.g., "Simbiocity"). It is noteworthy that 14 of the 20 games promote interdisciplinary elements (economic, ecological, social, cultural, etc.); by contrast, only 4 games provide educational guidelines or an educational guide to help the teacher in the development of the objectives and contents of the game, such as: "Where the River Meets the Sea". Finally, most of the games do not favour mechanisms to evaluate the learning acquired by the students, nor do they facilitate group work.

6. Conclusions

This paper provides a useful tool for teachers and serious game designers to carry out a quality assessment based on evidence from narrative, gameplay and education. The validation process by consensus and traits evaluated in each of the dimensions through a triple round cycle with a panel of experts provides methodological rigor to the decisions that a teacher can adopt on the resources to be used in learning scientific-ecological content about themes like water. The final normalization of scores carried out with this sample of games allows us to follow a procedure to classify the games separating the excellent ones from those of lowest quality according to the learning purposes pursued by the teacher when integrating this type of resource in their classes.

6.1. Evaluation of the Dimensions

Despite some authors acknowledge that the existing common game formats lack clarity as well as consistency and thus cannot serve as a solid reference to inform research on digital educational games which are increasingly used as learning tools, the study of the characteristics of educational videogames on water has allowed us to confirm that the format of simulation supports to a greater extent learning about water: it facilitates reflection, development of critical thoughts, and the contribution of solutions through the analysis of information and creativity in decision making. The format of simulation, understood here as those games simulating aspects of a real or fictional reality, has been found to be an appropriate tool for learning by other authors [89,93]. However, the literature on simulation for teaching emphasizes that "the effect of gaming exercises crucially depends on a subsequent debriefing, as processing of experience is necessary to provide insight" [93].

It is still necessary to make an effort in relation to how to favour the change of attitudes and behaviours through the virtual game. This is a great challenge, since the change of attitudes and behaviours is a gradual process that requires time and scenarios that allow taking actions in the daily sphere or in the social sphere. On the other hand, it seems that protagonist figures with responsibility in managing a city or territory could favour a deeper learning about water (from a more global and complex perspective), dealing with real and everyday situations in the subject of water. In the world of narrative construction, whether it is in books or in games, characters are seen as "key" to identification and immersion [94]. Psychologically, immersion can be explained as "a state where the primary driver to engage is intrinsic motivation; this state is extended where the player's attention to stimuli is exclusive, and awareness of the other stimuli in the environment loses awareness" [95]. Intrinsic motivation is, therefore, a key element in engaging players to learn the contents of the game. Other aspects to be improved would be to incorporate further the use of scientific terms and official sources of information, and to place greater emphasis on solutions or possible measures to be carried out to face the challenges and problems related to water in the different thematic areas addressed by the game (city, territory, watershed, ecosystem).

We can also conclude that in relation to the gameplay, those games whose main dynamics are decision making and the profile of the explorer or creator involve more interactivity, allowing the player a greater incidence in the capacity of action and modification of the reality or the history of the game. Regarding the educational aspect of the games, it is important to emphasize the lack of educational orientations that favour the achievement of the objectives. It would also be interesting to advance in games that promote group work, since this method of work supports the exchange of knowledge between a diversity of students, promoting values of mutual support, solidarity and understanding. Lastly, the study shows the limited capacity of games to incorporate cultural and artistic competence, fundamental to foster the development of creativity, in order to face current and future challenges.

6.2. *Quality Indicator*

With the ranking of games, we have noticed that the format of the game (simulation, adventures, platforms or questions) does not determine the quality of the game, although generally speaking, simulations and adventure games are placed in a range of medium or high quality. In relation to the theme, it is not possible to clarify if a certain theme can obtain a higher score, although it could be concluded that games that pursue objectives related to design and management of a territory in a sustainable way are also located between medium- and high-quality scenarios. In this sense, it deserves special attention that those games that support participatory learning versus theoretical learning are those games that have obtained a better score, as we saw in the four games that obtained a high-quality ranking. In relation to narrative, it can be determined that those games placed in a high-quality scenario must include almost all the elements of the matrix identified as high scores, as occurs with the examples like Simbiocity or Citizen Science, which obtained the maximum score. Regarding the curricular integration of the games, according to the typology of teaching methodologies it is observed that the games that encourage active/interactive learning seem to gain higher quality scores than those that have more traditional/passive methods.

Author Contributions: Conceptualization and literature review, L.G.-P., T.O. and J.G.-P.; Investigation, L.G.-P. and T.O.; Methodology, M.T.P.-L. and J.G.-P.; Delphy design, T.O. and M.T.P.-L; Formal analysis, L.G.-P. and J.G.-P.; Resources, L.G.-P. and T.O.; Data curation, L.G.-P. and T.O.; Writing—original draft preparation, L.G.-P., T.O. and J.G.-P.; Writing—review and editing, L.G.-P., T.O., M.T.P.-L. and J.G.-P.

Funding: This research received funding of the Excelence Unit of Research, Faculty of Education & Vicerectorate of Research and Transference, University of Granada.

Acknowledgments: We would like to thank the two anonymous reviewers for their suggestions and comments.

Conflicts of Interest: The authors declare no conflict of interest.

References

1. Crutzen, P.J.; Stoermer, E.F. The 'Anthropocene'. *Glob. Change Newslett.* **2000**, *41*, 17–18.
2. Crutzen, P.J. Geology of Mankind. *Nature* **2002**, *415*. [CrossRef] [PubMed]
3. Sanders, R. Was First Nuclear Test the Start of New Human-Dominated Epoch, The Anthropocene? Available online: http://news.berkeley.edu/2015/01/16/was-first-nuclear-test-dawn-of-new-human-dominated-epoch-the-anthropocene/ (accessed on 7 December 2017).

4. Trichler, H. El Antropoceno, ¿un concepto geológico o cultural, o ambos? *Desacatos: Rev. C. Sociales* **2017**, *54*, 40–57. [CrossRef]
5. Climate Change 2014: Mitigation of Climate Change. Available online: http://www.ipcc.ch/report/ar5/wg3/ (accessed on 10 May 2018).
6. United Nations. International Decade for Action on Water for Sustainable Development, 2018–2028. Available online: http://www.un.org/en/events/waterdecade/ (accessed on 10 May 2018).
7. Resolution Adopted by the General Assembly on 25 September 2015. Available online: http://www.un.org/en/development/desa/population/migration/generalassembly/docs/globalcompact/A_RES_70_1_E.pdf (accessed on 10 May 2018).
8. Muros, B.; Aragón, Y.; Busto, A. La ocupación del tiempo libre de jóvenes en el uso de videojuegos y redes. *Comunicar* **2013**, *40*, 31–39.
9. Garrido, D.C. El papel de los videojuegos en el desarrollo cognitivo. In *Contenidos Digitales en la era de la Sociedad Conectada*; Sánchez, J.S., Parente, D.R., Eds.; Fragua: Madrid, Spain, 2014; pp. 163–178.
10. Zhou, Q.Q.; Mayer, I.S. Models, simulations and games for water management: A comparative Q-method study in the Netherlands and China. *Water* **2018**, *10*, 10. [CrossRef]
11. Yang, J.C.; Lin, Y.L.; Liu, Y.C. Effects of locus of control on behavioral intention and learning performance of energy knowledge in game-based learning. *Environ. Educ. Res.* **2017**, *23*, 1–14. [CrossRef]
12. Magnuszewski, P.; Królikowska, K.; Koch, A.; Paja, M.; Allen, C.; Chraibi, V.; Giri, A.; Haak, D.; Hart, N.; Hellman, M.; et al. Exploring the role of relational practices in water governance using a game-based approach. *Water* **2018**, *10*, 346. [CrossRef]
13. McGonigal, J. *Reality is Broken: Why Games Make Us Better and How They Can Change the World*; Penguin Press: New York, NY, USA, 2011.
14. Haste, H. Citizenship Education: A Critical Look at a Contested Field. In *Handbook of Research on Civic Engagement in Youth*; Sherrod, L.R., Torney-Purta, J., Flanagan, C.A., Eds.; John Wiley & Sons: New Jersey, NJ, USA, 2010; pp. 161–188.
15. Aragón, Y. Desarmando el poder antisocial de los videojuegos. *REIFOP* **2011**, *14*, 97–103.
16. Del Castillo, H.; Herrero, D.; García, A.; Checa, M.; Mojelat, N. Desarrollo de competencias a través de los videojuegos deportivos: Alfabetización digital e identidad. *Rev. Ed. a Distancia.* **2012**, *33*, 1–22.
17. Cremers, A.; Stubbé, H.; van der Beek, D.; Roelofs, M.; Kerstholt, J. Does playing the serious game B-SaFe! make citizens more aware of man-made and natural risks in their environment? *J. Risk Res.* **2014**, *18*, 1280–1292. [CrossRef]
18. Tahir, R.; Wang, A.I. State of the art in game based learning: Dimensions for evaluating educational games. In Proceedings of the 15th European Conference on Knowledge Management (ECKM 2014), Santarém, Portugal, 4–5 September 2014; p. 75.
19. Prensky, M. Digital natives, digital immigrants. *On The Horizon* **2001**, *9*, 1–6. [CrossRef]
20. Prensky, M.H. Sapiens digital: From digital immigrants and digital natives to digital wisdom. *IJOE* **2009**, *5*, 1–11.
21. González-Lorenzo, J.M.; Quero-Gervilla, M. Los videojuegos como recurso educativo. In *Recursos Tecnológicos en Contextos Educativos*; Cacheiro-González, M.L., Romero, C.S., González-Lorenzo, J.M., Eds.; UNED: Madrid, Spain, 2015; pp. 357–398.
22. Anderson, L.; Krathwohl, D.; Airasian, P.W.; Cruiksank, K.A.; Mayer, M.R.; Pintrch, P.R.; Wittrock, M.C. *A Taxonomy for Learning, Teaching and Assesing: A revision of Bloom´s Taxonomy of Educational Objectives*; Logman: New York, NY, USA, 2001.
23. Connolly, T.M.; Boyle, E.A.; MacArthur, E.; Hainey, T.; Boyle, J.M. A systematic literature review of empirical evidence on computer games and serious games. *Comput. Educ.* **2012**, *59*, 661–686. [CrossRef]
24. Victor, D. Climate change: Embed the social sciences in climate policy. *Nature* **2015**, *520*, 27–29. [CrossRef] [PubMed]
25. Meira-Cartea, P.A.; González-Gaudiano, E.; Gutiérrez-Pérez, J. Climate crisis and the demand for more empiric research in social sciences: Emerging topics and challenges in environmental psychology. *Psyecology* **2018**, *9*, 1–12.
26. Rodriguez-Hoyos, C.; Gomes, M.J. Videojuegos y educación: una visión panorámica de las investigaciones desarrolladas a nivel internacional. *Profesorado* **2013**, *17*, 479–494.

27. Heintz, S.; Law, E.L.-C. The game genre map: A revised game classification. In Proceedings of the Annual Symposium on Computer-Human Interaction in Play (CHI PLAY 2015), London, UK, 5–7 October 2015; pp. 175–184.
28. De Freitas, S. Are games effective learning tools? A review of educational games. *J. Educ. Technol. Soc.* **2018**, *21*, 74–84.
29. Heintz, S.; Law, E.L.-C. Digital educational games: methodologies for evaluating the impact of game type. *ACM Trans. Comput.-Hum. Interact.* **2018**, *25*, 1–47. [CrossRef]
30. Lamb, R.L.; Annetta, L.; Firestone, J.; Etopio, E. A meta-analysis with examination of moderators of student cognition, affect, and learning outcomes while using serious educational games, serious games, and simulations. *Comput. Hum. Behav.* **2018**, *80*, 158–167. [CrossRef]
31. Frasca, G. Play the Message: Play, Game and Video Game Rhetoric. Ph.D. Thesis, IT University of Copenhagen, Copenhagen, Denmark, January 2007.
32. Bogost, I. *Persuasive Games: The Expressive Power of Videogames*; The MIT Press: Cambridge, MA, USA, 2007.
33. Courbet, D.; Bernard, F.; Joule, R.; Halimi-Falkowicz, S.; Gueguen, N. Small clicks, great effects: The immediate and delayed influence of websites containing serious games on behavior and attitude. *Int. J. Advert.* **2016**, *35*, 949–969. [CrossRef]
34. Mouaheb, H.; Fahli, A.; Moussetad, M.; Eljamal, S. The Serious Game: What Educational Benefits? *Procedia Soc. Behav. Sci.* **2012**, *46*, 5502–5508. [CrossRef]
35. National Research Council. *Learning Science through Computer Games and Simulations*; National Academies Press: Washington, DC, USA, 2011.
36. Soekarjo, M.; van Oostendorp, H. Measuring effectiveness of persuasive games using an informative control condition. *Int. J. Serious Games* **2015**, *2*, 37–53. [CrossRef]
37. Reeves, B.; Cummings, J.J.; Scarborough, J.K.; Yeykelis, L. Increasing energy efficiency with entertainment media: An experimental and field test of the influence of a social game on performance of energy behaviors. *Environ. Behav.* **2015**, *47*, 102–115. [CrossRef]
38. Ouariachi, T.; Gutiérrez-Pérez, J.; Olvera-Lobo, M.D. Can serious games help to mitigate climate change? Exploring their influence on Spanish and American teenagers' attitudes. *Psyecology* **2018**, *9*, 1–31. [CrossRef]
39. Medina, L. Tecnologías emergentes al servicio de la educación. In *Aprender y educar con las tecnologías del Siglo XXI*; Orduz, R., Ed.; Colombia Digital: Bogotá, Colombia, 2012; pp. 35–47.
40. Sušnik, J.; Chew, C.Z.; Domingo, X.; Mereu, S.; Trabucco, A.; Evans, B.; Vamvakeridou-Lyroudia, L.; Savic, D.A.; Laspidou, C.; Brouwer, F. Multi-Stakeholder Development of a Serious Game to Explore the Water-Energy-Food-Land-Climate Nexus: The SIM4NEXUS Approach. *Water* **2018**, *10*, 139. [CrossRef]
41. Gunter, A.G.; Kenny, F.R.; Vick, H.E. Taking educational games seriously: Using the retain model to design endogenous fantasy into standalone educational games. *Educ. Technol. Res. Dev.* **2008**, *56*, 511–537. [CrossRef]
42. Fu, F.L.; Su, R.C.; Yu, S.C. EGameFlow: A scale to measure learners enjoyment of e-learning games. *Comput. Educ.* **2009**, *52*, 101–112. [CrossRef]
43. Neville, D.; Shelton, B. Literary and Historical 3D Digital game-based learning: Design guidelines. *For. Lang. Ann.* **2010**, *41*, 607–629. [CrossRef]
44. Ouariachi, T.; Olvera-Lobo, M.D.; Gutiérrez-Pérez, J. Evaluación de juegos online para la enseñanza y aprendizaje del cambio climático. *Rev. Enseñ. Cienc.* **2017**, *35*, 193–214.
45. Gee, J.P. *Good Video Games and Good Learning*; Peter Lang: New York, NY, USA, 2007.
46. Shchiglik, C.; Barnes, S.J.; Scornavacca, E. The development of an instrument to measure mobile game quality. *J. Comput. Inf. Syst.* **2016**, *56*, 97–105. [CrossRef]
47. Rankin, Y.; Shute, M.; Gooch, B. User centered game design: Evaluating massive multiplayer online role playing games for second language acquisition. In Proceedings of the 2008 ACM SIGGRAPH Symposium on Video Games, Los Angeles, CA, USA, 9–10 August 2008; pp. 43–49.
48. Williams, D.; Lai, G.; May, Y.; Prejean, L. Using an Educational Computer Game to Teach History in a Pedagogical Laboratory. In Proceedings of the Society for Information Technology & Teacher Education International Conference, Las Vegas, NV, USA, 3 March 2008; pp. 1847–1852.
49. Kim, B.; Park, H.; Baek, Y. Not just fun, but serious strategies: Using meta-cognitive strategies in game-based learning. *Comput. Educ.* **2009**, *52*, 800–810. [CrossRef]

50. Barbour, M.; Evans, M.; Toker, S. Making sense of video games: Pre-service teachers struggle with this new medium. In Proceedings of the SITE 2009—Society for Information Technology and Teacher Education International Conference, Charleston, SC, USA, 2–6 March 2009; pp. 1367–1372.
51. Cuenca, J.M.C.; Martín, M.J.M. Virtual games in social science education. *Comput. Educ.* **2010**, *55*, 1336–1345. [CrossRef]
52. Marín, V. La Gamificación educativa. Una alternativa para la enseñanza creativa. *Dig. Educ. Rev.* **2015**, *27*, 5–8.
53. Cano, I. La contaminación del agua: una propuesta para trabajar de forma funcional y significativa en la educación secundaria. *Rev. Inv. Esc.* **2007**, *63*, 47–63.
54. Taiwo, A.; Ray, H.; Motswiri, M.J.; Masene, R. Perceptions of the water cycle among primary school children in Botswana. *Int. J. Sci. Educ.* **2010**, *21*, 413–429. [CrossRef]
55. Fernández, J. Los procesos de construcción del conocimiento significativo del agua en bachillerato. *Rev. Enseñ. Cienc.* **2012**, *30*, 177–194.
56. Marcén, C. Argumentos educativos para enseñar-aprender el agua en la enseñanza obligatoria. *Rev. Ser. Geogr.* **2012**, *18*, 65–75.
57. Robinson, M.; Kaleta, P. Global environmental priorities of secondary students in Zabrze, Poland. *Int. J. Sci. Educ.* **2010**, *21*, 499–514. [CrossRef]
58. Gutiérrez-Pérez, J.; Galván-Pérez, L. Educational guidance on water under the paradigm of complexity as result of a comparative study between Spain and Mexico. In Proceedings of the ERPA International Congresses on Education 2015 (ERPA 2015), Athens, Greece, 4–7 June 2015; pp. 1–7.
59. Fernández, J.; Marín, F. Los procesos de enseñanza–aprendizaje relacionados con el agua en el marco de las hipótesis de transición. *Rev. Enseñ. Divulg. Cienc.* **2017**, *14*, 227–243. [CrossRef]
60. Smith, P.; Sawyer, B. Serious Games Taxonomy. Presented at the Serious Games Summit at the Game Developers Conference, San Francisco, CA, USA, 19 February 2008.
61. Mehlenbacher, B.; Kampe, C. Expansive genres of play: Getting serious about game genres for the design of future learning environments. In *Emerging Genres in New Media Environments*; Palgrave Macmillan: Cham, Switzerland, 2017; pp. 117–133.
62. Morales, E.M. El videojuego y las nuevas tendencias que presentan al mercado de la comunicación. *Anu. Electron. Estudios Comun. Social Disert.* **2011**, *4*, 36–54.
63. Katsaliaki, K.; Mustafee, N. Edutainment for sustainable development: A survey of games in the field. *Simul. Gaming* **2015**, *46*, 647–672. [CrossRef]
64. Mitchell, A.; Savill-Smith, C. *The Use of Computer and Video Games for Learning: A Review of the Literature*; Learning and Skills Development Agency: London, UK, 2004.
65. Emblen-Perry, K. Enhancing student engagement in business sustainability through games. *Int. J. Sustain. Higher Educ.* **2018**, *19*, 858–876. [CrossRef]
66. Rahayu, R.; Edy, I. Modeling analysis effect of online game to individual quality. *Eur. J. Bus. Manag.* **2015**, *7*, 218–229.
67. Shoukry, L.; Göbel, S. Story play multimodal: A research tool for the multimodal evaluation on serious games. In Proceedings of the 11th European Conference on Games Based Learning (ECGBL 2017), Graz, Austilia, 5–6 October 2017; p. 68.
68. Boomsma, C.; Hafner, R.; Pahl, S.; Jones, R.V.; Fuertes, A. Should we play games where energy is concerned? Perceptions of serious gaming as a technology to motivate energy behaviour change among social housing resident. *Sustainability* **2018**, *10*, 1729. [CrossRef]
69. Johnson, R.B.; Onwuegbuzie, A.J.; Turner, L.A. Toward a definition of mixed methods research. *J. Mix. Methods Res.* **2007**, *1*, 112–133. [CrossRef]
70. Creswell, J.W.; Plano-Clark, V.L. *Designing and Conducting Mixed Methods Research*, 2nd ed.; SAGE: Thousand Oaks, CA, USA, 2011.
71. Garfield, E. How ISI selects journals for coverage: Quantitative and qualitative considerations. *Curr. Contents* **1990**, *22*, 5–15.
72. IEGC. Available online: https://www.academic-conferences.org/conferences/ecgbl/ecgbl-international-educational-games-competition/ (accessed on 15 May 2018).
73. BSW. Available online: http://www.bestofswissweb.ch/ (accessed on 15 May 2018).
74. Prix Europa. Available online: https://www.prixeuropa.eu/ (accessed on 15 May 2018).

75. Available online: http://www.businessdictionary.com/definition/quality.html (accessed on 15 May 2018).
76. ISO/IEC 9126-1:2001. Available online: https://www.iso.org/standard/22749.html (accessed on 15 May 2018).
77. ISO/IEC 25010:2011. Available online: https://www.iso.org/standard/35733.html (accessed on 15 May 2018).
78. Nielsen, J. Heuristic evaluation. In *Usability Inspection Methods*; Nielsen, J., Mack, R.L., Eds.; John Wiley & Sons: New York, NY, USA, 2004.
79. Federoff, M.A. Heuristic and Usability Guidelines for the Creation and Evaluation of Fun Video Games. Master's Thesis, Indiana Univserity, Bloomington, IN, USA, December 2002.
80. Desuvire, H.; Capñan, M.; Toth, A.J. Using heuristic to evaluate the playability in games. In Proceedings of the CHI 2004 Conference on Human Factors in Computing Systems, Vienna, Austria, 24–29 April 2004.
81. Korhonen, H.; Koivisto, E. Playability Heuristic for Mobile Games. In Proceedings of the 8th Conference on Human-Computer Interaction with Mobile Devices and Services, Helsinki, Finland, 12–15 September 2006.
82. Desurvire, H.; Wiberg, C. Game usability heuristics (play) for evaluating and designing better games: The next iteration. In *Online Communities and Social Computin*; Ozok, A.A., Zaphiris, P., Eds.; Springer-Verlag: Berlin/Heidelberg, Germany, 2009.
83. Shoukry, L.; Göbel, S. Reasons and Responses: A Multimodal Serious Games Evaluation Framework. *IEEE Trans. Emerg. Top. Comput.* **2017**. [CrossRef]
84. González, J.L. Jugabilidad. caracterización de la experiencia del jugador en videojuegos. Ph.D. Thesis, Universidad de Granada, Granada, Spain, 2010.
85. Ouariachi, T.; Gutiérrez-Pérez, J.; Olvera-Lobo, M.D. Criterios de evaluación de juegos online sobre cambio climático: aplicación del método Delphi para su identificación. *Rev. Mex. Investig. Educ.* **2017**, *22*, 445–474.
86. Pérez-Latorre, O. Consorci de Serveis Universitaris de Catalunya, Barcelona, Spain. Unpublished work. 2010.
87. Garfield, E.; Welljams-Dorof, A. Citation data: Their use as quantitative indicators for science and technology evaluation and policy-maker. *Curr. Contents* **1992**, *24*, 5–13.
88. Churches, A. Available online: http://www.eduteka.org/TaxonomiaBloomDigital.php (accessed on 10 May 2018).
89. Crookall, D. Serious games, debriefing, and simulation/gaming as a discipline. *Simul. Gaming* **2010**, *41*, 898–920. [CrossRef]
90. Gee, J.P. *What Video Games Have to Teach Us about Learning and Literacy*; Palgrave Macmillan: New York, NY, USA, 2007.
91. Prensky, M. Digital game-based learning. *Comput. Entertain.* **2003**, *1*. [CrossRef]
92. Bellotti, F.; Kapralos, B.; Lee, K.; Moreno-Ger, P.; Berta, R. Assessment in and of serious games: An overview. *Adv. Hum-Comput. Inter.* **2013**, *2013*. [CrossRef]
93. Eisenack, K. A climate change board game for interdisciplinary communication and education. *Simul. Gaming* **2012**, *44*, 328–348. [CrossRef]
94. Bizzocchi, J. Games and narrative: An analytical framework. *Loading—J. Can. Games Stu. Assoc.* **2007**, *1*, 1–10.
95. Burns, C.; Fairclough, S. Use of auditory event-related potentials to measure immersion during a computer game. *Int. J. Hum. Comput. Stu.* **2015**, *73*, 107–114. [CrossRef]

water

MDPI

Article

Stakeholder Engagement in Maritime Spatial Planning: The Efficacy of a Serious Game Approach

Xander Keijser [1,2,*], Malena Ripken [3], Igor Mayer [4], Harald Warmelink [4], Lodewijk Abspoel [5], Rhona Fairgrieve [6] and Crawford Paris [7]

1 Environmental Policy Group, Wageningen University, Hollandseweg 1, 6706 KN Wageningen, The Netherlands
2 Rijkswaterstaat, Zuiderwagenplein 2, 8224 AD, Lelystad, P.O. Box 2232, 3500 GE Utrecht, The Netherlands
3 COAST—Centre for Environment and Sustainability Research, Carl von Ossietzky University of Oldenburg, P.O. Box 2503, 26111 Oldenburg, Germany; malena.ripken@uni-oldenburg.de
4 Breda University of Applied Sciences, Academy for Digital Entertainment, Monseigneur Hopmansstraat 1, 4817 JT Breda, The Netherlands; i.s.mayer@hotmail.com (I.M.); warmelink.h@nhtv.nl (H.W.)
5 Ministry of Infrastructure and Water Management, Rijnstraat 8, 2515 XP The Hague, P.O. Box 2090, 2500 EX The Hague, The Netherlands; lodewijk.abspoel@minienm.nl
6 Scottish Coastal Forum, Area 1-A South, Victoria Quay, Edinburgh EH6 6QQ, UK; rhona.fairgrieve@gov.scot
7 School of Geosciences, University of Edinburgh, Drummond Street, Edinburgh EH8 9XP, UK; c.paris@ed.ac.uk
* Correspondence: xander.keijser@rws.nl

Received: 6 May 2018; Accepted: 29 May 2018; Published: 2 June 2018

Abstract: The 2014 EU Directive on Maritime Spatial Planning (MSP) lays down obligations for the EU Member States to establish a maritime planning process, resulting in a maritime spatial plan by 2020. Consultation should be carried out with local, national and transnational stakeholders. Stakeholder engagement in MSP is complex because of the great number and diversity of maritime stakeholders and the unfamiliarity of some of these stakeholders with MSP and its potential impact. To facilitate stakeholder engagement in MSP, the 'MSP Challenge' table top strategy game was designed and played as part of several stakeholder events in different European countries. The authors study the efficacy of the game for stakeholder engagement. Background and evaluation data of nineteen game sessions with a total of 310 stakeholders with different backgrounds were collected through post-game surveys. Furthermore, the efficacy of the game for stakeholder engagement processes, organised by competent MSP authorities in Scotland and Belgium, is studied in more detail. The results show that the board game, overall, has been a very efficient and effective way of familiarising a great diversity of stakeholders with MSP and to create meaningful interaction and learning among stakeholders in formal planning processes. However, the case studies also show that contextual factors—the level of familiarity with MSP and participants' perception to sustainability—influences the efficacy of the game.

Keywords: Maritime Spatial Planning (MSP); stakeholder participation; serious game; Blue Growth; Good Environmental Status

1. Introduction

The diversity and intensity of human activities at sea are staggering. This is due to the rich ecosystem services and resources that are provided by the sea, combined with rapid technological innovation and economic globalization [1,2]. Human activities at sea involve, but are not exclusively reserved to, traditional maritime sectors such as shipping, fishing, dredging, recreation and mineral extraction (oil and gas) along with emerging economic sectors such as blue energy (offshore wind, tidal, wave), aquaculture (fish farming) and blue biotechnology [3].

Maritime sectors are increasingly getting into each other's way, not only within an Exclusive Economic Zone (EEZ) but also at a sea basin level, i.e., across borders. The ambitions of one country for offshore wind energy, can easily become the problem of another sector, such as fishing, in a nearby country [4]. National economic interests are at stake. Linear infrastructures—such as cables, pipelines, shipping routes—are transboundary by definition [5]. Sectoral interests and ambitions are voiced by maritime stakeholders, who form intricate webs of actor-networks in countries around a sea basin and try to influence policies at all administrative levels [3,6]. The spatial allocation of maritime activities, therefore, has to be coordinated effectively among the countries that share a sea basin [1,7].

An important constraint to the ambitions of maritime sectors and countries is the health status of marine ecosystems. Globally, marine and coastal ecosystems are under enormous pressure [3,8]. Human uses are having a cumulative effect on ecosystems that is not yet fully known. Different international treaties and agreements, such as the Convention on Biological Diversity (CBD), including the Aichi targets [9], and the United Nations' Sustainable Development Goal (SDG) 14 call upon nations and stakeholders to 'conserve and sustainably use the oceans, seas and marine resources for sustainable development' [10]. The European Union (EU) adopted the European Marine Strategy Framework Directive (MSFD) which aims to achieve clean, healthy and productive seas (e.g., Good Environment Status) [11]. The Integrated Maritime Policy [12,13] aims to find a balance between Blue Growth (BG) and Good Environmental Status (GES) by using, among other methods, Maritime Spatial Planning.

1.1. Maritime Spatial Planning

In 2014, the EU Parliament and Member States agreed on the Maritime Spatial Planning (MSP) Directive (2014/89/EU) [14]. In short, this Directive lays down obligations for the EU Member States to establish a 'maritime planning process', resulting in a 'maritime spatial plan, or plans' (Art 9.) by 2020. MSP is defined as 'a process by which the relevant Member State's authorities analyse and organise human activities in marine areas to achieve ecological, economic and social objectives [14]'. Some key principles underpinning the EU MSP Directive (2014/89/EU) [14] are:

- Integrated planning of all spatial uses and possible conflicts.
- Evidence-based i.e., guided by best available knowledge.
- Ecosystem-based e.g., taking into account monitoring of the cumulative impact of human activities on the ecology.
- Transboundary consultation and co-ordination at sea basin level.
- Information of the general public and consultation of all relevant stakeholders.

At the time of writing, EU Member States are in different implementation stages of the MSP directive. Some countries, such as Denmark, are in the first planning cycle, while other countries have recently implemented a Maritime Spatial Plan or already started a second or third iteration after a review process has been completed [15,16]. This makes transnational cooperation even more challenging as countries need to reconcile differences in approaches, mandates and capacity. In light of the above, the EU is actively supporting transboundary co-ordination and stakeholder consultation in the Baltic, Mediterranean, Celtic Seas and North Sea regions, amongst others, via projects supported by European funding mechanisms see [17] for an overview of these projects.

1.2. Stakeholder Engagement

Stakeholder engagement, or public participation in relation to policy-making and implementation, is required by both national legislation and international resolutions, such as the Aarhus Convention 1998 [18]. Furthermore, art. 9 of the MSP Directive [14] states: "Member States shall establish means of public participation by informing all interested parties and by consulting the relevant stakeholders and authorities, and the public concerned, at an early stage in the development of maritime spatial plans, in accordance with relevant provisions established in Union legislation".

Stakeholder engagement and involvement processes in MSP and other areas is relatively well studied in the academic literature [19]. There are specific handbooks and guidelines about MSP and stakeholder engagement [20], best practices examples from transnational projects [19] and more critical literature on, for instance, stakeholder empowerment [21,22] and inclusivity [23]. In the context of MSP, stakeholders have been defined as 'individuals, groups or organizations who are, in one way or another, interested, involved or affected (positively or negatively) by a particular project or action toward resources' [21]. Requirements for effective stakeholder participation are, among others: willingness and motivation of stakeholders to participate [24]; inclusivity of all possible interests [25]; equal access to information and knowledge [24,26]; timeliness of participation [19,20,23]; safety and fairness in the process [24]. However, studies of (regional) experiences have demonstrated that there are many barriers to effective stakeholder involvement in MSP [19], including the following non-exclusive list of factors that play a role:

- Stakeholder involvement in MSP often means tokenistic participation or cosmetic engagement in a very early stage of the planning process or at the end of process, whereas stakeholders should be 'on board' during the whole planning process [20,21,23,26].
- Not all stakeholders are familiar with MSP, so they are unaware of its possible impact. The sense of urgency among specific type of stakeholders can be low.
- MSP stakeholders include often other States, or government bodies or agencies therein [27].
- Stakeholders have great diversity in terms of capacity and areas of interest: compare national-scale oil and gas companies, offshore wind industry and fishing industries with wreck and recreational divers, recreationists, local fishing communities or indigenous tribes [24,26,28].
- Maritime communities—local fishermen, shippers—have deeply rooted value and belief systems and often reason and talk from 'tacit knowledge', a form of experiential and everyday understanding about the ocean.
- Trust levels in maritime sectors—e.g., local (fishing) communities—have often been adversely affected by previous negative participation experiences and/or political decisions (e.g., discussion on pulse fishing, discard ban) [4].

Because of the great number and diversity of maritime stakeholders in a transboundary context with high and (potentially) conflicting interests, combined with a significant unfamiliarity of these stakeholders with MSP and its potential impacts, stakeholder engagement in MSP is as important as it can be difficult. Several authors have called for innovative approaches and tools for stakeholder involvement in MSP [29]. They should be effective to help identify and reach out to new stakeholders; engage stakeholders; stimulate them to participate; increase the general level of understanding on MSP; create a level playing field; build a common language; give a sense of safety and openness [30]. Personal experience and interactive practice in MSP is essential, as planning can hardly be taught and learned individually or by books [31].

1.3. Serious Gaming

Collaborative tools, such as serious game/simulation games (SG) can be used to facilitate stakeholder engagement, to provide opportunities to deepen mutual understanding and to explore and integrate new ideas and solutions. 'Serious games' make use of the technologies of games and the principles of play to achieve objectives that are valued not for their intrinsic value—i.e., merely for the sake of entertainment—but for the extrinsic value achieved by the consequences, such as the fact that engagement and feedback in play is a good condition for learning and change [32].

Since 2011, the authors have been developing and using serious games to support MSP in various ways for education, social learning, stakeholder engagement and decision-support [33]. At the time of writing, the MSP Challenge brand—a not for profit initiative—holds several types of serious games; a role-playing game, advanced digital games, a table top strategy board game and a playful discussion method called the Living Q. The MSP Challenge board game has been designed to let stakeholders experience some of the dynamic and complex interactions in ecosystem-based MSP and to start 'thinking and talking' about the interrelations among different marine uses and objectives [34,35].

1.4. Research Problem

The aim of this paper is to study the efficacy—the ability in reaching its designed objectives—of the MSP Challenge board game for stakeholder engagement in various planning contexts. We use the following indicators to measure efficacy: the reported perception of learning about MSP, players' satisfaction ratings and the uptake of the game in the MSP community. The main research questions are:

- How and to what extent does playing the MSP Challenge board game raise stakeholder's understanding of MSP?
- How and to what extent do different types of stakeholders enjoy playing the MSP challenge board game?
- How and to what extent do background factors—age, gender, prior level of MSP involvement, sector affiliation and sustainability perception—affect the engagement and learning of stakeholders?
- What are the insights for real world MSP from stakeholder interactions in the game?

2. Materials and Methods

2.1. The MSP Challenge Board Game

At the end of 2015, the MSP Challenge board game Short Sea Shipping edition, was developed. It was first played in Amsterdam in February 2016 with around 30 stakeholders at a high-level meeting on Short Sea Shipping as part of The Netherlands' EU Presidency [34]. The idea behind the board game was to communicate in an engaging manner the emerging concepts of ecosystem-based MSP, the European Blue Growth Agenda (BG), and Good Environmental Status (GES) to the Short Sea Shipping community [34,35]. The simplicity of the game made it a powerful tool for engaging a wider range and greater diversity of people.

Since 2016, several revised editions—with translations into other languages, customisation of icons and symbols and adaptation of planning scales—have been made. The MSP Challenge: Scottish Marine Region edition has been developed to consider transboundary marine planning issues at a sub-national level [36], while the MSP Challenge: Blue Development edition widened the focus from shipping to other economic sectors [37]. At the time of writing, the MSP Challenge board game has been played in twenty countries with hundreds of people in the global MSP community (for an overview of the sessions, see Table 1). Dozens of institutions have been involved in its further development and dissemination.

Figure 1. Impression of the MSP Challenge board game.

Table 1. Overview of MSP Challenge sessions.

No. of Events	Name	No. of Sessions	Date	Duration	No. of Part.	No. of Resp.	Context
1	High level meeting on Short Sea Shipping; Amsterdam, The Netherlands	2	15 February 2016	1.5 h	Circa 30	N.a.	Generic
2	20th anniversary event of Scottish Coastal Forum; Edinburgh, Scotland	2	10 March 2016	1.5 h	Circa 40–50	N.a.	Generic
3	NorthSEE kick-of meeting; Hamburg, Germany	1	4 April 2016	1.5 h	Circa 20	N.a.	Generic
4	World Maritime University; Malmö, Sweden	1	13 July 2016	3 h	Circa 11	N.a.	Generic
5	3rd Atlantic Stakeholder platform conference; Dublin, Ireland	1	26 September 2016	1 h	Circa 50–60	32	Generic
6	North Sea Days Conference; Scheveningen, The Netherlands	1	6 October 2016	1 h	Circa 35	22	Generic
7	Scottish Coastal Forum; Millport, Scotland	1	15 October 2016	3 h	9	9	Generic
8	2nd Baltic MSP Forum; Riga, Latvia	1	24 November 2016	1 h	Circa 35	13	Generic
9	Van Hall Larenstein; Leeuwarden, The Netherlands	1	12 January 2017	2 h	Circa 25	N.a.	Generic
10	Stakeholder start event Belgian MSP; Bruges, Belgium	5	8 February 2017	1 h	Circa 150	89	MSP process
11	Seas at Risk; Brussels, Belgium	1	21 February 2017	1.5 h	Circa 15	N.a.	Generic
12	Kick start Erasmus+ strategic partnership; Liverpool, UK.	1	23 February 2017	1.5 h	Circa 15	N.a.	Generic
13	MapSIS Conference; Las Palmas, Gran Canaria	2	27/28 April 2017	75 min.	Circa 35	31	Generic
14	2nd International Conference on MSP; Paris, France	1	14 March 2017	1 h	Circa 20	N.a.	Generic
15	EUSBSR Forum; Berlin, Germany	1	13 June 2017	1 h	Circa 10	6	Generic
16	Memorial's Master of Marine Studies; St Johns, Canada	1	15 June 2017	1.5 h	Circa 20	N.a.	Generic
17	Riverside Museum, Glasgow, Scotland	1	4 July 2017	1.5 h	7	7	MSP process
18	Arran High School, Lamlash, Scotland	1	5 July 2017	1.5 h	10	10	MSP process
19	MARE Conference; Amsterdam, The Netherlands	1	6 July 2017	1.5 h	Circa 15	14	Generic
20	Royal West of Scotland Amateur Boat Club, Greenock, Scotland	1	7 July 2017	1.5 h	19	19	MSP process
21	Blue Growth Summer School; Ostend, Belgium	1	21 September 2017	2 h.	Circa 25	21	Generic
22	ArtReefs workshop; Varna, Bulgaria	1	10 October 2017	1.5 h	Circa 25	11	Generic
23	SimCelt closing conference; Liverpool, UK	1	28 November 2017	1.5	Circa 15	N.a.	Generic
24	Van Hall Larenstein; Leeuwarden, The Netherlands	1	30 November 2017	2 h	26	26	Generic
	Total	32	N.a.	N.a.	Circa 650	310	N.a.

N.a. means not applicable.

The Maritime Spatial Planning Challenge board game is a table top strategy game, based on a constructionist learning philosophy [38], which suggests that players learn about the world and their relation to it, through creation and construction. The game was originally developed in a period of three months with a small development team consisting of game developers, designers and MSP experts. It has been designed to run for one to three hours, depending on the setting, and involve 12 to 30 players. The game is played on a 2.8 × 1.6 m plywood game board, printed with a map of a fictional sea, showing a grid with holes drilled through the middle of each square (See Figure 1). The fictional sea, the 'RICA Sea' is shared by three countries: Bayland, Peninsuland and Island'; each having their own country profile and objectives about the future development of the RICA Sea. At the start of the game, players stand around the board and a simple narrative introduces them to the game: 'Jointly develop the RICA Sea so that at the end of the game, you and others feel comfortable with the state of the RICA Sea and how you developed it'. Players are assigned different roles—maritime spatial planners, nature conservationists or a representative of a marine-related industry—in one of the three countries. Players can develop the RICA Sea by placing various tokens and threads on the game board. Many colourful acrylate squares (i.e., the tokens), with various symbols representing marine environment and human activities, are designed to fit onto the grid in order to build up the picture of ecosystem services and human activities above, on and below the sea surface. Different colours of threads indicate forms of linear infrastructure (e.g., cables, pipelines) and shipping lanes relating to various types of vessels. During the game, the players may gradually find out that they get into each other's way and this should start players 'thinking and talking' about the interrelations among the different activities and objectives. At chosen intervals during the game and at the end of the game, the Moderator, together with the Game Overall Director (G.O.D.), facilitates a discussion with the players, asking questions like 'How did it go?', 'What happened and why?' and 'How does this resemble real life?'. The G.O.D. is an MSP policy expert and he/she has the authority to give additional information, decide on, or intervene in, all matters that are unclear or on situations that arise out of the game scenario. Rules of the game can be tailored to session objectives or invented on the spot to give the gameplay a certain twist. The board game is designed to be an impressionistic, fast and hectic, even chaotic experience; much like reality. Over everything, however, it is intended to be fun. For a more details, see www.mspchallenge.info.

2.2. Study Design

2.2.1. Questionnaires

Background and evaluation data of nineteen game sessions have been analysed for this study. A questionnaire, based on an evaluation framework [39], was developed to capture participants' feedback. It uses validated questions and constructs for game evaluation to elicit information on demographics, sector of employment, pre-existing MSP involvement, game play experience and appreciation and understanding of Maritime Spatial Planning. The survey is based on self-reporting, with answer categories ranging from 'strongly agree' (5) to 'strongly disagree (1)'. Respondents can also write down a short comment.

A total of 310 stakeholders filled in a post-game survey. The questionnaire is intentionally brief to ensure maximum cooperation. Not all game sessions have been surveyed as the questionnaire was not developed until after success of the game became apparent. Table 1 details the sessions where questionnaires were deployed and the context of the game play.

2.2.2. Observations

The majority of game sessions were moderated by the development team with considerable involvement of the different organising hosts. All but two of the sessions in Scotland were moderated by the organising authority only, without involvement of the original developers. Moderators noted down observations for many game sessions. On a few occasions, participants were interviewed on camera to capture useful footage for promoting the game and short time lapse videos were made for a couple of sessions to demonstrate how the players developed their maritime world [40]. All sessions had intermediate and post-game reflective discussions to transfer MSP related knowledge and stories and to stimulate and capture learning. The case study in Scotland was systematically observed as part of a Master of Science thesis [41] and the results of the recorded observations illustrate and sustain the conclusions noted here.

2.2.3. Sessions

The game sessions differed in terms of duration, number and experience of participants and context. Generic sessions typically took place—with stakeholders or students—as part of an event, such as a conference, a workshop, or learning sessions (e.g., a MSP summer school). These generic game sessions varied from one to three hours and their objectives were to create more attention, awareness and understanding of MSP. Two other types of sessions, identified as 'Belgium' and 'Scotland' were organised by the relevant competent authority to kick start a formal MSP process with stakeholders. These sessions were between 90 min and 2 h and are more comparable with each other, allowing analysis in more detail. Table 2 gives a short overview of the session characteristics.

Table 2. MSP Challenge board game sessions—context and characteristics.

	Generic	Belgium	Scotland
Host organisation	Conferences, workshops, education	Belgium Federal government	Clyde Marine Planning Partnership
MSP level	N.a.	National	Regional
MSP phase	N.a.	Revision process of marine plan	Pre-planning (initial stage)
Planning area	N.a.	Belgium North Sea, Ca 3500 km^2	Firth of Clyde, Ca 3650 km^2
Type of stakeholders	Students, experts, intermediaries.	Sector stakeholders	Recreational stakeholders
Number of respondents	185	89	36
MSP Challenge Edition	Various	Blue Development Edition	Scottish Marine Region Edition
Duration	1–3 h. per session	1.5 h. per session	2 h. per session
Nr. of sessions	11	5	3

MSP Sessions in Scotland

In Scotland, the Marine (Scotland) Act 2010 provides the framework for marine spatial planning. This Act requires the creation of a national marine plan and enables the production of regional marine plans under the authority of Scottish Ministers [42]. On 25 March 2015, the Scottish National Marine Plan was adopted. The National Marine Plan will be augmented at the regional level by local Marine Planning Partnerships (MPP), which will develop a suite of regional marine plans in territorial waters (12 nautical miles) [43]. The Clyde Marine Planning Partnership (CMPP) is the first of eleven regional partnerships in Scotland that have been delegated the task to prepare a statutory policy document for the Clyde Marine Region and, as part of the process, will implement a series of consultation sessions with local marine stakeholders over a period of three years [41].

In Scotland, the CMPP organised three separate sessions with the MSP Challenge board game in the summer of 2017. The purpose of these sessions was to develop a dialogue between policy experts and target marine users about the complexities of MSP issues. A special Scottish Marine Region edition of the MSP board game was created for the purpose, overseen by the Scottish Coastal

Forum and the development team, and supported under the auspices of the EU-funded SIMCelt initiative, which was testing novel ways of stakeholder engagement to facilitate cross-border and transboundary marine planning. These sessions were held to encourage engagement from sectors that had been under-represented in marine planning in the Clyde area up until that point, including a group of stakeholders from the marine recreational sector (coastal rowers, divers, kayakers and coastal swimmers) [41].

MSP Sessions in Belgium

The first legally binding Marine Spatial Plan for the Belgian Part of the North Sea was approved by Royal Decree in March 2014 [44]. The plan needs to be revised at least every six years. In the spring of 2017, the Ministry for the Environment started with the revision procedure of the Belgian Maritime Spatial Plan which, under Belgian law, must lead to a new Plan by 2020. For the start of this process, a workshop with all competent authorities and stakeholders was organised on 8 February 2017 in Bruges, Belgium [45]. The purpose of the workshop was to inform the public about the revision procedure and to receive ideas from the participants about the future development of the Belgian part of the North Sea. In an event with circa 150 participants from different maritime sectors including scientists and civil servants, the MSP Challenge: Blue Development edition game was used to kick start this revision procedure [45]. The MSP Challenge development team has facilitated five MSP board game sessions together with MSP experts from Belgium.

2.2.4. Description of the Data Set

The players were asked about their background; results are shown in Table 3. A total of 310 participants of the nineteen game sessions filled in the post-game questionnaires. The male/female distribution of the total respondent group is relatively balanced with a slight overrepresentation of male participants. Persons of all age categories participated, from slightly younger than eighteen to above sixty-five. About forty percent of all respondents are active in the non-profit sector (NGOs, research, education) and about thirty percent in the public sector.

Table 3. Player backgrounds.

	Generic (n = 185)	Belgium (n = 89)	Scotland (n = 36)	Total (n = 310)
Age	0–18: 0 (0%) 18–25: 37 (20%) 25–35: 65 (35%) 35–45: 49 (27%) 45–55: 16 (9%) 55–65: 14 (8%) 65+: 3 (2%)	0–18: 0 (0%) 18–25: 3 (3%) 25–35: 14 (16%) 35–45: 23 (26%) 45–55: 23 (26%) 55–65: 18 (20%) 65+: 8 (9%)	0–18: 1 (3%) 18–25: 1 (3%) 25–35: 6 (17%) 35–45: 1 (3%) 45–55: 11 (31%) 55–65: 11 (31%) 65+: 3 (8%)	0–18: 1 (0%) 18–25: 41 (13%) 25–35: 85 (28%) 35–45: 73 (24%) 45–55: 50 (16%) 55–65: 43 (14%) 65+: 14 (5%)
Male/Female	Male: 91 (50%) Female: 92 (50%)	Male: 55 (62%) Female: 34 (38%)	Male: 18 (50%) Female: 16 (44%)	Male: 164 (53%) Female: 142 (46%)
Involvement *	1: 70 (38%) 2: 57 (31%) 3: 22 (12%) 4: 16 (9%) 5: 17 (9%)	1: 6 (7%) 2: 32 (36%) 3: 27 (30%) 4: 17 (19%) 5: 5 (6%)	1: 23 (82%) 2: 2 (7%) 3: 2 (7%) 4: 0 (0%) 5: 1 (4%)	1: 99 (33%) 2: 91 (30%) 3: 51 (17%) 4: 33 (11%) 5: 23 (8%)
Public sector	49 (27%)	44 (51%)	9 (26%)	102 (34%)
Private sector	23 (13%)	28 (32%)	16 (46%)	67 (22%)
Non-profit	104 (58%)	15 (17%)	5 (14%)	124 (41%)

* 1 = not or hardly involved; 2 = from time to time; 3 = quite some involvement; 4 = strong involvement; 5 = deeply involved. Missing values in questionnaire and/or rounding off explain differences in the summation of numbers and/or percentages between rows and columns.

Table 3 also shows that the three cases—Generic, Belgium, Scotland—differ significantly in terms of backgrounds. In the Generic case, the players are internationally mixed (country background not presented here) with an equal gender balance. On average, the players in this category tend to be early or mid-career professionals (around 20–35 years of age), with some previous involvement in MSP, mainly in the non-profit sector (education, NGO). Typical players in this category were students or graduates from different countries, young researchers or policy officers, who played the game at an MSP conference or workshop.

In the Belgian case, there are more male players with a moderate to high level of involvement in MSP. On average, they have a higher age and are working in the public sector. The players in the Belgian case are already part of a national MSP actor-network, where many of them have previously been involved in the first national Maritime Spatial Planning process.

In the Scottish case, the players on average come from the private sector and have a higher age. They have hardly any or no pre-involvement or knowledge of MSP. For these local and regional players, it was a first encounter with MSP. The intermediate conclusion is that in the analysis, the significant differences among the case studies, need to be taken into account.

3. Results

3.1. MSP Understanding

According to the participants, the MSP Challenge board game reflects the challenges in real life MSP accordingly and playing the board game has made respondents, in general, more interested in MSP (see Table 4). Furthermore, the players on average report that they have learned about MSP as a result of playing the game. By and large, and based on a great many player debriefings, and observations, as well as player comments in the survey, we conclude that the game is considered meaningful and insightful.

Table 4. MSP understanding—key descriptives.

Statement/Mean (Standard dev.)	Generic (n = 185)	Belgium (n = 89)	Scotland (n = 36)	Total (n = 310)
The issues in the game represent the challenges in MSP	4.1 (0.8)	3.9 (0.8)	4.4 (0.7)	4.1 (0.8)
I have become more interested in MSP	3.8 (0.9)	3.4 (1.1)	4.1 (0.7)	3.7 (1.0)
I know better what MSP is	3.8 (1.0)	3.4 (1.0)	4.2 (0.7)	3.7 (1.0)
I can better imagine the different viewpoints on MSP	4.0 (0.9)	3.6 (0.9)	4.2 (0.7)	3.9 (0.9)
I gained more insight into what the important factors in MSP are and how they (can) influence each other	3.9 (0.8)	3.6 (0.8)	4.1 (0.5)	3.8 (0.8)
I gained insights on how different planning scales (local, regional, national, international, etc.) can influence decisions made	3.7 (1.0)	3.6 (0.8)	N.a.	3.6 (1.0)
General understanding of MSP * (Cronbach's Alpha = 0.85)	3.9 (0.8)	3.5 (0.8)	4.2 (0.5)	3.8 (0.8)

* General MSP understanding is based upon 3 statements: "I know better what MSP is", "I can better imagine the different viewpoints on MSP", and "I gained more insight into what the important factors in MSP are and how they (can) influence each other".

> "It has been [an] interesting game, because you can obtain a global idea about MSP and the conflict resolution between sectors, environmental issue and neighbouring countries."
>
> —(participant of MapSiS session).

There are marked differences among the three cases. Table 4 shows that the respondents who have played the MSP Challenge board game as part of the Clyde Marine Planning Partnership report a significantly stronger perception of learning about MSP than the respondents who played the game at a conference or workshop. Participants who played the game as part of the revision of the Belgian Maritime Spatial Plan reported a learning effect that was not as high as the players in the other cases. This might be explained by the level of pre-existing familiarity with MSP of the players.

> "As I was familiar with this topic, the game itself did not help me personally to understand the issue better. I can imagine, however, that the MSP Challenge will be very helpful in providing stakeholders (your average layman) in understanding the complicated task of planning and policy implementation in the coastal area."
>
> —(participant of ArtReefs workshop).

More important than the players' reviews or testimonials are the dialogues between the players on various MSP aspects. All players—students, professionals, novices or proficient planners—struggle with words and symbols during the game. This can be exacerbated as the players come from different countries and cultural backgrounds. Through the gameplay players get to know and understand one another better.

3.2. *Engagement*

Based upon observations of the facilitators, video recordings, and the results of the questionnaires, we can safely conclude that the participants of the MSP Challenge board game sessions get, in general, deeply engaged and immersed in the game. After the short introduction, the players jump into it, and the moderator often has a hard time stopping the players for intermediate or post-game reflections. The following quotes are an illustration of player-stakeholder engagement.

> "Really interesting and innovative way [for] getting people to think about the different interests and viewpoints of the various stakeholders."
>
> —(participant of Atlantic Strategy Stakeholder Conference).

> "The game shows clearly how governance and communication links work in reality. Everyone undermines the planners and mind[s] their own interest."
>
> —(participant of Baltic MSP forum).

The players take their role as planners or stakeholder very seriously. They show all kinds of emotions, signs of pleasure, team work as well as argument and conflict. In general, across almost all sessions, the experience is that all stakeholders pursue their own interests and try to realise their ambitions; the Marine Planners have a difficult task getting everyone aligned. The game tends to be hectic and chaotic, but extremely immersive. The half way reflection challenges the players to improve and solve their conflicts through better national, sectoral and transboundary planning in the second part of the game.

Table 5 shows that the participants enjoy playing the game although the responses differ per case, with a very high enjoyment score for Scotland. The respondents strongly recommend the game to others. This is further evidenced by the uptake of the MSP Challenge board game by the MSP community.

Table 5. Gameplay experience—key descriptives.

Statement/Mean (Standard dev.)	Generic (n = 185)	Belgium (n = 89)	Scotland (n = 36)	Total (n = 310)
I think it is easy to learn how to play the game.	4.0 (0.8)	3.8 (0.8)	4.1 (0.8)	3.9 (0.8)
I think the information provided in the game is clear.	3.8 (0.8)	3.7 (0.8)	N.a.	3.7 (0.8)
I think the game is visually appealing.	4.6 (0.6)	4.2 (0.7)	N.a.	4.4 (0.6)
I feel creative while playing the game.	4.1 (0.9)	3.9 (0.8)	N.a.	4.0 (0.9)
I think the game is fun.	4.5 (0.6)	4.0 (0.9)	N.a.	4.3 (0.8)
I enjoy playing the game.	4.4 (0.7)	4.0 (0.9)	4.7 (0.6)	4.3 (0.8)
I am likely to recommend this game to others.	4.2 (0.8)	4.0 (0.8)	4.4 (0.6)	4.2 (0.8)

3.3. Mediating Factors

In order to understand better how the game works for different player backgrounds, we examined the possible intermediating effects of age, gender, country of origin, MSP involvement, sector of professional affiliation and sustainability perception, by performing various statistical analysis.

Table 6 gives an overview of the results of the correlation between different background factors (age, gender, level of involvement, sector affiliation) on MSP understanding and gameplay enjoyment, for all sessions and for the three cases separately. The analysis shows that younger players significantly enjoy the gameplay a little more than older players while females seem to enjoy playing the game more than men. There are also significant differences with regard to self-reported learning on MSP. The analysis shows a significant, negative relationship between previous involvement in MSP and self-reported learning. Players who have previously been engaged in MSP seem to pick up less new things from the game. The analysis also shows a significant difference in terms of learning based on the country of origin; with regard to enjoyment this relationship is less strong. Possibly, players with different cultural backgrounds do not respond the same to the gameplay.

Table 6. Correlation results MSP understanding and enjoyment.

		Generic	Belgium	Scotland	Total
'MSP understanding' vs 'Age'	Spearman's rho (sig)	−0.11 (0.15)	0.04 (0.74)	0.18 (0.32)	−0.09 (0.11)
'MSP understanding' vs 'Gender'	Kruskal Wallis sign.	0.02	0.28	0.92	0.07
'MSP understanding' vs 'Sector'	Kruskal Wallis sign	0.03	0.05	0.92	0.87
MSP understanding' vs 'MSP involvement'	Spearman's rho (sig)	−0.04 (0.56)	−0.23 * (0.03)	-0.05 (0.82)	−0.19 ** (0.00)
'MSP understanding' vs 'Country of origin'	Kruskal Wallis sign.	0.00	N.a.	N.a.	0.00
'MSP understanding' vs 'Soft sustainability'	Spearman's rho (sig)	0.20 ** (0.01)	0.29 ** (0.01)	N.a.	0.16 ** (0.01)
'MSP understanding' vs 'Hard sustainability'	Spearman's rho (sig)	0.03 (0.70)	−0.02 (0.88)	N.a.	0.05 (0.44)
'Enjoyment' vs 'Age'	Spearman's rho (sig)	−0.03 (0.69)	−0.18 (0.09)	−0.26 (0.13)	−0.15 ** (0.01)
'Enjoyment' vs 'Gender'	Kruskal Wallis sign.	0.00	0.59	0.82	0.00
'Enjoyment' vs 'Sector'	Kruskal Wallis sign	0.23	0.47	0.39	0.39
'Enjoyment' vs 'MSP involvement'	Spearman's rho (sig)	0.15 * (0.04)	−0.10 (0.36)	0.29 (0.15)	−0.03 (0.61)
'Enjoyment' vs 'Country of origin'	Kruskal Wallis sign.	0.10	N.a.	N.a.	0.00
'Enjoyment' vs 'Soft sustainability'	Spearman's rho (sig)	−0.02 (0.80)	−0.05 (0.64)	N.a.	−0.08 (0.21)
'Enjoyment' vs 'Hard sustainability'	Spearman's rho (sig)	0.01 (0.93)	0.23 * (0.04)	N.a.	0.12 * (0.05)

* means result is significant at 0.05; ** means result is significant at 0.01, Note: only Spearman's rho results are presented; Kendall's tau analysis showed similar results.

Table 7 provides an overview of the results of respondents with regard to two statements on sustainability ('hard' vs 'soft' sustainability). Hard sustainability indicates an attitude that ecology should be given priority over economic growth. Soft sustainability indicates the attitude that economic growth and sustainability can go hand in hand, for instance through technological innovation. Player-stakeholders from the Belgium group who agree more with the statement that measures 'hard sustainability', significantly report more enjoyment in the game, while player-stakeholders who agree more with the statement measuring 'soft sustainability', report a significantly higher level of self-reported learning (see Table 6). This is an interesting result, although the reasons behind it are largely speculative. It may relate to the level of trust of the participants in whether this game, or a game-based approach in general, is able to address the real problems of MSP.

Table 7. Sustainability perspective—key descriptives.

Statement/Mean (Standard dev.)	Generic (n = 181)	Belgium (n = 86)	Scotland	Total (n = 267)
Depletions in natural resources and decline of biodiversity can be compensated for through economic growth and improvements in technology (Soft sust. Factor).	2.7 (1.3)	3.2 (1.1)	N.a	2.9 (1.2)
Human activity and economic development should not be allowed to undermine natural systems and processes that are vital to the existence of humans. (Hard sust. Factor).	4.2 (0.9)	3.9 (1.0)	N.a	4.1 (0.9)

Note: Mann-Whitney test shows significance differences in response between the Generic and Belgium case with regard to soft sustainability statement (p = 0.00) and hard sustainability statement (p = 0.01).

3.4. Insights and Lessons for Real World MSP

From the observations of the facilitators during the different MSP Challenge sessions the following insights and lessons can be drawn.

Players engage deeply in discussions about 'the ecosystem-based approach', 'land-sea interaction' or 'cumulative effects'. Participants ask all sorts of questions, such as "What is blue biotech?", "Which functions can be combined (co-exist)? and 'How does one re-route an international shipping lane?" The MSP Challenge board game allows players to make mistakes or act in an inconsistent way, such as planning wind farms in water deeper than 50 m despite being told not to do so at first. The game uses this as a learning experience and allows for future innovations to change game conditions and introduce new situations that must be taken into account (e.g., floating wind farms that are not restricted to certain water depths).

The MSP Challenge board game also lets players learn about different roles, responsibilities, and limitations. The role of the maritime spatial planner, including their authority and planning instruments, is quite challenging to comprehend and is especially heavily debated. Planners deploy different planning styles as in real life. In the game, most planners try a pro-active, top-down approach, although others plan in reaction to proposed developments and a few are able to plan in an adaptive mode. Real world stakeholders—shippers, fishers, scientists—are often ask to step into the role of the planner to experience first-hand what it means to be a marine spatial planner and consider multiple different interests all at once. Some of them immediately take the lead by nature, while others let themselves be overrun by the stakeholders (e.g., other players). The players also experience the consequences of their decisions through the response of other players in the board game.

Another aspect that plays an important role in the game as well as in real life is resources (e.g., man power, time, tokens). In one session, there was no fisheries representative in a country and, as a result, other stakeholders were putting claims (e.g., tokens of other economic functions) on important fishing grounds without anyone to stop them. The important lesson for the stakeholder-players is that it is important to be 'at the table' and participating actively to avoid finding yourself being served up 'on it'.

The game also brings many realistic conflicts and tensions to the surface. They revolve around the disputed boundaries of an Exclusive Economic Zone, the development of shared energy infrastructure such as an offshore power grid and energy island, and the development of short sea shipping connections, such ferries and cruise lines. Negotiations take place among the stakeholders and planners within the same country, as well as among the three countries. Transnational cooperation, however, is not a given even when players know that they share the RICA Sea with others. The focus of the players is first to develop their own part of the RICA Sea and only afterwards start to think about transnational cooperation. Often, at the end of the session, participants describe the RICA Sea as 'chaotic'. If they could play again, they would do it differently.

"Nice to see how you experience that you first go for your own interests, then national, and only then international in this game. Actually, it isn't good, but this is how it worked. Also: everyone wants to plan as much as possible."

— (participant of North Sea Days session).

4. Discussion

In terms of the MSP stakeholder engagement purpose, indicators for a good efficacy of the board game are: the reported perception of learning about MSP, players' satisfaction ratings and the uptake of the game in the MSP community. Our analysis shows that, in general, all participants enjoyed playing the MSP Challenge board game but that there are differences between the three cases relating to the self-reported perception of learning about MSP and gameplay enjoyment. These differences can be explained by key session characteristics (such as quality of facilitation, duration of the session, number of participants, personality of the participants and group dynamics), but also on key characteristics of the participants (such as age, gender, country of origin, level of involvement and perspective on sustainability).

Each MSP challenge session is as unique as the number and combination of players, the interaction between them, the context of the session and the facilitation of the game by the moderator and G.O.D. The players, together with the facilitators, make the game; the discussions between them and the facilitators on different aspects of MSP are key to the gameplay and learning experience. Although the design of the game is similar across the sessions, different facilitators and the way they run the events might have an influence on the way the players perceive the game and the subsequent reported learning about MSP. Some participants might benefit from greater explanation about their roles or possible interactions between activities, while others prefer a looser gameplay. To what extent these factors play a role is a topic for further research. Based upon our experience and feedback received from the players, it is recommended to play the MSP Challenge board game for 90–120 min with 15 to 21 players. Experience has shown that sessions with fewer players might miss out on interesting discussions as some stakeholder roles are not being played. More than 7 players per country makes the gameplay more chaotic, which might influence transnational cooperation as planners and stakeholders have a hard time in realising national objectives.

The MSP Challenge board game works very well as an introductory game (i.e., entry level game) for participants with a limited knowledge of MSP as it provides a build-up of information in an informal manner. From the Scottish case, it became evident that the participants' general lack of previous involvement in MSP did not appear to affect their ability to learn and enjoy playing the game. Instead, it could be argued that selecting and characterising these stakeholders because of their lack of involvement actually enhances learning and knowledge exchange. The interaction in the game among players who are proficient in MSP and those who are novices can of course be very beneficial at a group or team level, where both might learn from each other. The experts, for instance, form the questions and explain things that are unclear to the incumbents and the novices learn from the contributions of the experts. The game is not only an individual learning experience but also a shared social learning experience.

Additionally, sustained debate and enthusiasm for MSP, coupled with a likeliness to recommend the game to others, also hinted at the potential for wider learning within the broader stakeholder community. Participants continuing debates and demonstrating enthusiasm over MSP issues after the MSP Challenge sessions had finished alluded to the potential comparative effectiveness of gaming over more traditional methods of engagement. Although this cannot be decisively concluded here, further studies may wish to examine this comparison, perhaps investigating the extent to which post-workshop knowledge of MSP is demonstrated within broader stakeholder groups after playing a 'serious game' such as the MSP Challenge board game. Future research will focus on the role of advanced (digital) gaming for other forms of stakeholder engagement in the MSP process, such as in later stages of the development of Maritime Spatial Plans, for scenario-analysis and for transboundary co-operation among stakeholders and planners.

5. Conclusions

We can conclude that, in general, MSP stakeholders from any sector and background can enjoy and appreciate the experience of the Maritime Spatial Planning Challenge board game. Whether participants have a better understanding of MSP after playing the board game depends, amongst others, on key characteristics of the participant's own professional background, notably their country of origin, level of involvement in MSP and their perspective on sustainability.

Author Contributions: The original idea of the MSP Challenge board game is from L.A. The concept of the board game has been further developed by a small team, including the authors L.A., I.M. and X.K. The Scottish Marine Region edition has been developed together with R.F. All authors have moderated and/or facilitated various MSP Challenge board game sessions in which data collection for this paper has taken place. C.P. has analysed the Scottish board game sessions into detail for his master thesis and provided input on the Scottish case study for this paper. The conceptualisation of the paper has been done by I.M., H.W. and X.K. Data analysis has been done by X.K. and H.W. The first drafts of the paper have been written by X.K. with input of M.R. and I.M, I.M. supervised the process. All authors have provided input on different sections of the paper. R.F. and C.P. have performed a final editing and language check.

Acknowledgments: This research is part of the PhD thesis by the first author X.K. on the use of Serious Gaming in (transboundary) Maritime Spatial Planning at Wageningen University, The Netherlands, with support of Rijkswaterstaat. The board game editions have been made possible by the Dutch Ministry of Infrastructure and Water Management, the Scottish Government (Marine Scotland, with financial support from the EU-funded SIMCelt project (www.simcelt.eu), the University of Liverpool in the UK, the Federal Ministry of Transport and Digital Infrastructure in Germany and the University of Oldenburg in Germany. The authors are grateful to Gesine Meissner, chair of the Searica Intergroup European Parliament for giving us permission to use 'RICA Sea' as playground of the board game. The authors thank all other contributors, supporters and players, too many to mention by name.

Conflicts of Interest: The authors declare no conflict of interest.

References

1. Jay, S.; Alves, F.L.; O'Mahony, C.; Gomez, M.; Rooney, A.; Almodovar, M.; Gee, K.; de Vivero, J.L.S.; Gonçalves, J.M.S.; da Luz Fernandes, M.; et al. Transboundary dimensions of marine spatial planning: Fostering inter-jurisdictional relations and governance. *Mar. Policy* **2016**, *65*, 85–96. [CrossRef]
2. Douvere, F.; Ehler, C.N. New perspectives on sea use management: Initial findings from European experience with marine spatial planning. *J. Environ. Manag.* **2009**, *90*, 77–88. [CrossRef] [PubMed]
3. Kannen, A. Challenges for marine spatial planning in the context of multiple sea uses, policy arenas and actors based on experiences from the German North Sea. *Reg. Environ. Chang.* **2014**, *14*, 2139–2150. [CrossRef]
4. Jentoft, S.; Knol, M. Marine spatial planning: Risk or opportunity for fisheries in the North Sea? *Marit. Stud.* **2013**, *12*, 13. [CrossRef]
5. Ehler, C.N.; Douvere, F. An International Perspective on Marine Spatial Planning Initiatives. *Environments* **2010**, *37*, 9–20.
6. Weible, C.M. An Advocacy Coalition Framework Approach to Stakeholder Analysis: Understanding the Political Context of California Marine Protected Area Policy. *J. Public Admin. Res. Theory* **2006**, *17*, 95–117. [CrossRef]
7. Van Tatenhove, J.P.M. Transboundary marine spatial planning: A reflexive marine governance experiment? *J. Environ. Policy Plan.* **2017**, *7200*, 783–794. [CrossRef]
8. Tamis, J.E.; de Vries, P.; Jongbloed, R.H.; Lagerveld, S.; Jak, R.G.; Karman, C.C.; Van der Wal, J.T.; Slijkerman, D.M.; Klok, C. Toward a harmonized approach for environmental assessment of human activities in the marine environment. *Integr. Environ. Assess. Manag.* **2016**, *12*, 632–642. [CrossRef] [PubMed]
9. United Nations Convention of Biological Diversity. Available online: https://www.cbd.int/sp/targets/ (accessed on 4 May 2018).
10. United Nations. *Transforming Our World: The 2030 Agenda for Sustainable Development*; General Assembly Seventieth Session (A/70/L.1); United Nations: New York, NY, USA, 2015; pp. 1–35.

11. European Union. Directive 2008/56/EC of the European Parliament and of the Council of 17 June 2008 establishing a framework for community action in the field of marine environmental policy (Marine Strategy Framework Directive). *Off. J. Eur. Union* **2008**, *L164*, 19–40. Available online: http://eur-lex.europa.eu/LexUriServ/LexUriServ.do?uri=OJ:L:2008:164:0019:0040:EN:PDF (accessed on 4 May 2018).
12. European Union. *Communication from the Commission to the European Parliament, the Council, the European Economic and Social Committee and the Committee of the Regions—An Integrated Maritime Policy for the European Union*; COM (2007) 574 Final; European Commission (EC): Brussels, Belgium, 2007; Available online: http://eur-lex.europa.eu/legal-content/EN/TXT/PDF/?uri=CELEX:52007DC0575&from=EN (accessed on 4 May 2018).
13. European Union. *European Commission Progress of the EU's Integrated Maritime Policy*; COM 46 Final; European Commission: Luxembourg, 2012; p. 15. Available online: https://ec.europa.eu/maritimeaffairs/sites/maritimeaffairs/files/docs/publications/imp-progress-report_en.pdf (accessed on 4 May 2018).
14. European Union. Directive 2014/89/EU of the European Parliament and of the Council of 23 July 2014 Establishing a Framework for Maritime Spatial Planning. Available online: http://eur-lex.europa.eu/legal-content/EN/TXT/PDF/?uri=CELEX:32014L0089&from=EN (accessed on 4 May 2018).
15. Olsen, E.; Fluharty, D.; Hoel, A.H.; Hostens, K.; Maes, F.; Pecceu, E. Integration at the Round Table: Marine Spatial Planning in Multi-Stakeholder Settings. *PLoS ONE* **2014**, *9*, e109964. [CrossRef] [PubMed]
16. The United Nations Educational, Scientific and Cultural Organisation (UNESCO). MSP around the World. Available online: http://msp.ioc-unesco.org/world-applications/overview/ (accessed on 4 May 2018).
17. European Union. Maritime Affairs Maritime Spatial Planning. Available online: https://ec.europa.eu/maritimeaffairs/policy/maritime_spatial_planning_en (accessed on 4 May 2018).
18. United Nations Economic Commission for Europe Convention on Access to Information, Public Participation in Decision-Making and Access to Justice in Environmental Matters. Aarhus Convention, 25 June 1998. Available online: http://www.unece.org/fileadmin/DAM/env/pp/documents/cep43e.pdf (accessed on 4 May 2018).
19. Pentz, A.T.-A. *BaltSeaPlan Report 24 Stakeholder Involvement in Maritime Spatial Planning*; BaltSeaPlan: Hamburg, Germany, 2012; pp. 1–23.
20. Ehler, C.; Douvere, F. *Marine Spatial Planning: A Step-by-Step Approach toward Ecosystem-Based Management*; Intergovernmental Oceanographic Commission and Man and the Biosphere Programme; IOC Manual; UNESCO: Paris, France, 2009; pp. 1–98.
21. Pomeroy, R.; Douvere, F. The engagement of stakeholders in the marine spatial planning process. *Mar. Policy* **2008**, *32*, 816–822. [CrossRef]
22. Tafon, R.V. Taking power to sea: Towards a post-structuralist discourse theoretical critique of marine spatial planning. *Environ. Plan. C Politics Space* **2018**, *36*, 258–273. [CrossRef]
23. Flannery, W.; Healy, N.; Luna, M. Exclusion and non-participation in Marine Spatial Planning. *Mar. Policy* **2018**, *88*, 32–40. [CrossRef]
24. Gunton, T.; Rutherford, M.; Dickinson, M. Stakeholder Analysis in Marine Planning. *Environments* **2010**, *37*, 95–110.
25. Reed, M.S. Stakeholder participation for environmental management: A literature review. *Biol. Conserv.* **2008**, *141*, 2417–2431. [CrossRef]
26. Gopnik, M.; Fieseler, C.; Cantral, L.; McClellan, K.; Pendleton, L.; Crowder, L. Coming to the table: Early stakeholder engagement in marine spatial planning. *Mar. Policy* **2012**, *36*, 1139–1149. [CrossRef]
27. Smith, G.; Jentoft, S. Marine spatial planning in Scotland. Levelling the playing field? *Mar. Policy* **2017**, *84*, 33–41. [CrossRef]
28. Fletcher, S.; McKinley, E.; Buchan, K.C.; Smith, N.; McHugh, K. Effective practice in marine spatial planning: A participatory evaluation of experience in Southern England. *Mar. Policy* **2013**, *39*, 341–348. [CrossRef]
29. Pınarbaşı, K.; Galparsoro, I.; Borja, Á.; Stelzenmüller, V.; Ehler, C.N.; Gimpel, A. Decision support tools in marine spatial planning: Present applications, gaps and future perspectives. *Mar. Policy* **2017**, *83*, 83–91. [CrossRef]
30. Gazzola, P.; Onyango, V. Shared values for the marine environment–developing a culture of practice for marine spatial planning. *J. Environ. Policy Plan.* **2018**, 1–14. [CrossRef]

31. Morf, A.; Perus, J.; Steingrimsson, S.A.; Ekenger, M. *Results 2nd Nordic Workshop on MSP—Results of the 2nd Nordic Workshop on Marine Spatial Planning and an Update for 2014—Use and Management of Nordic Marine Areas: Today and Tomorrow*; Nordisk Ministerråd: Copenhagen, Denmark, 2014; 96p.
32. Mayer, I.S. *Playful Organisations & Learning Systems*; NHTV Breda University of Applied Science: Breda/The Hague, The Netherlands, 2016; ISBN 978-90-825477-0-2.
33. Mayer, I.S.; Zhou, Q.; Abspoel, L.; Keijser, X. Gaming the Future of the Ocean: The Marine Spatial Planning Challenge 2050. In *SGDA 2014: Serious Games Development and Applications*; LNCS 8778; Springer: Cham, Switserland, 2014; pp. 150–162.
34. Keijser, X.; Ripken, M.; Warmelink, H.; Abspoel, L.; Fairgrieve, R.; Mayer, I.S. Maritime Spatial Planning—A Board Game for Stakeholder Involvement. In *ISAGA 2017: Simulation Gaming. Applications for Sustainable Cities and Smart Infrastructures*; Springer: Berlin/Heidelberg, Germany; Delft, The Netherlands, 2017.
35. Government of The Netherlands; Breda University of Applied Sciences. *Maritime Spatial Planning Challenge: Short Sea Shipping Edition February 2016 Handbook*; Government of The Netherlands: The Hague, The Netherlands, 2016.
36. Government of The Netherlands; Breda University of Applied Sciences. *Maritime Spatial Planning Challenge: Scottish Marine Region Edition September 2016 Player's Handbook*; Government of The Netherlands: The Hague, The Netherlands, 2016.
37. Government of The Netherlands; Breda University of Applied Sciences. *Maritime Spatial Planning Challenge: Blue Development Edition January 2017 Handbook*; Government of The Netherland: The Hague, The Netherlands, 2017; pp. 1–13.
38. Papert, S.; Harel, I. *Constructionism*; Ablex Publishing Corporation: Norwood, NJ, USA, 1991.
39. Mayer, I.S.; Bekebrede, G.; Harteveld, C.; Warmelink, H.; Zhou, Q.; van Ruijven, T.; Lo, J.; Kortmann, R.; Wenzler, I. The research and evaluation of serious games: Toward a comprehensive methodology. *Br. J. Educ. Technol.* **2014**, *45*, 502–527. [CrossRef]
40. MSP Challenge. Available online: www.mspchallenge.info (accessed on 22 February 2018).
41. Paris, C. Using Serious Games in Regional Marine Planning. An Innovative Stakeholder Engagement Pilot Project in the Clyde Marine Region. Master's Thesis, University of Aberdeen, Aberdeen, UK, 2017.
42. The Scottish Government. *Marine (Scotland) Act 2010*; 2010 Asp 5; The Scottish Government: Edinburgh, 2010; pp. 1–112. Available online: https://www.legislation.gov.uk/asp/2010/5/pdfs/asp_20100005_en.pdf (accessed on 2 March 2018).
43. The Scottish Government. *Scotland's National Marine Plan: A Single Framework for Managing Our Seas*; The Scottish Government: Edinburgh, UK, 2015; ISBN 9781785442148.
44. FOD Volksgezondheid, Veiligheid van de Voedselketen en Leefomgeving. Royal Decree Marine Spatial Plan for the Belgian Part of the North Sea. 2014. Available online: https://www.health.belgium.be/sites/default/files/uploads/fields/fpshealth_theme_file/19094275/Summary%20Marine%20Spatial%20Plan.pdf (accessed on 2 March 2018).
45. FOD Volksgezondheid, Veiligheid van de Voedselketen en Leefomgeving. Available online: https://www.health.belgium.be/nl/news/startschot-herziening-marien-ruimtelijk-plan (accessed on 4 May 2018).

water

MDPI

Article

Exploring the Role of Relational Practices in Water Governance Using a Game-Based Approach

Piotr Magnuszewski [1,2,*], Karolina Królikowska [2,3], Anna Koch [2], Michal Pająk [2,4], Craig Allen [5], Victoria Chraibi [6], Anil Giri [7], Danielle Haak [8], Noelle Hart [8], Michelle Hellman [8], Donald Pan [9], Nathan Rossman [10,11], Jan Sendzimir [12], Maggi Sliwinski [8], Joanna Stefańska [2], Tharsi Taillieu [13], Denise Marie Weide [10,14] and Ilonka Zlatar [8]

1 International Institute for Applied Systems Analysis, A-2361 Laxenburg, Austria
2 Centre for Systems Solutions, Wroclaw 50-305, Poland; karokrolikowska@gmail.com (K.K.); anna.koch@crs.org.pl (A.K.); michal.pajak@crs.org.pl (M.P.); jlstefanska@gmail.com (J.S.)
3 Faculty of Finance and Accounting, WSB University in Wroclaw, Wroclaw 53-609, Poland
4 Faculty of Economics, Wroclaw University of Economics, Wroclaw 53-345, Poland
5 School of Natural Resources, Nebraska Cooperative Fish & Wildlife Research Unit, U.S. Geological Survey, University of Nebraska, Lincoln, NE 68583, USA; callen3@unl.edu
6 Department of Biological Sciences, Tarleton State University, Stephenville, TX 76049, USA; chraibi@tarleton.edu
7 Department of Biology and Agriculture, University of Central Missouri, Warrensburg, MO 64093, USA; giri@ucmo.edu
8 School of Natural Resources, University of Nebraska, Lincoln, NE 68583, USA; danielle.haak@gmail.com (D.H.); noelle.m.hart@gmail.com (N.H.); shelli.hellman@gmail.com (M.H.); maggi.sliwinski@gmail.com (M.S.); ilonka.zlatar@gmail.com (I.Z.)
9 Department of Subsurface Geobiological Analysis and Research, Japan Agency for Marine-Earth Science & Technology, Yokosuka, Kanagawa 237-0061, Japan; donald.pan@jamstec.go.jp
10 Department of Earth and Atmospheric Sciences, University of Nebraska, Lincoln, NE 68588, USA; nrossman@huskers.unl.edu (N.R.); dmarieweide@gmail.com (D.M.W.)
11 HDR, Inc., Omaha, NE 68114, USA
12 University of Natural Resources and Life Sciences, Vienna 1180, Austria; sendzim@gmail.com
13 Faculty of Psychology & Educational Sciences, KU Leuven, 3000 Leuven, Belgium; tharsi.taillieu@ppw.kuleuven.be
14 Department of Geography, Kansas State University, Manhattan, KS 66506, USA
* Correspondence: piotr.magnuszewski@gmail.com

Received: 22 January 2018; Accepted: 13 March 2018; Published: 20 March 2018

Abstract: The growing complexity and interdependence of water management processes requires the involvement of multiple stakeholders in water governance. Multi-party collaboration is increasingly vital at both the strategy development and implementation levels. Multi-party collaboration involves a process of joint decision-making among key stakeholders in a problem domain directed towards the future of that domain. However, the common goal is not present from the beginning; rather, the common goal emerges during the process of collaboration. Unfortunately, when the conflicting interests of different actors are at stake, the large majority of environmental multi-party efforts often do not reliably deliver sustainable improvements to policy and/or practice. One of the reasons for this, which has been long established by many case studies, is that social learning with a focus on relational practices is missing. The purpose of this paper is to present the design and initial results of a pilot study that utilized a game-based approach to explore the effects of relational practices on the effectiveness of water governance. This paper verifies the methods used by addressing the following question: are game mechanisms, protocols for facilitation and observation, the recording of decisions and results, and participant surveys adequate to reliably test hypotheses about behavioral decisions related to water governance? We used the "Lords of the Valley" (LOV) game, which focuses on the local-level management of a hypothetical river valley involving many stakeholders. We used an observation protocol to collect data on the quality of relational practices and compared this data with the quantitative outcomes achieved by participants in the game. In this pilot study, we ran the game three times with different groups of participants, and here we provide the outcomes within the context of verifying and improving the methods.

Keywords: serious games; social simulation; social learning; relational practices; river basin management; water governance; multi-party collaboration; stakeholders; experimental social research

1. Introduction

Water governance involves many complex issues beyond the conventional scope of technical and environmental aspects. Water catchments can be characterized by connectedness, complexity, uncertainty, conflict, multiple stakeholders, and multiple perspectives [1]. Berg [2] identifies seven crucial elements that affect water governance: institutions, interests of stakeholders, information, incentives, ideas, ideals (i.e., priorities placed on objectives), and individuals (meaning leadership). In order to integrate all these aspects in water governance, traditional top-down and technocratic approaches must be replaced by new management paradigms [3]. By this account, collaborative or participatory and multi-scale or multi-level modes of governance have been widely recommended to reconcile environmental, economic, and societal goals in the water sector [4–7].

Crucial factors for successful collaborative governance include face-to-face dialogue, trust building, commitment, and development of a shared understanding [8]. Dialogue enhances stakeholder empowerment and network building, and fosters learning [9]. Previous studies have found that learning processes based on social relations are essential for sustainable water resources management [10–12] since they provide positive outcomes, providing an increased understanding of key issues, reframing, building trust, improving relations, developing new organizations, and producing substantive results [13]. These outcomes of social learning may be both technical (e.g., effectiveness, sustainability, and integration) and relational or normative, such as a sense of ownership of solutions by different stakeholders, active citizenship, inclusive governance, and self-governing capacities [14]. Social learning occurs in multi-party collaboration processes that take place in the actors' networks or in "communities of practice" [3]. Multi-party collaboration is an emerging work system of two or more legally independent parties formed to address a certain problem while still retaining their autonomy within the new entity, e.g., network.

Multi-party collaboration involves a process of joint decision-making among key stakeholders in a problem domain directed towards the future of that domain. However, the common goal is not apparent from the beginning but rather is created through a process of collaboration [14,15]. According to the relational theory of multi-party collaboration processes, organizing can be defined as the process of making new meaning from the given diverse inputs while building communities of practice, and relational practices constitute the core of the social learning process (Figure 1). Relational activities (we use term interchangeably with relational practices in this paper) include getting the attention of stakeholders, committing to collaborate, legitimating stakeholders, fostering dialogue, connecting stakes and interests, negotiating roles and identities, guaranteeing commitment of constituencies, and aligning efforts and agreements [14]. They are often neglected at the expense of the focus on technical problem-solving. All too often such neglect of relational practices turns out to be highly detrimental to the outcomes. Social learning for interdependence among the parties is regarded as a central aim of multi-party collaboration, of which the main goal is to acknowledge and accommodate interdependent interests through the process of continuous negotiations. Unfortunately, most multi-party projects do not reliably develop improvements in policy or practice [14]. This might be caused by stakeholder assumptions about intended outcomes. Quaghebeur et al. [16] document that the emerging participation may be very different from what the convening party originally intended or what the financing institution expected. A greater challenge arises in regard to framing and reframing issues, e.g., when different actors, by definition, take different perspectives on issues. This is often the core problem in interdisciplinary projects related to natural resources [17]. Difficulties in creating co-ownership of the solution among all parties may occur even if the interactions among the parties are very intensive and well-guided [18].

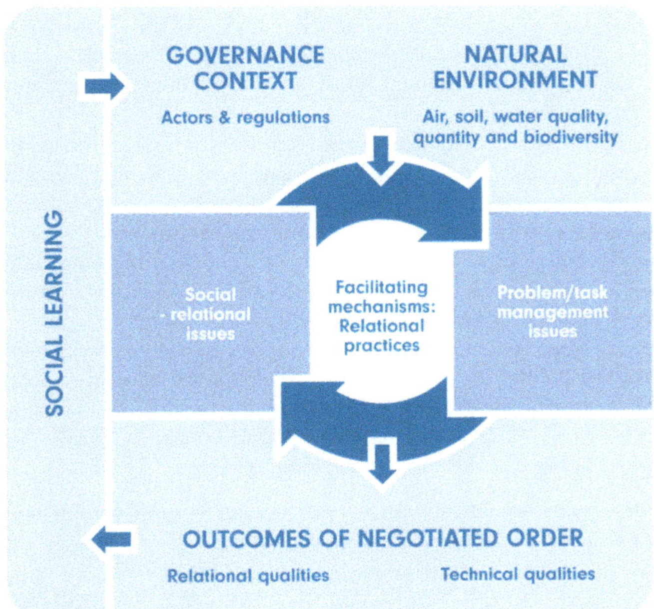

Figure 1. Social learning cycle through multi-party collaboration on natural resources [14] (p.149).

When conflicting interests of different actors are at stake in environmental projects, social learning that is centered on relational practices may prove to be crucial in overcoming these problems. Water management can serve as a perfect reflection of multi-party collaboration issues. Water itself might be perceived as a strong metaphor for the interconnectedness among many stakeholders since it connects geographical regions and diverse groups of providers and users, and as such symbolizes ecological and social interdependencies [14]. The purpose of this paper is to consider the abovementioned issues within the context of a multi-party collaboration experiment based on a serious game devoted to the challenges of water management in a river valley.

Environmental management games have long been used to support learning and to promote awareness of sustainability challenges [19] in a broad range of policy domains, such as climate change mitigation and adaptation [20,21], flood risk management [22–24], and land use and urban planning [25]. Environmental management games have been used both at different levels of education systems as well as in public policy to support social learning and encourage collective action [26–28]. Educational uses of serious games are further expanded into support for participatory policy analysis and strategic management [29,30], and it has been found that almost one-quarter of these games relate to water governance [31]. According to Mayer [32], the success of gaming for policymaking derives from the power of gaming to tackle both the technical-physical and the social-political complexities of policy problems.

Parallel to gaming applications in education and policy making, serious games have also been used to support experimental research. The application of laboratory experiments in the social and political sciences has continued to increase since the 1980s; however, there are still many objections regarding the lack of realism and generalizability. However, Falk and Heckman [33] argue that many of these objections are misguided, and laboratory experiments provide the possibility to control variation, which is basic for empirical scientific knowledge. According to these authors, such experiments, including games, can be used to study behaviors and institutions under the conditions of strong control of decision environments (e.g., payoffs, information that parties possess, and rules of action). There is an established tradition of experiments in the field of experimental economics regarding policy-relevant issues, such as cooperation and public goods [34] or the importance of reciprocity and social approval [33,35]. Progress in experimental games has been achieved by evoking the essence of complex dilemmas in simple settings, though it leaves many questions about social processes in more realistic policy situations untouched.

Regarding complex policy questions, games seem to be a research method with potentially high external validity, especially when it is not possible (or very difficult) to test alternative policies in real-life situations. Games may be designed in a manner that is less artificial than laboratory experiments, and thus, they may more adequately depict real-life scenarios [36]. Empirical studies in the organization and policy sciences might have a form of "research with gaming" [30] or "gaming for pure research" [36], i.e., when games are used as an observation context to develop and test hypotheses on the policy-relevant behavior of individuals, groups, and organizations. The strongest advantage of games as quasi-experiments relates to the controlled variation that is crucial for causal knowledge [37]. Games provide the possibility of systematically manipulating variables in complex organizational environments and measuring the effects in systematic ways, including statistical methods, even in mostly qualitative projects [36]. The analysis of available experiments with games provided several conclusions on the advantages of "research with gaming" [30,36]:

- identical policy environments for experimental and control groups, though the stimuli for all groups can be structured in a controlled and desired way;
- the possibility of complete observation and reconstruction (by writing down all actions in observation protocols) of all decision-making processes, which is hard to achieve in real-life situations;
- opportunities to observe situations (raised by simulations) that are rare, hidden, risky, and socially unacceptable or in other ways that are difficult to access in reality.

Questionable correspondence between the experimental environment and reality can be checked among participants [30,36], who should include experienced policy makers (strongly recommended), not just students. This "member check" provides a serious opportunity to assess the psychological and procedural validity of the game. However, there will always be difficult-to-avoid trade-offs between the control of experimental conditions and the real-life resemblance of experimental environments. Furthermore, a trade-off exists between representing real-life experiences and playability since participants often undertake certain actions within a game in order to "have fun" that they would not initiate in the real world. Next, an issue that always arises with social research is the effect of the observer's presence on participants' behavior. However, the same problem occurs in interviews, so this problem is not exclusive to gaming research.

The LOV game has been used repeatedly as an educational serious environmental management game. Because it is based on a real-world situation in the Tisza River Valley of Hungary [38], it also has the potential to serve as an experimental environment for research in the field of water governance [22]. With a more complex representation than experimental economic games, LOV allows researchers to examine the social interactions that emerge from the role-playing situation linked with causal mechanisms from the environment. The main outcome from this paper is the verification of the methods used by addressing the following question: are game mechanisms, protocols for facilitation and observation, the recording of decisions and results, and participant surveys adequate to reliably test hypotheses about behavioral decisions related to water governance?

2. Materials and Methods

We developed the LOV game as a tool to support a multi-party collaboration quasi-experiment. The original version of the game was created as a part of the NeWater project [39]. The goal was to create a simulation, based on interviews with stakeholders and experts, which enables participants to explore alternative river management regimes and introduces the concept of river-landscape connectivity. After the finalization of the project, the game was further developed in cooperation with the University of Leuven (Belgium) to expand on the existing biophysical model by introducing new roles for the participants, thus exploring social aspects of floodplain management. The goal was to represent the problems connected with engaging multiple stakeholders and thereby emphasize the importance of relational activities. This development created the social simulation that is currently used regularly in the context of educating about river basin management and social aspects of creating multi-lateral agreements. Furthermore, it has been implemented as a supplementary tool in real-world processes to engage participants and give them an opportunity to explore the problem, similar to the real one, in a safe, simulated environment (for example, sessions with Dutch Water Sector Intelligence [40]. The current version of the game is available to be played online using tablets and computers for all operations and calculations [41].

The biophysical model in the game refers to a short river reach. It is limited and does not include any details on water quality, aquatic ecology, or groundwater. Since the game focuses on a small community in a river valley, upstream–downstream relationships are not included. The game is not intended to recreate biophysical complexity—rather its goal to represent a minimalistic representation with only basic feedbacks. Even such a minimalistic representation is hard to handle for participants. The main purpose of our quasi-experimental approach is to study the social complexity of water governance systems. Many experimental studies follow an experimental economics approach that limits game features to bare essentials. Adding game features has to be done carefully in order to match the cognitive limits of participants within the assigned time (which is rarely longer than a 1-day workshop). Hence, the scope of our biophysical model must have been limited. At the other end of the game complexity spectrum, role-playing simulations are rarely assessed with quantitative methods. The contribution of this study lies in bringing more structure to qualitative studies of social behavior relevant to water governance.

We applied a "research with gaming" approach in order to examine the impact of relational activities on the outputs in the game setting. It is not the goal of this experiment to reproduce real life social learning, although the game itself might be used as a support for multi-stakeholder processes (we provide some comments on such a use of this game in the discussion section). In other words, we use the game for empirical research on complex systems in a laboratory situation.

Using the LOV game as an observational context [30], we have designed structured observation protocols and used the record of game results to measure the impact of relational activities on the game's outputs. Relational activities such as leadership, sharing information, stereotyping, and ground rules were selected for this study from the longer list based on research and training experience [42].

In this article, we report our initial results from three case studies. While it is obvious that these results are not statistically significant, these results represent a pilot study of the research method that allows one to explore relational activities and social learning in a controlled environment. Therefore, the contribution of this article is methodological in nature, and there is no claim made regarding the validity of the results. Rather, we demonstrate how this type of research can be conducted; additionally, although the actual results in the three case studies support our hypothesis, we are aware that a larger sample is needed to develop valid conclusions. Despite the modesty of these assumptions, we believe the challenges inherent in the empirical research of complex social-ecological systems make this pilot study novel and relevant to both the use of serious games for empirical research the understanding of social processes in complex environments.

Our main research direction focuses on determining whether relational activities act as facilitating mechanisms that affect the game's two main outcomes: the players' economic status and the state of the environment. Based on the existing research in the area of multi-party collaboration [14,15], we propose the following hypotheses:

1. Groups that allow/stimulate all members to share information, express their understanding of the problem, and build a shared reality are better able to cope with inherent system complexity; thus, they achieve better content outcomes.
2. Having a leader (leaders) that is (are) both process- and outcome-oriented facilitates the processes of defining the problem and agreeing upon solutions, which, therefore, leads to better content outcomes.
3. Stereotyping and/or lack of ground rules to facilitate interactions make conflict management more difficult; as a result, it lowers the chance of developing common solutions and negatively affects content outcomes.
4. The ability of the group to formulate and agree upon a common solution leads to better content outcomes.

We will attempt to verify these hypotheses in the subsequent research. In this article, based on the pilot study, we will verify if the methods that we use are appropriate for testing these hypotheses. In the LOV game, participants take the roles of inhabitants of a river valley threatened with droughts and floods. They are asked to identify with their roles; however, no decisions are imposed on them—they can make them freely. Some assume the roles of farmers, and their task is to manage the farms they own. They make decisions about the type of production on a particular plot. There is also the possibility to buy land, so they can expand their farms. Free plots belong to the Local Government, which determines the sale conditions and taxation levels of specific farmers. Money acquired in this way can be transferred in the form of subsidies to other players.

The Water Board main task is to build and maintain elements of the flood-protection infrastructure, e.g., dykes and the water steering system. Dykes protect the valley from flooding, while the water distribution system helps reduce flood losses or increases soil moisture in the valley during a drought. The farmers' financial performance depends on external factors, such as the precipitation, the water levels in the river, the associated soil humidity, and the performance of different types of production on the individual plots under different moisture regimes. Farmers can decide to set up crop cultivation (with options to use high-yield grain), animal husbandry, orchards, and fish ponds, or they can decide on the commercial use of pre-existing forests. Each production type is characterized by a different resilience to environmental conditions as well as by losses caused by floods or excess irrigation from the water steering system. Therefore, it is crucial to match each production type with the anticipated conditions on the board. These are, in turn, the result of both external factors, independent of the participants and the investments made by the Water Board. Additionally, all activities undertaken on plots by farmers as well as the extent of the water steering system affect the biodiversity in the valley. The intensification of production may damage the local ecosystem, which elevates crop sensitivity to pest attacks and may cause additional losses.

There are also two other important institutional roles in the game, i.e., the Bank and the Environmental Non-Governmental Organization (NGO). In the first case, as Banker, a player has the option of granting loans to other entities in the valley, but he must also ensure that they are recovered, including the interest charged. An environmental NGO has information on the impacts of particular methods and types of production on biodiversity. This knowledge, together with the possibility of making complaints about other players to the Central Government, which is represented by the Moderator, can help protect the valley's ecosystem.

The game presents two layers of complexity. One comes from the general complexity of the simulation model, i.e., multiple interactions between environmental variables that affect production within the valley and its economic condition. That, in turn, results in further investments that create feedbacks, which can affect the environmental conditions. The second layer of complexity comes from interactions among participants. Therefore, the process can be described as encompassing social complexity. The economic dependencies between specific roles and the asymmetry of information mean that communication and cooperation between and among players are the key aspects of the game. Game facilitator applies the protocols of gameplay to impose restrictions on conversational possibilities between individual participants. These restrictions are introduced to represent real-life problems related to the dispersion of information. For example, representatives of more than three roles cannot talk with each other at the same time. This simulates real-life challenges connected with engaging other parties. As in real life, the only opportunity to assemble all inhabitants of the valley and reach a broad agreement is through a so-called "community meeting." Community meetings are convened twice throughout the entire game, and representatives of all roles can participate. The way they are carried and the topics discussed are regulated by the participants themselves. The overview of the game physical model is attached in Appendix A.

The game was played over 7 rounds with the total time between 4 and 5 h, followed by 1–2 h debriefing. It is a long and intensive game designed to allow players to immerse in its narrative and for the social relations to develop within the game world (Figures 2 and 3).

Figure 2. Social interactions in progress—Lords of the Valley (LOV) game workshop, Wrocław (photo Anna Zemlak).

During the game, participants have to deal not only with both layers of complexity but also with their own preconceptions and ideas about the effective management of the river valley. The underlying model simulating environmental conditions (flood, drought) tests these ideas in terms of feasibility and provides feedback about the consequences of their decisions. This process shows that some actions may have unforeseen long-term consequences that can affect the environment of the entire valley. Yet, for participants, the greatest challenge can be the social element within the simulation. Creating a common understanding of the situation, let alone some management agreements, can be difficult when all participants have their own perspectives on the problems being addressed, their own goals and personal values, as well as different levels of access to the in-game information. The setting, as well as the emphasis on the role-playing element of the game, enables each participant to observe how real-life communication problems emerge and how they affect the outcomes. The safe environment of social simulation creates a "laboratory" where participants can not only learn about the challenges connected with multi-party collaboration but also test their ideas and potential solutions in a safe environment.

Figure 3. Financial performance in progress—LOV game workshop, Wroclaw (photo Anna Zemlak).

The game was played during workshops in three cities: Wrocław, Rzeszów, and Kraków. The number of players participating in these workshops varied from 15 to 25. Observations involved 11 observers trained in observation procedures (Appendix B), who were assigned to different roles; one observer was assigned to follow general interactions at the game board. The results of the observations include both quantitative and qualitative parts. However, due to the pilot study character of this research, the quantitative results are not statistically significant and are used here only for testing the methodology. There were two observation protocols applied: one for the rounds of the game and one for the community meetings (Appendixs C and D). They both contained multiple choice as well as open-ended questions regarding communication and interactions between teams, sharing information, leadership, conflicts, and overall group dynamics. Moreover, in the community meetings, where representatives of all groups could participate, the observation protocol questions referred to defining, understanding, and solving problems. Questions about exclusion (are some parties ignored by others?), a common understanding of the problem, establishing agreed solution, and ground rules are relevant only for community meetings. Results concerning internal dynamics come from participants' evaluation feedback, while the rest of the results come from observation protocols.

The questions in the observation protocol were related to specific categorizations (relevant for the stated hypotheses) that were in turn narrowed down into specific variables. Observations were recorded as data and transformed to indices (added) to measure variables that were later compared between the cases (cities) to verify our hypotheses.

For binary variables, observations were recorded as "yes" if that response was the majority, "no" if that response predominated, and "hard to say" (HTS) if the distribution between the two was roughly equal. In case of conflicts, answers indicate how many of the observers indicated "yes", e.g., 7 from 10. For the variables measured on ordinal scales, values presented in the table are the average of the scores.

In the case of stereotypes, the number indicates how many times they were noticed. Sharing information specific to a player's role was measured on a scale from—3 (*No, actors actively refuse to share information very much*) to 3 (*Yes, actors share information very much*) with no neutral value. Both competition and cooperation were measured on a scale from 3 (very much) to 1 (a little), and observers were able to estimate both scales in each round. Common understanding of the problem was measured on a scale from 0 (not at all) to 3 (very much). Establishing and agreed solution is measured on a scale from 1 (*Yes, and the solution is made explicit*) to 4 (*No, actors explicitly set individual solutions*). For all questions represented with ordinal scales, the values shown in the table are the average of the scores.

Additionally, over the course of the game, participants completed surveys (3 times during the game and before and after the game), intended to gather information on their subjective assessment of cooperation and competition levels (Appendix E). In order to understand the time dynamics of the sequence of game events round after round, both quantitative and qualitative data (including participants direct quoted) have to be used jointly.

Recruitment for the games was open, advertised mostly in the academic community (mostly students, doctoral students, and researchers) and resulted in a rather random group composition, the impact of which is later discussed. Again, the study should be treated only as a pilot study.

3. Results

The results of observations carried out by observers during rounds and community meetings, as well as from evaluation protocols, are presented in Tables 1–3.

Each single workshop (gameplay) created its own history of multi-party collaboration related to managing natural resources in the face of threats. Multiple actors collaborated to address the issues of water and soil management, nature preservation, land use, farming practices, introduction of technology, innovation, and related problem domains. Participants in different cities had different attitudes and different definitions of the problem domain. Each story was unique in terms of participants decisions, results, and frames applied by participants about what happened in the valley and why.

In Wrocław, participants initially focused on "deal-making" among themselves. They were mainly engaged in bilateral conversations and agreements. Actors shared information specific for their roles and goals; however, there was primarily "lots of talking between groups to figure out other team's roles and how to operate" (Quotes in italics come from observation protocols). In Rzeszów, there was almost no communication at the beginning; some players openly refused to share or "held information close and shared little with other teams." In Kraków, actors were looking for general solutions. They "seemed to be looking for the best solutions; not entirely self-focused." In addition, there were some bilateral conversations without open communication, but actors shared information, especially organizations (i.e., Water Board, Bank, and NGO). These differences in attitudes at the beginning of the game influenced further development of trajectories in different cities gameplays.

In Wrocław, leadership functions were rotating between selected teams (playing their roles): NGO, Local Government (LG), and Water Board (WB). The participant playing the LG was perceived as a natural leader. The participant playing the NGO was focused on both technical content (problem-solving) and on group processes. The NGO emerged as a group facilitator and "was accepted by all". It was the only group (Wrocław) where the function of the facilitator was clearly identified and enacted. The first round began in a climate of competition. However, better outcomes became possible as other factors appeared: the emergence of a leader and facilitator, much smaller number of conflicts, and sharing information during the first community meeting. At the time when the drought occurred, competition reappeared (the second Community Meeting) and even conflict (9/10), which clearly shows the effect of the crisis. However, due to the collaborative practices established earlier as well as effective leadership and facilitation, the group was able to recover from the crisis and, already in the 5th round, information sharing was very intensive and cooperation more pronounced than competition. This led to higher profits (visible in the game results). Although participants indicated a high level of competition in their evaluation sheets, a summary of observational data of the participants provides a broader view than the participants' perceptions of competition and cooperation by revealing the effect of a facilitator or leader and of sharing information on the outcomes. This also demonstrates that conflict can be constructive when relational practices are positively established.

In Rzeszów, responsibility for the group was dispersed among different actors in each round. There was no clear leader identified until the seventh round. As a consequence, there was "more chaos than conflicts or cooperation." Actors were "not sure who's leading the circus." Institutions like the LG or the WB were recognized as leaders responsible for the valley, but "not all players paid attention to them."

In Kraków, there was no unambiguous leader. Rather, responsibility was dispersed. The LG and the WB were recognized as leaders a few times, but they were mainly focused on technical content. Cooperation was less intensive than in Wrocław and Rzeszów. There was much more competition and rivalry, with obvious arguments and conflicts.

In each gameplay, stereotyping emerged over the course of interactions, and it was probably transferred from actors' experiences in the real world. Expressions of stereotyping affected mainly the LG but also the NGO, WB, and sometimes (rarely) farmers:

> "Government was initially distrusted by farmers."
>
> "Role of the government as a money source."
>
> "Thought that the government is greedy."
>
> "Farmers think government should subsidize poor farmers. NGO thinks farmers want money for silly things. Farm 3 called NGO the "green guy." NGO did not trust in the farmers' abilities to change their land use, since they claimed at the meeting that they do not have the knowledge. Farmers are poor and not environmentally friendly."
>
> "I don't like the Bank, they are bloodsuckers (...) Those blood suckers, those bank people ... you like them?"

The data collected indicate that every single multi-party collaboration process was characterized by different internal dynamics. Behaviors linked with some variables, such as exclusion or stereotyping, were common in each gameplay, but there was significant variation with respect to most of the variables observed. For example, some groups were poorly organized. Some had a history of conflict, and there were disparities in power and resources. In addition, there were differences in access to expertise and information. In general, the Wrocław group was better organized than the other communities. In Wrocław, the leadership functions were shared successively between different roles not only during game rounds but also during the community meetings. The final outcome was much more satisfactory because parties were able to cooperate effectively, share information, and agree on which solutions to implement. This is confirmed by both qualitative and quantitative data. In Rzeszów and Kraków, information sharing was low, and there were fewer leadership attempts and no facilitation function. Exclusion and conflicts occurred much more often than they did in Wrocław. Observations from community meetings indicate that, in all cities, participants were not able to achieve a shared understanding of the problem (nevertheless, Wrocław has the highest rate). This is often the case in the real world, where stakeholders quickly jump to solutions without deeper reflection on how problem perception differs among them.

Table 1. Observation and evaluation data analysis from Wroclaw.

WROCŁAW	Sharing Information	Exclusion	Common Understanding	Solution	CLEAR Leader	Facilitator	Rules	Stereotypes	Conflicts	Competition	Cooperation	Internal Dynamics
Observation data Round 1	1.13	–	–	–	YES	HTS	–	4	7/10	1.38	1.40	Competition
Evaluation data Round 1	–	–	–	–	–	–	–	–	–	2.00	1.83	Cooperation and competition
Observation data Round 2	1.22	–	–	–	YES	NO	–	5	7/10	1.40	2.10	More cooperation than competition
						FLOOD						
Observation data #1CM	1.22	5	1.33	1.22	YES	YES	NO	1	5/9	1.25	2.50	More cooperation than competition
Evaluation data #1CM	–	–	–	–	–	–	–	–	–	2.00	2.71	More cooperation than competition
Observation data Round 3	2.13	–	–	–	YES	NO	–	4	2/10	1.40	2.14	More cooperation than competition
Observation data Round 4	1.83	–	–	–	YES	NO	–	0	3/10	1.33	2.00	More cooperation than competition
						DROUGHT						
Observation data #2CM	1.63	7	0.85	3	YES	NO	NO	4	9/10	2.38	1.60	Competition
Evaluation data #2CM	–	–	–	–	–	–	–	–	–	2.00	2.14	Cooperation and competition
Observation data Round 5	2.33	–	–	–	YES	NO	–	1	2/10	1.00	2.00	More cooperation than competition
Observation data Round 6	1.13	–	–	–	YES	NO	–	1	4/10	2.00	2.14	More cooperation than competition
						FLOOD						
Observation data Round 7	1.83	–	–	–	HTS	NO	–	3	6/10	1.40	1.78	More cooperation than competition
Evaluation data Round 7	–	–	–	–	–	–	–	–	–	2.00	1.86	Cooperation and competition
Evaluation data debriefing	–	–	–	–	–	–	–	–	–	2.00	2.50	Cooperation and competition
Summary	14.44							23	45	23.53	28.71	

Table 2. Observation and evaluation data analysis from Rzeszów.

RZESZÓW	Sharing Information	Exclusion	Common Understanding	Solution	CLEAR Leader	Facilitator	Rules	Stereotypes	Conflicts	Competition	Cooperation	Internal Dynamics
Observation data Round 1	0.50	–	–	–	NO	NO	–	4	6/11	1.60	1.60	Cooperation and competition
Evaluation data Round 1	–	–	–	–	–	–	–	–	–	2.00	2.14	Cooperation and competition
Observation data Round 2	0.78	–	–	–	NO	NO	–	1	6/11	1.40	1.25	More cooperation than competition
						FLOOD						
Observation data #1CM	0.60	8	1.28	2.71	NO	NO	NO	1	8/11	1.43	1.75	More cooperation than competition
Evaluation data #1CM	–	–	–	–	–	–	–	–	–	1.86	2.57	Cooperation and competition
Observation data Round 3	0.57	–	–	–	YES	NO	–	2	4/10	1.50	1.67	More cooperation than competition
Observation data Round 4	1.00	–	–	–	HTS	NO	–	1	3/10	2.00	1.80	Cooperation
						DROUGHT						
Observation data #2CM	0.83	8	1	2,4	NO	NO	NO	1	5/9	1.50	1.60	Cooperation and competition
Evaluation data #2CM	–	–	–	–	–	–	–	–	–	1.43	2.29	Cooperation and competition
Observation data Round 5	1.33	–	–	–	YES	NO	–	3	3/10	1.40	2.00	More cooperation than competition
Observation data Round 6	1.75	–	–	–	YES	NO	–	3	5/10	1.67	2.33	Cooperation
						FLOOD						
Observation data Round 7	1.56	–	–	–	YES	NO	–	1	2/10	1.50	2.13	More cooperation than competition
Evaluation data Round 7	–	–	–	–	–	–	–	–	–	1.43	2.57	Cooperation and competition
Evaluation data debriefing	–	–	–	–	–	–	–	–	–	1.67	2.43	More cooperation than competition
Summary	8.92							17	42	22.38	28.13	

Table 3. Observation and evaluation data analysis from Kraków.

KRAKÓW	Sharing Information	Exclusion	Common Understanding	Solution	CLEAR Leader	Facilitator	Rules	Stereotypes	Conflicts	Competition	Cooperation	Internal Dynamics
Observation data Round 1	1.50	–	–	–	YES	NO	–	4	3/11	1.50	1.67	Cooperation
Evaluation data Round 1	–	–	–	–	–	–	–	–	–	2.40	2.29	More cooperation than competition
Observation data Round 2	1.80	–	–	–	YES	NO	–	0	2/11	1.17	2.00	Cooperation and competition
						FLOOD						
Observation data #1CM	1.80	9	1	3.12	NO	NO	NO	5	10/10	1.67	1.78	Cooperation and competition
Evaluation data #1CM	–	–	–	–	–	–	–	–	–	1.75	2.17	More cooperation than competition
Observation data Round 3	1.36	–	–	–	YES	NO	–	2	4/11	1.00	2.09	Cooperation
Observation data Round 4	0.67	–	–	–	NO	NO	–	1	3/11	1.33	2.11	More cooperation than competition
						DROUGHT						
Observation data #2CM	1.22	5	0.6	2.87	NO	NO	NO	4	10/10	2.11	1.86	Conflict
Evaluation data #2CM	–	–	–	–	–	–	–	–	–	1.50	2.57	More cooperation than competition
Observation data Round 5	1.10	–	–	–	HTS	HTS	–	0	6/11	1.60	1.50	More cooperation than competition
Observation data Round 6	1.00	–	–	–	YES	NO	–	1	6/11	1.40	2.14	More cooperation than competition
						FLOOD						
Observation data Round 7	1.18	–	–	–	YES	NO	–	2	8/10	1.00	1.55	More cooperation than competition
Evaluation data Round 7	–	–	–	–	–	–	–	–	–	1.43	2.25	Cooperation and competition
Evaluation data debriefing	–	–	–	–	–	–	–	–	–	1.50	2.29	More cooperation than competition
Summary	11.63							19	52	21.36	28.25	

Besides observations, game decisions and results over the course of the gameplay were collected to support the key hypotheses. Economic outcomes in the context of events in the river valley ecosystem (flood and drought) are presented in Figure 4. The Wrocław group achieved better economic results. Natural disasters (floods and droughts) occurred similarly in every scenario, but in Wrocław, there was the greatest variability in outcomes related to events (i.e., floods and droughts); however, this variability did not decrease accumulated profits. The most difficult economic situation was in Kraków, where multi-party collaboration finished with conflict and economic bankruptcy. In Rzeszów, a leader emerged in the seventh round, and they finished the game with better economic results than those observed in Kraków. In Kraków, the greatest economic losses occurred due to drought, while losses in Rzeszów were due to flooding. Regarding biodiversity, both Kraków and Rzeszów resulted in a decreasing value, in contrast with Wrocław. Obviously, one can achieve high outputs in both accumulated profit and biodiversity. This is because the underlying game model includes feedback mechanisms that undermine economic outcomes in the long term when ecosystem services are reduced due to negative environmental impacts.

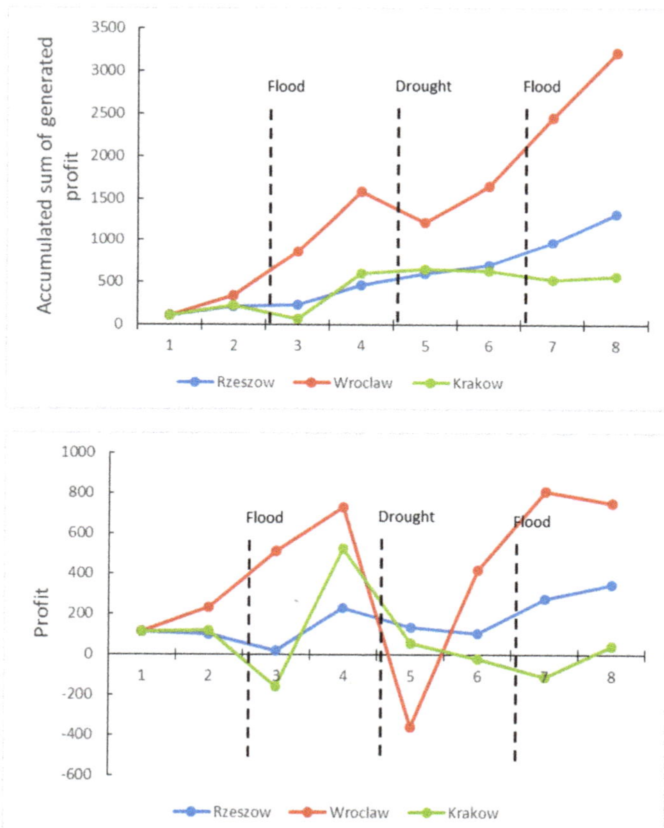

Figure 4. *Cont.*

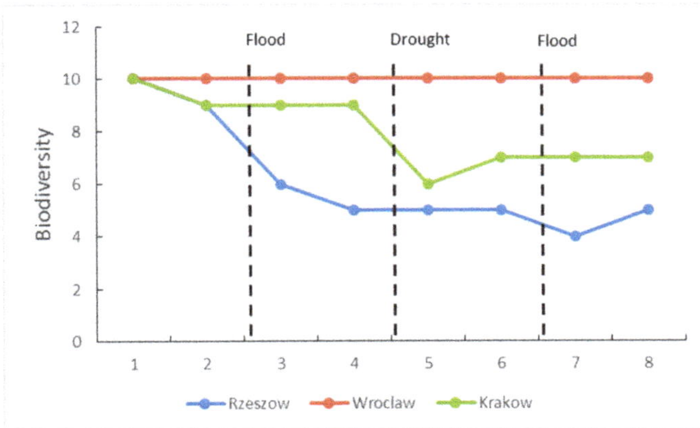

Figure 4. Economic outcomes and biodiversity levels in the context of the events in the ecosystems of the valleys.

The results and their analysis presented above (Figure 4) provide the basis for the assessment of our hypotheses. As indicated earlier in the article, verification of the stated hypotheses is beyond the scope of this paper because our sample is too small. However, we can discuss the findings from our case studies (gameplays in different cities) and note to what extent observed regularities contribute to the stated hypotheses, which is necessary to verify our "research with gaming" approach. Subsequent discussion of the pilot results should be seen in this light.

The Wrocław group achieved the best game outcomes both in terms of profits and biodiversity. Table 4 presents observations indicating potential causes of this advantage, linked to specific hypotheses. There is supporting evidence for Hypotheses 1, 2, and 4 but not for Hypothesis 3.

Table 4. Summary of observations highlighting the Wrocław group that achieved the best outcomes.

Hypothesis	Summary of Observations
Hypothesis 1. *Groups that allow/stimulate all members to share information, express their understanding of the problem, and build a shared reality are better able to cope with inherent system complexity, and thus achieve better content outcomes.*	The Wrocław group scored high on information sharing, much better than other groups. The difference between the groups with respect to achieving the common problem definition was not pronounced.
Hypothesis 2. *Having a leader (leaders) that is (are) both process- and outcome-oriented facilitates the processes of defining the problem and agreeing on solutions, therefore leading to a better content outcomes.*	The leadership effect was observed throughout the game in the Wrocław case. It emerged naturally (NGO) or was discussed, agreed and transferred between the roles. It allowed the whole group to get through the crisis caused by floods and droughts.
Hypothesis 3. *Stereotyping and/or lack of ground rules facilitation of interactions makes conflict management more difficult; as a result, it lowers the chance of developing common solutions and negatively affects content outcomes.*	Wrocław case was the only one where facilitation was present and this group was definitely better organized than other cases; however, establishment of the ground rules was not recorded. The intensity of stereotyping was similar between the cases.
Hypothesis 4. *The ability of the group to formulate and agree on a common solution leads to better content outcomes.*	An agreement on a common solution was reached only in the Wrocław case.

Above we have presented all the steps necessary to analyze the results of the pilot study with the game using data gathered with the observations protocol and participants' surveys. These data provide us with measures to draw the conclusions with respect to the stated research goal. Once we have a larger dataset, we can test stated hypotheses statistically. The method is also rich enough for qualitative research of smaller samples.

4. Discussion and Conclusions

The LOV game was designed in such a way that relational practices were as important as problem-solving aspects, and this is a novelty that combines game mechanics with role-playing (identifying with and role taking in a party/organization from the game world). We aimed to provide a space to observe how players behaved both in terms of the openness of interactions and the structured observation protocol. Since this was a pilot study, the results are not statistically significant, and the results are presented to demonstrate the method rather than to emphasize specific findings. The limitations of this approach are similar to those in other experimental settings in social sciences and include the artificiality effect, as well as the issues of psychological and process validity. The real-life resemblance might be evaluated by participants themselves if participants were real stakeholders, and the artificiality effect is difficult to avoid in laboratory experiments. On the other hand, during the game, we may ensure a similar experimental environment for different groups in order to achieve a level of controlled variability that is not possible in real-life situations. In order to make comparisons possible, a larger study with more LOV sessions is necessary. Moreover, additional data about participants should be collected and included in the analysis. For example, it is not possible to retroactively determine why the group in Wrocław was better at embracing relational activities, especially leadership. In future research experiments, participants should be examined with regard to their background, including domain knowledge, previous experience in public policy making and serious gaming, and social or public activities in real life. The authors are aware of their own cultural background, including the current emphasis on participatory approaches in water governance research. In spite of our effort to create a neutral environment (a "sandbox"), where participants could try different governance styles, there are unavoidable cultural biases that may be present in game and research design. Different governance regimes can produce different outcomes in different cultures [43,44]. In order to verify the stated hypotheses, this cultural background should also be controlled in the experimental design, especially regarding the nature of relational activities under observation, since this might differ depending on the cultural context. This issue might be the subject of further research.

Indices constructed from observations can be developed with greater complexity or greater specificity to identify relevant processes. For instance, conflicts may be focused on relations or tasks. Several task conflicts can be a sign of diversity, so the balance of task conflict to relational conflict is important. In this case, an index can be based on the ratio of task/relational conflict. In terms of ground rules, they concerned procedural issues, task issues, and relationship issues. We usually observe process interventions by some actors when they handle these issues, and gradually, these become accepted (i.e., written) rules. Improved indices may take the nature of these interventions under consideration, e.g., the effort to reframe the common ground to ensure that the concerns and interests of all members are recognized in the problem formulation.

The application of the LOV game to do research with gaming described in this article is not the only possible option for using this game. It can also be used to support real-world multi-party processes related to the management of river valleys. Often the soft, relational activities do not receive enough attention in the actual processes of formulating policy and making decisions. By participating in a "social simulation" based on this game, stakeholders can become aware of the importance of relational practices, and based on this experience, discuss how they can change their real world relations, interactions, rules, or procedures.

By considering all limitations, we find the initial results obtained from this pilot study provide a solid verification of the proposed research method used to study the effects of relational activities and social learning for improving water governance.

Acknowledgments: The Nebraska Cooperative Fish and Wildlife Research Unit is jointly supported by a cooperative agreement between the U.S. Geological Survey, the Nebraska Game and Parks Commission, the University of Nebraska–Lincoln, the United States Fish and Wildlife Service, and the Wildlife Management Institute. This work was supported in part by the National Science Foundation's Integrative Graduate Education and Research Traineeship (IGERT) program (NSF #0903469). Any opinions, findings, and conclusions or recommendations expressed in this material are those of the authors and do not necessarily reflect the views of the NSF. International Institute for Applied Systems Analysis encourages and actively supports its researchers to publish their research in journal articles or books that are made available for free to all users (gold open access). We would like to express our gratitude to Magdalena Liszka, Elzbieta Szlauer, Anna Dubel, and Agata Pierscieniak, who greatly supported the organization of the game sessions for this study.

Author Contributions: The concept of the research and the article: Piotr Magnuszewski; Research design (including observation protocol): Joanna Stefańska, Piotr Magnuszewski, Tharsi Taillieu, Jan Sendzimir; Research implementation (including observations and processing of results): Victoria Chraibi, Anil Giri, Danielle Haak, Noelle Hart, Michelle Hellman, Donald Pan, Nathan Rossman, Maggi Sliwinski, Denise Marie Weide, Ilonka Zlatar, Craig Allen; Analysis of Results: Anna Koch, Michal Pająk, Karolina Królikowska, Piotr Magnuszewski; Writing and editing: Karolina Królikowska, Anna Koch, Jan Sendzimir, Danielle Haak, Victoria Chraibi, Nathan Rossman, Piotr Magnuszewski.

Conflicts of Interest: The authors declare no conflicts of interest.

Appendix A. Overview of the Biophysical Model for the Lords of the Valley Game

The game uses a simple stylized model defining critical relationships and processes connected to the small-scale management of the floodplain valley. The area is restricted to the segment of the river with surrounding water infrastructure and local farms. Furthermore, all values presented to the participants are treated as yearly averages (rainfall, soil moisture, the water level in the river, etc.). What is more, due to the limited size of the modeled area and the scope of the game, biodiversity is treated as an aggregated value representing average biodiversity within the valley. Introducing such simplifications to the system was conscious design choice as limiting the physical complexity of the simulation was required to conduct the workshop within the sensible timeframe along with the debriefing. Figure A1 presents the general schematics of the system as well as the feedbacks between simulation elements. The model transforms the participants' decisions into the changes of food production and of water infrastructures, such as dikes. It also combines the environmental scenario (rainfall and the water level in the river) with choices made by the players (farmers and the Water Board) to produce the concrete values of farmers' production. Together the model and the players' decisions form an integrated social-ecological system with many interacting feedback loops.

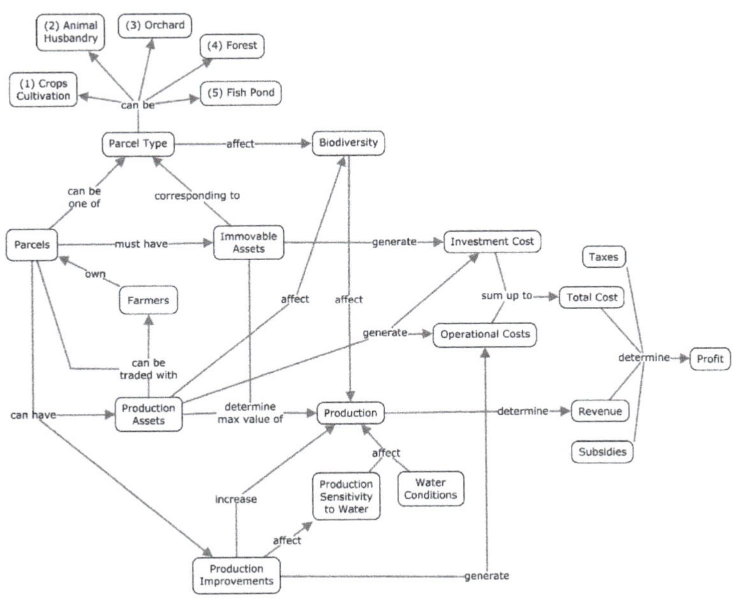

Figure A1. *Cont.*

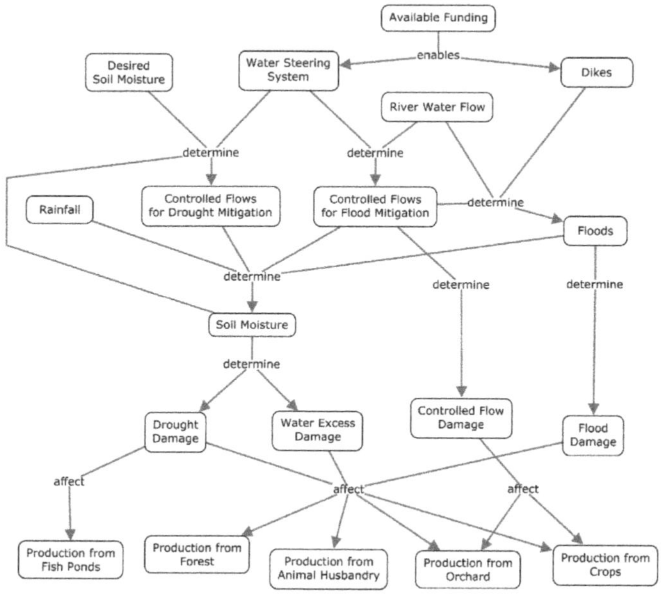

Figure A1. General schematics of the game system.

Appendix B. Instructions for Observers

Appendix B.1. Observation of Multi-Party Processes—Selection of Behaviors to be Recorded

- Communication between teams
- Conflicting goals and interests at different levels
- Dealing with unequal distribution of information
- Shared reality (common understanding of the problem/the system)
- Inter-group relations-conflict, cooperation, competition
- The role of trust (inter-group and intragroup)
- Setting interaction ground rules
- Commitments, responsibilities, control
- Learning about interdependence of actors
- Leadership style and facilitation
- Dealing with risk and uncertainty

Appendix B.2. Main Directives for the Observers

1. Stay "invisible"—do not talk to participants. In case they have questions—gently but firmly redirect them to the main moderator. Do not to express your feelings and opinions about whatever is going on—do not comment, nod, smile, etc. Do not laugh at their jokes.
2. Whatever you hear is confident during the game—you must not share any information with other players.
3. Focus on relevant issues—do not listen/make notes if players are discussing topics unrelated to the game. Respect their privacy!
4. Try not to distract players but make sure you can hear what they are saying.
5. Remember to write down interesting, vivid quotes, they will be useful also for debriefing.

6. After the game, go through your notes and make a short list of the 3 most important observations you had—report them to Piotr.

Appendix B.3. If You are Assigned to Observe One Team in the Game

1. In case "your group" splits—follow one player for a while to find out what she/he is doing, then try to find the second player and figure out what they are doing, etc. Your information will not be complete but you need to know more or less about all players in your group.
2. During community meetings stay focused on "your group" and try to understand (and record) how they relate to other groups. Be as specific as you can.

Appendix B.4. If You are Assigned to Observe the Whole Game at Once

1. Try to observe where people congregate, walk around the room, try to listen to their conversations, and record who talked to whom, about what and how. You should visit all actors in each round to see what they are up to—even if you cannot capture everything, it will give you a good overall impression.
2. During the community meeting, try to observe group-level dynamics. Identify dominant actors and observe how the groups decide the mode of discussion (process-related issues) and topics of discussion (content-related issues).

Appendix B.5. Examples

Below, you will find examples of behaviors/quotes that correspond to each item of the observation protocol. This is NOT a complete list, just a guide, to give you the sense of what we are looking for. Participants' behaviors are likely to be slightly different, so use your common sense to qualify them. For each item of the observation protocol, we list several possible behaviors that we've see during our pilots (with examples of what participants might actually say in italics)—please code this behavior based on occurrence of any of these examples. For example, you code that participants share information if they talk about anything related to the current situation in the valley with other role representatives.

Appendix B.5.1. Example 1

Actors share information specific for their roles

- Comments about the state of the situation in the Valley
    - *In this round we have no money for ...*
    - *We have cut all the trees now ...*
- Comments about changes
    - *We will have a problems with low level of water*
    - *The prices of parcels will raise*

Appendix B.5.2. Example 2

Participants declare adopting common norms/rules

- Announcing common rules
    - *Next round everybody in the group can only ... (at least several participants nod or confirm verbally)*
- Announcing sanctions for disobedience
    - *If anyone ... , we will all sanction him/her*

Appendix B.5.3. Example 3

Can you identify a leader among participants in this round? Please write who it is. What is the main focus of the leader?

- Who do you think is the most influential participant?
- Who can affect other participants' decisions/behaviors?
- Who is asked for advice or guidance? (Technical content and problem solving)
- Who can stop fights and solve conflicts? (Interpersonal/group process)

Other interesting behaviors (please describe)

- Active resistance
    - *"I am not going to let you order me around."*
    - *"I don't have to tell you how much I earned"*
- Aggressive behaviors
    - *I don't trust some of you, I think we should check each other.*
    - *Visible conflicts—squabbling, yelling . . .*
- Direct statements suggesting lack of trust
    - *I don't trust that bastard, he always cheats.*
    - *I want to see how much money you earned*
- Nonverbal behaviors suggesting lack of trust (depending on the context)
    - *Remaining silent during the general discussion or when asked questions*
    - *Trying to check the content of other peoples' boxes*
- Confusion about rules of the activity
    - *I don't understand how . . . works.*
    - *What happens if I . . . . . . ?*
- And more . . .

Appendix C. Observation Protocol for the Game

1. How would you evaluate communication between teams regarding identifying, understanding, and solving the problem of the Valley?

○(very much)	○(moderately)	○(a little)	○(a little)	○(moderately)	○(very much)	○
Actors are mainly engaged in bilateral conversations and agreements			Actors are looking for solutions that will work for the whole group			It's hard to say

2. Who is taking responsibility for the future of the whole valley? (multiple choice possible)

- Local Government
- Water Board
- NGO
- some farmer(s)
- all farmers
- Bank
- It's hard to say

3. Do actors share information specific for their roles?

○(very much)	○(moderately)	○(a little)	○(a little)	○(moderately)	○(very much)	○
No, actors actively refuse to share information			Yes, actors share information			It's hard to say

4. Under what conditions and for whom do they provide the information? (please describe below)
5. How would you evaluate interactions between teams? (you may tick on both sides of the scale)

○(very much)	○(moderately)	○(a little)	○(a little)	○(moderately)	○(very much)	○
Competition/rivalry/conflicts			Cooperation			It's hard to say

6. Can you observe conflicts?

- Yes, there are conflicts (please describe how they are dealt with)
- No, there seems to be no conflict

7. Can you observe stereotyping?

- Yes, certain actors get stereotyped (please describe)
- No, there seems to be no use of stereotypes

8. Is there a clear leader?

   Yes, there is clearly a leader. Choose one:

- Local Government
- Water Board
- NGO
- Farmer
- Bank

   Vaguely yes, there seems to be a few actors aspiring to leadership. Choose all applicable:

- Local Government
- Water Board
- NGO
- Farmer
- Bank

   No, there seems to be no leader.

9. What is the main focus of the leader?

○(very much)	○(moderately)	○(a little)	○	○(a little)	○(moderately)	○(very much)	○
Technical content and problem solving			Both equally	Interpersonal/group process			It's hard to say

10. Describe the internal dynamics of the team you are observing—how are they making decisions, sharing information, and delegating tasks?
11. How were the actors dealing with uncertainty and risk?
12. Other interesting observations

Appendix D. Observation Protocol for Community Meetings

1. Can all actors give their view on the definition of the problem of the Valley?

- Yes, they all said how they see the problem.
- Most actors are able to say how they see the problem but some actors are ignored.

- Discussion is dominated by one or few actors, and many actors are ignored.
- There is no discussion about the problem of the Valley.

2. Which actor(s) are repeatedly ignored by others when discussing the problem of the Valley?

- Local Government
- Water Board
- NGO
- Some farmer(s)
- All farmers
- Bank

3. Do the actors seem to achieve common understanding of the problem?

○(not at all)	○(a little)	○(moderately)	○(very much)	○It's hard to say

4. Did the group establish an agreed solution?

- Yes, and the solution is made explicit (please write it down below)
- Vaguely yes, the solution is implicit
- Vaguely no, it seems that actors have different solutions and these solutions are not articulated
- No, actors explicitly set individual solutions (please write them down below)
- It's hard to say

5. Who is taking responsibility for the future of the whole valley? (multiple choice possible)

- Local Government
- Water Board
- NGO
- Some farmer(s)
- All farmers
- Bank
- It's hard to say

6. Do actors share information specific for their roles?

○(very much)	○(moderately)	○(a little)	○(a little)	○(moderately)	○(very much)	○
No, actors refuse to share information			Yes, actors share information			It's hard to say

7. How would you evaluate group-level interactions? (you may tick on both sides of the scale)

○(very much)	○(moderately)	○(a little)	○(a little)	○(moderately)	○(very much)	○
Competition/rivalry/conflicts			Cooperation			It's hard to say

8. Can you observe conflicts?

- Yes, there are conflicts (please describe how they are dealt with)
- No, there seems to be no conflict

9. Can you observe stereotyping?

- Yes, certain actors get stereotyped (please describe)
- No, there seems to be no use of stereotypes

10. Is there a leader?

Yes, there is clearly a leader. Choose one:

- Local Government
- Water Board
- NGO
- Farmer
- Bank

Vaguely yes, there seems to be a few actors aspiring to leadership. Choose all applicable:

- Local Government
- Water Board
- NGO
- Farmer
- Bank

No, there seems to be no leader.

11. What is the main focus of the leader?

○(very much)	○(moderately)	○(a little)	○	○(a little)	○(moderately)	○(very much)	○
Technical content and problem solving			Both equally	Interpersonal/group process			It's hard to say

12. Are there any ground rules created by the group? (multiple choice possible)

- rules about making decisions
- rules about dealing with conflict
- rules about sharing information
- rules about mutual respect
- other rules (please describe)
- the group does not have any rules established

Other interesting observations

Appendix E. Evaluation Questionnaires for Farmers

FARMER No . . . .
Beginning of Round 1
Please recall the moment AT THE BEGINNING OF THE GAME, JUST BEFORE ROUND 1.
How did you expect your interactions to develop with the following teams?

Local Government	a little	moderately	very much	hard to say
cooperation				
competition				
Water Board	a little	moderately	very much	hard to say
cooperation				
competition				
ecoNGO	a little	moderately	very much	hard to say
cooperation				
competition				
Bank	a little	moderately	very much	hard to say
cooperation				
competition				
Other farmers	a little	moderately	very much	hard to say
cooperation				
competition				

Additional Comments:

The whole valley community	a little	moderately	very much	hard to say
cooperation				
competition				

After community meeting #1

Please recall the moment IMMEDIATELY AFTER THE FIRST COMMUNITY MEETING.

Considering your experience of the previous rounds and what happened at the first community meeting, how did you evaluate your interactions with the following teams at that moment?

Local Government	a little	moderately	very much	hard to say
cooperation				
competition				
Water Board	a little	moderately	very much	hard to say
cooperation				
competition				
ecoNGO	a little	moderately	very much	hard to say
cooperation				
competition				
Bank	a little	moderately	very much	hard to say
cooperation				
competition				
Other farmers	a little	moderately	very much	hard to say
cooperation				
competition				

Additional Comments:

The whole valley community	a little	moderately	very much	hard to say
cooperation				
competition				

After community meeting #2

Please recall the moment IMMEDIATELY AFTER THE SECOND COMMUNITY MEETING.

Considering your experience of the previous rounds and what happened at the first community meeting, how did you evaluate your interactions with the following teams at that moment?

Local Government	a little	moderately	very much	hard to say
cooperation				
competition				
Water Board	a little	moderately	very much	hard to say
cooperation				
competition				
ecoNGO	a little	moderately	very much	hard to say
cooperation				
competition				
Bank	a little	moderately	very much	hard to say
cooperation				
competition				
Other farmers	a little	moderately	very much	hard to say
cooperation				
competition				

Additional Comments:

The whole valley community	a little	moderately	very much	hard to say
cooperation				
competition				

Immediately after the last round

Considering your experience during the game, how do you evaluate your interactions with the following teams?

Local Government	a little	moderately	very much	hard to say
cooperation				
competition				
Water Board	a little	moderately	very much	hard to say
cooperation				
competition				
ecoNGO	a little	moderately	very much	hard to say
cooperation				
competition				
Bank	a little	moderately	very much	hard to say
cooperation				
competition				
Other farmers	a little	moderately	very much	hard to say
cooperation				
competition				

Additional Comments:

The whole valley community	a little	moderately	very much	hard to say
cooperation				
competition				

After the debriefing

Considering all you have heard and experienced during the debriefing session, how do you evaluate your interactions with the following teams?

Local Government	a little	moderately	very much	hard to say
cooperation				
competition				
Water Board	a little	moderately	very much	hard to say
cooperation				
competition				
ecoNGO	a little	moderately	very much	hard to say
cooperation				
competition				
Bank	a little	moderately	very much	hard to say
cooperation				
competition				
Other farmers	a little	moderately	very much	hard to say
cooperation				
competition				

Additional Comments:

The whole valley community	a little	moderately	very much	hard to say
cooperation				
competition				

References

1. Blackmore, C.; Ison, R.; Jiggins, J. Social learning: An alternative policy instrument for managing in the context of Europe's water. *Environ. Sci. Policy* **2007**, *10*, 493–498. [CrossRef]
2. Berg, S.V. Seven elements affecting governance and performance in the water sector. *Util. Policy* **2016**, *43*, 4–13. [CrossRef]
3. Pahl-Wostl, C.; Craps, M.; Dewulf, A.; Mostert, E.; Tabara, D.; Taillieu, T. Social learning and water resources management. *Ecol. Soc.* **2007**, *12*, 5. [CrossRef]
4. Mostert, E. The challenge of public participation. *Water Policy* **2003**, *5*, 179–197.
5. Neef, A. Transforming rural water governance: Towards deliberative and polycentric models? *Water Altern.* **2009**, *2*, 53–60.
6. Parkes, M.W.; Morrison, K.E.; Bunch, M.J.; Hallström, L.K.; Neudoerffer, R.C.; Venema, H.D.; Waltner-Toews, D. Towards integrated governance for water, health and social-ecological systems: The watershed governance prism. *Glob. Environ. Chang.* **2010**, *20*, 693–704. [CrossRef]
7. Wiek, A.; Larson, K.L. Water, people, and sustainability—A systems framework for analyzing and assessing water governance regimes. *Water Res. Manag.* **2012**, *26*, 3153–3171. [CrossRef]
8. Ansell, C.; Gash, A. Collaborative governance in theory and practice. *J. Pub. Adm. Res. Theory* **2008**, *18*, 543–571. [CrossRef]
9. Kochskämper, E.; Challies, E.; Newig, J.; Jager, N.W. Participation for effective environmental governance? Evidence from Water Framework Directive implementation in Germany, Spain and the United Kingdom. *J. Environ. Manag.* **2016**, *181*, 737–748. [CrossRef] [PubMed]
10. Ison, R.; Röling, N.; Watson, D. Challenges to science and society in the sustainable management and use of water: Investigating the role of social learning. *Environ. Sci. Policy* **2007**, *10*, 499–511. [CrossRef]
11. Pahl-Wostl, C.; Mostert, E.; Tàbara, D. The growing importance of social learning in water resources management and sustainability science. *Ecol. Soc.* **2008**, *13*, 24. [CrossRef]
12. Jones, R.M. Leadership and Public Learning. In *Adaptive Governance and Water Conflict: New Institutions for Collaborative Planning*; Scholz, J.T., Stiftel, B., Eds.; Routledge: Abington-on-Thames, UK, 2010.
13. Mostert, E.; Pahl-Wostl, C.; Rees, Y.; Searle, B.; Tàbara, D.; Tippett, J. Social learning in European river-basin management: Barriers and fostering mechanisms from 10 river basins. *Ecol. Soc.* **2007**, *12*, 19. [CrossRef]
14. Bouwen, R.; Taillieu, T. Multi-party collaboration as social learning for interdependence: Developing relational knowing for sustainable natural resource management. *J. Community Appl. Soc. Psychol.* **2004**, *14*, 137–153. [CrossRef]
15. Schruijer, S.; Vansina, L. The dynamics of multiparty cooperation and leadership. In *The Emerging World of Chains and Networks: Bridging Theory and Practice*; Camps, T., Diederen, P., Vos, G.C.J.M., Eds.; Reed Business: Amsterdam, The Netherlands, 2004.
16. Quaghebeur, K.; Masschelein, J.; Nguyen, H.H. Paradox of participation: Giving or taking part? *J. Community Appl. Soc. Psychol.* **2004**, *14*, 154–165. [CrossRef]
17. Dewulf, A.; Craps, M.; Dercon, G. How issues get framed and reframed when different communities meet: A multi-level analysis of a collaborative soil conservation initiative in the Ecuadorian Andes. *J. Community Appl. Soc. Psychol.* **2004**, *14*, 177–192. [CrossRef]
18. Pahl-Wostl, C.; Hare, M. Processes of social learning in integrated resources management. *J. Community Appl. Soc. Psychol.* **2004**, *14*, 193–206. [CrossRef]
19. Games4Sustainability, Gamepedia. Available online: https://games4sustainability.org/gamepedia/ (accessed on 14 January 2018).
20. Mendler de Suarez, J.; Suarez, P.; Bachofen, C.; Fortugno, N.; Goentzel, J.; Gonçalves, P.; Grist, N.; Macklin, C.; Pfeifer, K.; Schweizer, S.; et al. Games for a New Climate: Experiencing the Complexity of Future Risks. In *Pardee Center Task Force Report. Boston: The Frederick S. Pardee Center for the Study of the Longer-Range Future*; Boston University: Boston, MA, USA, 2012.
21. Juhola, S.; Driscoll, P.; de Suarez, J.M.; Suarez, P. Social strategy games in communicating trade-offs between mitigation and adaptation in cities. *Urban Clim.* **2013**, *4*, 102–116. [CrossRef]
22. Stefanska, J.; Magnuszewski, P.; Sendzimir, J.; Romaniuk, P.; Taillieu, T.; Dubel, A.; Flachner, Z.; Balogh, P. A Gaming Exercise to Explore Problem-Solving versus Relational Activities for River Floodplain Management. *Environ. Policy Gov.* **2011**, *21*, 454–471. [CrossRef]

23. Centre for Systems Solutions/IIASA. Flood Resilience Game. Available online: http://floodresilience.games4sustainability.org/ (accessed on 14 January 2018).
24. Harteveld, C.; Guimarães, R.; Mayer, I.S.; Bidarra, R. Balancing play, meaning and reality: The design philosophy of levee patroller. *Simul. Gaming* **2010**, *41*, 316–340. [CrossRef]
25. Krolikowska, K.; Kronenberg, J.; Maliszewska, K.; Sendzimir, J.; Magnuszewski, P.; Dunajski, A.; Slodka, A. Role-playing simulation as a communication tool in community dialogue: Karkonosze Mountains case study. *Simul. Gaming* **2007**, *38*, 195–210. [CrossRef]
26. Rumore, D.; Schenk, T.; Susskind, L. Role-play simulations for climate change adaptation education and engagement. *Nat. Clim. Chang.* **2016**, *6*, 745–750. [CrossRef]
27. Bachofen, C.; Suarez, P.; Steenbergen, M.; Grist, N. *Can Games Help People Manage the Climate Risks They Face? The Participatory Design of Educational Games*; Working Paper Series; Red Cross/Red Crescent Climate Centre: Leeghwaterplein, The Netherlands, 2012; p. 3.
28. United Nations Human Settlements Programme. Using Minecraft for Community Participation Manual. 2016. Available online: https://unhabitat.org/books/manual-using-minecraft-for-community-participation (accessed on 14 January 2018).
29. Geurts, J.L.; Joldersma, C. Methodology for participatory policy analysis. *Eur. J. Oper. Res.* **2001**, *128*, 300–310. [CrossRef]
30. Duke, R.D.; Geurts, J. *Policy Games for Strategic Management*; Rozenberg Publishers: Amsterdam, The Netherlands, 2004.
31. Madani, K.; Pierce, T.W.; Mirchi, A. Serious games on environmental management. *Sustain. Cities Soc.* **2017**, *29*, 1–11. [CrossRef]
32. Mayer, I.S. The gaming of policy and the politics of gaming: A review. *Simul. Gaming* **2009**, *40*, 825–862. [CrossRef]
33. Falk, A.; Heckman, J.J. Lab experiments are a major source of knowledge in the social sciences. *Science* **2009**, *326*, 535–538. [CrossRef] [PubMed]
34. Chaudhuri, A. Sustaining cooperation in laboratory public goods experiments: A selective survey of the literature. *Exp. Econ.* **2011**, *14*, 47–83. [CrossRef]
35. Gintis, H. Beyond Homo economicus: Evidence from experimental economics. *Ecol. Econ.* **2000**, *35*, 311–322. [CrossRef]
36. De Caluwé, L.; Geurts, J.; Kleinlugtenbelt, W.J. Gaming research in policy and organization: An assessment from the Netherlands. *Simul. Gaming* **2012**, *43*, 600–626. [CrossRef]
37. Guala, F. *The Methodology of Experimental Economics*; Cambridge University Press: Cambridge, UK, 2005.
38. Sendzimir, J.; Magnuszewski, P.; Flachner, Z.; Balogh, P.; Molnar, G.; Sarvari, A.; Nagy, Z. Assessing the resilience of a river management regime: Informal learning in a shadow network in the Tisza River Basin. *Ecol. Soc.* **2008**, *13*, 11. [CrossRef]
39. New Approaches to Adaptive Water Management under Uncertainty. Available online: http://www.newater.uni-osnabrueck.de/ (accessed on 20 February 2018).
40. Dutch Water Sector Intelligence. Available online: https://www.dwsi.nl (accessed on 20 February 2018).
41. Lords of the Valley, Games4Sustainabilit. Available online: https://lordsofthevalley.games4sustainability.org/ (accessed on 20 February 2018).
42. Ridder, D.; Mostert, E.; Cernesson, F.; Harmony, C.T. *Learning Together to Manage Together: Improving Participation in Water Management*; University of Osnabrück: Osnabrück, Germany, 2005.
43. Flores, C.C.; Özerol, G.; Bressers, H. "Governance restricts": A contextual assessment of the wastewater treatment policy in the Guadalupe River Basin, Mexico. *Util. Policy* **2017**, *47*, 29–40. [CrossRef]
44. Zhou, Q.; Mayer, I.S. Models, Simulations and Games for Water Management: A Comparative Q-Method Study in The Netherlands and China. *Water* **2017**, *10*, 10. [CrossRef]

water

MDPI

Article

Experiential Learning through Role-Playing: Enhancing Stakeholder Collaboration in Water Safety Plans

Giuliana Ferrero [1,*], Françoise Bichai [1,2] and Maria Rusca [3]

1 IHE Delft Institute for Water Education, 2611 AX Delft, The Netherlands
2 École Polytechnique de Montréal, P.O. Box 6079, Downtown Station, Montreal, QC H3C 3A7, Canada; fbichai@polymtl.ca
3 King's College London, Department of Geography, The Strand, London WC2R 2LS, UK; maria.rusca@kcl.ac.uk
* Correspondence: g.ferrero@un-ihe.org; Tel.: +31-15-21-51-773

Received: 13 December 2017; Accepted: 15 February 2018; Published: 23 February 2018

Abstract: Improved water safety management, as addressed by the Sustainable Development Goals, can be aided by Water Safety Planning, a risk-assessment and risk-management approach introduced by the World Health Organization and implemented to date in 93 countries around the globe. Yet, this approach still encounters some challenges in practice, including that of securing collaboration among the broad range of stakeholders involved. This paper presents a role-playing game designed to foster stakeholder collaboration in Water Safety Plans (WSP). In this role-play, participants take on different stakeholders' roles during a collective (team-based) decision-making process to improve water supply safety in a fictive town. The game is the result of a transdisciplinary initiative aimed at integrating knowledge across technical and governance aspects of WSPs into an active learning experience for water sector actors from diverse backgrounds. It exposes participants to the four phases of Kolb's experiential learning cycle: concrete experience, reflective observation, conceptualization and active experimentation. This paper discusses potential impacts of the WSP role-play, including skills and knowledge development among participants, which can support cross-sectoral integration and dealing with complexity in decision-making. These are capacity assets strongly needed to address water safety management challenges in a sustainable way.

Keywords: active learning; drinking water; role-play; stakeholder collaboration; Water Safety Plan; water supply

---

1. Introduction

The Sustainable Development Goals (SDGs) launched by the United Nations [1] aim at achieving, by 2030, "universal and equitable access to safe and affordable drinking water for all" (target 6.1). This poses a triple challenge: reaching unserved populations, improving existing service levels, and ensuring sustainability of existing and future services [2]. This target will be monitored by tracking the proportion of "safely managed drinking water services", which entails water that is free from contamination, available on premises and available when needed [3].

Water safety management may be aided by risk management approaches, such as Water Safety Plans (WSPs), a preventive risk assessment and management approach that encompasses all steps in water supply from catchment to consumers, including water treatment and distribution [4]. WSPs were originally introduced by the WHO in 2004 [5], following the 2001 Stockholm Framework [6] for the development of risk-based microbial water quality standards for all water uses. Since then, WSPs have been implemented in 93 countries around the world [7]. In the context of drinking water, WSPs require site-specific assessment of hazards impacting water quality from the watershed to the tap, and the identification and implementation of risk control measures, monitoring and management plans. Generally, the implementation of urban WSPs is led by the water supplier. Benefits related to WSP implementation include improved management and traceability, and a demonstrated reduction of the incidence of clinical cases of diarrhea [8].

Nevertheless, some challenges persist, such as securing executive buy-in (i.e., support from the top management), encountering a favorable organizational culture (i.e., a sense of collaboration towards a common goal of improved management practice and public health protection), and overcoming the perception of WSPs being an additional burden for utilities [9,10]. Further, one of the greatest challenges is to actively involve stakeholders other than the water supplier in the WSP process, despite the WSP manual reporting that favoring dialogue and collaboration among stakeholders is one of the benefits promoted by the WSP approach [11]. Improving water safety at the treatment and distribution stages may be manageable by the utility with limited commitment of other stakeholders, but a WSP cannot be fully implemented without the involvement of major stakeholders at catchment and at household scale. In fragmented water governance structures, cooperation among stakeholders in the implementation of WSPs, as encouraged by the Integrated Water Resources Management (IWRM) principles [12], generally doesn't occur.

Multiple approaches have been adopted for building capacities on WSP: these range from a variety of short courses (face-to-face or on-line) to Water Operators' Partnerships or courses at universities and higher education institutes. Most of the courses are based on teaching material developed by WHO & IWA [13]. Nevertheless, after over 10 years of capacity building and the implementation of WSPs worldwide, research highlights that in many cases there is little evidence of involving stakeholders from the catchment in WSPs [14]. Hence, there is a need to support implementers with tools that facilitate or enhance collaboration between stakeholders. Additionally, Parker and Summerill [15] reported that the ability of WSPs to involve all stakeholders was seen as a motivation to engage in WSPs; yet, poor relationships with other stakeholders formed a barrier in several countries.

In this paper, we argue that role-plays can significantly contribute to both enhancing experience-based learning on governance of water safety management and developing skills required for water professionals dealing with the design and management of WSPs. Role-plays are a form of simulation in which learners engage actively in problem solving, while the teacher acts as facilitator and uses his/her knowledge to develop realistic scenarios, creating a 'situational interest' by presenting a problem stimulus [16]. They address the demand for active learning in multiple ways. First, pedagogical theories suggest that knowledge is not transferred from the teacher to the learner, but rather constructed by learners whilst the teacher acts as a facilitator [17,18]. This has important implications on the role of the learner, who ultimately holds the responsibility of his/her learning process [19,20]. Learning, therefore, requires actively engaging with the subject matter, as 'students learn best when they engage with course material and actively participate in their learning' [21–24], active learning techniques are increasingly being encouraged. Through the role-play, learners are exposed to a real-life situation in a protected environment that allows them to experience, experiment and repeat (trial and error), thus putting the learner in the condition of both constructing and enhancing their knowledge [25]. Participants take on an active role (e.g., a given stakeholder involved in water management) during the simulation of an activity that involves interacting and making decisions. In this process, activation is intrinsic: "individuals assume roles, act out their characters, experience the interaction and see the outcome" [20]. Second, the flexible structure of the role-play generally allows the pursuit of different scenarios and, in turn, multiple learning objectives. Because of the trial and error pattern, these also allow testing and reinforcing knowledge by practicing in a variety of scenarios [26]. Role-plays often entail multidisciplinary scenarios, in which different dimensions of a problem or a process are addressed in the controlled and safe environment of the role-play [25].

We present an original role-playing game developed to complement existing training materials with the following objectives: (i) raising awareness of water professionals of the added value of WSPs and ensuring a safe water supply globally, and thus of achieving Sustainable Development Goal 6, target 6.1 [3]; (ii) creating a learning environment where water professionals can practice stakeholder engagement while making strategic decisions on financial investments in the rehabilitation of a water supply system; (iii) demonstrating the importance of effective collaboration and integration among stakeholders in decision making on water supply safety. After presenting the game, we discuss the potential impact of the game on water professionals and on water governance at large.

2. Materials and Methods

The context of this role-playing game is a multiple-stakeholder decision-making process to address public health protection in relation to drinking water safety. Participants are divided into groups to experiment with decision making in two consecutive game rounds: (i) the first round simulates a 'fragmented' institutional environment, where stakeholders are segregated and communication is limited (ii) the WSP 'integrated' institutional environment is evaluated in the second round, when communication between stakeholders is intensified, eventually leading to a different outcome of the decision-making process. The outcome of both rounds is evaluated in terms of water quality risk reduction, revealing how stakeholder engagement and cooperation in the WSP decision-making process could lead to improved water quality through more efficient investment planning.

The target group includes water professionals of different backgrounds (i.e., public administration, engineers, policy makers) who may either implement or advocate for the implementation of WSP in the future. At a broader level, the game can also be used with other practitioners within the water sector (including—but not limited to—all stakeholders represented in this game), and as an educational tool for any group of students taking courses related to water governance and water services management.

The game was developed in 2016–2017 as a result of a collaboration between the Water Supply Engineering and Water Services Management chair groups at IHE Delft. This game is part of a set of tools developed by IHE Delft to support capacity building for water utilities in Water Operators' Partnerships towards achieving a range of performance goals. The game is the result of a transdisciplinary initiative aimed at integrating technical knowledge with water management and governance of WSPs into an active learning experience for water sector actors from diverse training backgrounds and fields.

The development of this game was an iterative process based on initial brainstorming and review of other serious gaming tools used in water management. The game was tested with (i) students enrolled in the Master Programme in Urban Water and Sanitation at IHE Delft (n = 18), (ii) students enrolled in the Master Programme in Water Management at IHE Delft (n = 9), and (iii) researchers of the Dutch National Institute for Public Health and the Environment (RIVM) (n = 15). The game was part of a larger course or training on WSPs. Plenary feedforward sessions—that is, to listen to the suggestions for the future—were organized at the end of each game. The game developers listened to the suggestions, took notesk and integrated the suggested modifications into the subsequent revisions of the game. During the students' sessions, each team appointed a participant to take notes about potential improvements of the game. Additional guidance was sought from a WHO consultant experienced in WSP and role-play games. The game was finalized in August 2017 and is available open-access on the WSP Portal (http://www.wsportal.org) of the IWA/WHO and on the BEWOP project website (http://bewop.un-ihe.org). The final version of the game is intended to be played over a period of 4 h, but includes additional features to turn it into a longer version (suggested 6 h).

3. Results

In this section, the main features and stages of the game, i.e., the preparatory steps, round 1 and round 2, and the outcomes of the testing process are described.

3.1. Preparatory Steps

To begin with, the roles are assigned to participants and the case study is presented. The case study is a fictional town named BE, which is supplied with treated surface water. In BE town, agricultural activities in the catchment have significantly intensified over the past decades, moving from small-scale subsistence agriculture to larger-scale. A chemical factory has recently opened in the rural-urban fringe, and the quality of the river water has been drastically deteriorating. The water supply company is struggling to supply drinking water that meets basic water quality standards. In this scenario, the local government has received funds from a developing agency, which they must use to tackle multiple issues in the water and sanitation sector. In addition, the water company has its own budget to invest in rehabilitation. The total budget remains limited with respect to the range of issues faced by the utility, and therefore investment priorities need to be set.

Before diving into the game, participants are divided into teams, such that each team includes at least one participant representing each of the following stakeholders:

1. Catchment authority
2. Farmers' association
3. Industry
4. Water supply company
5. Local government
6. Ministry of Public Health
7. Consumers

3.2. Round 1

The first round of the game aims at allowing participants to experience a situation that is in many instances representative of typical real-life context, i.e., institutional fragmentation and limited communication among actors. The scenario of stakeholders' interconnections presented in Table 1 is given to participants, which use it to form the two sub-groups: one led by the local government (in grey in Table 1) and the other one led by the water supply company (in white in Table 1). Communication within the sub-groups is not allowed, and the information available to each sub-group differs, as detailed below.

Table 1. Stakeholders' interconnections and influence factors in decision making.

Stakeholder	Links	Influence Factor
Catchment authority	Weak relationship with farmers' association about land use and agricultural practices Weak relationship with industry about types of chemicals discharged in plant's effluent Weak (informative) relationship with city government as downstream user in the catchment	1/10
Farmers' association	Weak relationship with catchment authority	1/10
Industry	Weak relationship with catchment authority	2/10
Local government (municipality) *	Weak relationship with catchment authority	6/10
Water supply company *	Relationship with customers based on billing, customer complaints and satisfaction surveys Weak relationship with Ministry of Public Health which provides guidance on drinking water quality	5/10
Ministry of Public Health	Weak relationship with water supply company	3/10
Consumers	Necessary relationship with water supply company	2/10

Note: * Indicates the stakeholder leading the sub-group.

The water supply company has a budget of 1M BE$ and the local government of 3M BE$. Within each sub-group, stakeholders must discuss how to invest the money they have available over a period of 10 years, using the printed information available for each sub-group. Each stakeholder within the sub-group can influence investments of an amount corresponding to a fraction of the total budget managed by the local government or by the water supply company, based on the respective weight of each stakeholder. The amount is calculated as follows: amount of investment influenced by X = (influence factor) × (total budget sub-group). This relative influence of individual stakeholders with respect to collective investment decisions is referred to as the 'weight distribution' of stakeholders.

The sub-group led by the water supply company can use the WSP they recently developed (for the sake of the game, a summary of the WSP, focusing on the highest identified risks, is provided to the sub-group (i.e., it is assumed that this was developed by the water supply company through activities that are not part of the scope of the game)) (see an excerpt in Table 2), but they won't be able to invest on all the items listed in the improvement plan because of two reasons: (a) they have a limited budget, and (b) some interventions would require the collaboration of stakeholders from the other sub-group, with whom they cannot communicate (as exemplified in Table 2).

Table 2. Example of one high-risk hazard and possible control actions identified through the WSP.

Hazard	Likelihood	Severity	Risk	Control Action	Investment (M BE$)	Risk after Implementing the Control Action	Stakeholders Involved/Remarks
Pesticides from agricultural uses	4	3	12	1.A—Improved farmers' practices involving a reduced use of pesticides	0.8	9	Requires close collaboration with farmers and coordination through the catchment authority
				1.B—Investment in advanced treatment systems at the drinking water treatment plant	2	3	
				1.C—Improved treatment at the drinking water treatment plant through punctual dosing of powdered activated carbon during and after intense rainfall events	1.2	6	Requires close collaboration with catchment authority for accurate and timely information on rainfall events

Meanwhile, the sub-group led by the local government can only rely on a list of various issues that have been raised over the past few years of the government's mandate around water supply and water management, along with the estimated funding required to address the issue at the time it was raised. This sub-group therefore does not have access to the information contained in the WSP. The two sub-groups will eventually deliver different investment plans and it will become evident that both the presence of a WSP and stakeholders' collaboration could considerably improve the decisions to be taken.

3.3. Round 2

In this second round, stakeholders play as one group (i.e., both sub-groups are reunited into the original team). In addition, they have the opportunity to re-define the interconnections and the influence factor of each stakeholder. Participants reflect on what the consequences and limitations of such a weight distribution are, and how it can be reconsidered and improved in order to increase the benefits of coordinated decision-making on water supply.

The spending of the total budget (4M BE$ in total) is guided by an investment plan based on the prioritization of risks and control measures as informed by the WSP. The stakeholders can negotiate which investment they support, based on their own interest. If a given option requires collaboration from specific stakeholders, these stakeholders must agree to support at least part of the investment required to implement that control action. During the discussion, play-money can be placed on a chart that illustrates all possible investment decisions and their respective costs (Figure 1).

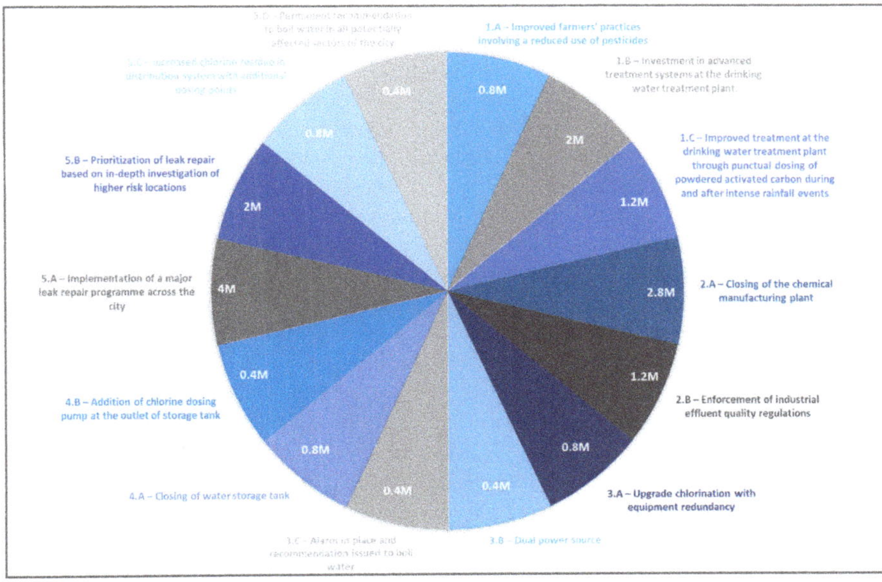

Figure 1. Chart used to illustrate investment decisions made by a team in round 2.

Stakeholders should aim at the set of investments that will result in the greatest risk reduction.

When playing in a large group with several teams, and/or if there is more time available for playing the game (~5–6 h), the trainer can distribute 'unexpected event cards', which include natural calamities that restrict the budget available or new risks that need to be immediately tackled, thereby influencing priorities and the set of investment decisions that the team settles on during this second round.

3.4. Feedback Session

The game ends with a discussion on various aspects of the game, particularly comparing the outcomes of the two rounds of the game. Additionally, the connections between stakeholders and their respective influence factor are discussed, as well as the rationale, interactions and processes behind the group's decisions on investment. Finally, participants are guided through assessing the 'take home' message from playing this game:

- Could participants relate the experience of stakeholders' fragmentation in the game's first round to the water supply sector situation in their own country?
- Could participants appreciate the value of the WSP as a support tool to protect public health in water provisioning?
- Could participants appreciate the value of stakeholders' collaboration while taking strategic decisions on investments during the second round?

3.5. Testing Process

A total of 27 Master students and 15 public health practitioners and researchers tested the game. Only 4 participants (15%) among the students group and 3 participants (20%) among the public health practitioners and researchers group had previous experience with WSP. Master students stated that they could clearly relate the stakeholders' fragmentation to the water sector in their own country, being mainly from developing and in-transition countries, but this was more difficult for the public health practitioners and researchers from The Netherlands. Moreover, the trainers observed that the group of Master students was easily acting out the roles throughout the game, whereas the group of public health practitioners and researchers had more difficulties in adhering to their roles.

Participants in both IHE student groups were asked to discuss their experience orally after playing the game. All participants in the student groups mentioned that, through the game, they gained a better understanding about the importance of stakeholders' engagement for the successful implementation of WSP. Many participants in the student group also mentioned that they would be likely to play this game in the future, to support training on WSP and stakeholder coordination on water management from catchment to tap. These last two points were not discussed with the public health practitioner and researcher groups. The latter was requested to fill in a simple evaluation form that included one question related to each part of a broader training workshop of which the game was part. Only partial results were obtained from this group: 7 out 15 participants filled in the evaluation; 5 out of 7 were extremely positive about the game and 2 out of 7 were neutral.

Another extremely important component of the testing process was the feedforward session. The main suggestions received from participants were related to the operationalization of the game. This included, for instance, adding a "fun element" to the game. As a result, the team developed play-money to be placed on a chart that illustrates all possible investment decisions (Figure 1). Additionally, Master students helped the trainers to improve the process for allocating money to each stakeholder by simplifying the influence factor approach (Table 1). Other suggestions included elucidating some minor clarity issues in the participant's and trainer's guidance manuals, which were addressed in the final version available online.

Finally, expert opinion on the game was sought from a WHO consultant who had extensive experience with role-playing games in relation to water safety. His most valuable suggestion was to let participants reflect on the risk minimization linked to each control action, in order to overcome the tendency to focus just on cost (Figure 1). His suggestions also enabled the improvement of operational elements of the game, such as color coding of stakeholders' tags, and adding guidance in the trainer's manual on how to organize the room and facilitate the role-play the when playing the game with larger or smaller groups.

4. Discussion and Conclusions

This section discusses the potential impact of the WSP role-play on knowledge and skills development among participants, as well as the potential impact on longer-term water quality planning and management. First, the game offers the opportunity to experience the importance of integrated approaches to water quality management. It is often argued that the so-called "water crisis" is, to a large extent, "a crisis of governance" [27–29]. Water governance is complex and involves a wide range of interconnected processes, embedded in fragmented and institutionally complex systems [30,31]. Managing water quality from catchment to tap and protection of public health require understanding water challenges as interconnected at the city, region or basin scale, as well as among users and across the water cycle. In experiencing the different scenarios (without and with collaboration of different stakeholders), participants become aware of the crucial need for stakeholder collaboration to achieve safe water supply.

Further, water quality management, just as suggested by IWRM principles [12], entails developing a set of skills and competencies for water professionals to deal with integration and complexity [25]. In fact, professionals in the water sector are increasingly required to have an in-depth knowledge in their specific field, whilst being well versed in adjacent and related ones [32]. Implementing WSPs requires not only an expertise on water quality and supply operations, but also an understanding of water governance processes and negotiation skills, such as dealing with allocation of resources or the right to pollute entails developing collaborations and handling potential conflicts [25,33,34]. During the game, participants are exposed to two different types of negotiations, where they must agree on budget allocations, establish priorities and reach an agreement with stakeholders that might have different positions and interests. The long version of the game also simulates unexpected events such as a sewage pipe break, unforeseen operational problems at the water treatment plant, new regulatory requirements for water quality, farmers going on strike, extreme weather conditions, etc. This scenario-based training forces participants to engage in new and more complex negotiations, and to think about how to address the event and return to business as usual or reach adapted, acceptable (or even improved) operational conditions. It exposes the participants to the reality of unexpected, dynamic, evolving conditions, and therefore increases their awareness to the context of uncertainty in decision-making processes. Moreover, understanding and addressing water governance challenges, which constitute one of the bottlenecks in effective water safety management, require interdisciplinary analyses encompassing both the institutional (i.e., socio-political, cultural, ethical and legal) and the technical dimensions (e.g., geo-hydrological and engineering assessments, water quality monitoring) of water management [35,36]. By simulating a real decision-making context involving multiple stakeholders, this game allows participants to experience some of these various dimensions of water management, as conflicting interests and priorities of the different stakeholders will emerge from the teams' discussion and decision processes.

Moreover, this role-play represents an important tool for enhancing participants' knowledge on water safety management from catchment to tap, i.e., the WSP approach, including associated challenges and benefits, and on the impact of various stakeholders on water supply safety. Participants build this knowledge through active problem solving by making decisions on investments on water safety and by conciliating different stakeholders' interests (especially at the second round). In fact, role-playing games have the potential to be especially effective when the desired performance objective of the training involves problem solving [16].

According to Kolb, experience plays into the learning process [37]. The game exposes participants to different learning moments that cater to different learning styles. At different points in the game, participants are exposed to the four stages of Kolb's experiential learning cycle (Figure 2). In the first round of the role-play exposes students to what Kolb defines a Concrete Experience (1): here, participants experience the behavior, different levels of influence and potentially conflicting interests of stakeholders involved or affected by water safety management. They experience this in a simulated 'fragmented' system, representative of real-life context in many instances. The concrete experience leads to Reflective Observation (2): before starting the second round, participants in both sub-groups reflect on the outcome of the first round and on the challenges they faced. Through the plenary feedback discussion, some level of Conceptualization (3) can be derived from the previous stages, e.g., on WSPs as tools for strengthening the decision-making process on water safety investment and on the need for stakeholders' collaboration. This stage also works as a tool for facilitators to test participants' knowledge on water safety management from catchment to tap. In this phase they identify and describe the challenges faced in the implementation of WSP and the impact of various stakeholders on water supply safety. Finally, by playing the second round of the game, participants can engage in Active Experimentation (4): they practice the concepts they have just approached through another round of role-play, in a simulated 'integrated' system, with the potential to further practice later in real-life situations as water professionals.

As a result, the game has the potential to provide both short-term impacts (e.g., achieving the learning objectives during the training) and medium-term impacts (real-life WSP implementation). Beyond this, the set of skills and the type of knowledge that the game seeks to develop among water professionals aspires to longer-term, broader impact on transformative change: skills and knowledge that help support cross-sectoral integration and deal with complexity in decision-making are increasingly needed in order to address sustainability challenges. The active learning material presented in this study aims at contributing towards a better understanding of the WSP approach, as a tool for water utilities to engage with a broader network of stakeholders, but also to further challenge-based training of water professionals on social and interactional skills' development, which are critical to solve the pressing issues underpinned by the SDGs.

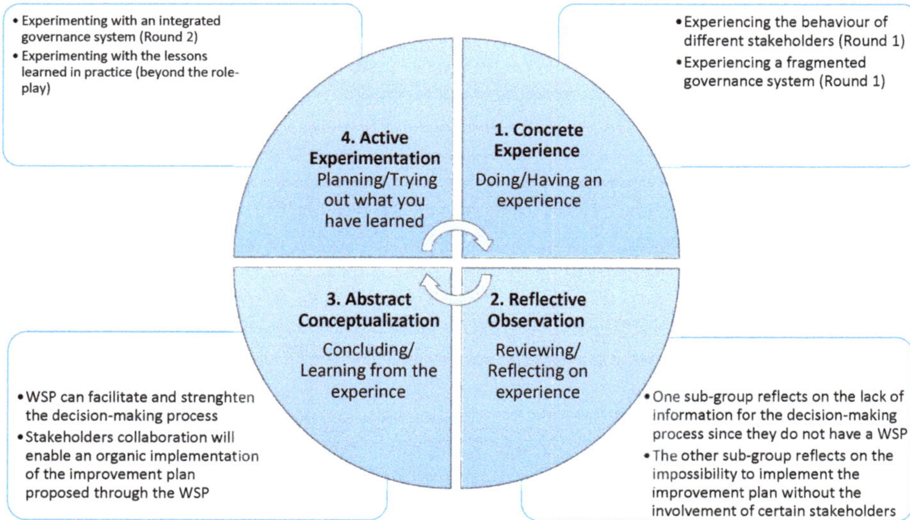

Figure 2. Kolb's experiential learning cycle stages through the WSP role-play.

Acknowledgments: The game has been developed within the framework of the project "Boosting the efficiency of Water Operators Partnerships (BEWOP)", sponsored by the Dutch Government (DGIS).

Author Contributions: Françoise Bichai and Giuliana Ferrero developed the game. Giuliana Ferrero tested the game. Darryl Jackson reviewed the game. Giuliana Ferrero, Françoise Bichai and Maria Rusca wrote the paper.

Conflicts of Interest: The authors declare no conflict of interest. The founding sponsors had no role in the design of the study, in the writing of the manuscript, and in the decision to publish the results.

References

1. UN. Transforming Our World: The 2030 Agenda for Sustainable Development. Resolution Adopted by the General Assembly on 25 September 2015, A/70/1. Available online: http://www.un.org/en/development/desa/population/migration/generalassembly/docs/globalcompact/A_RES_70_1_E.pdf (accessed on 3 December 2017).
2. World Bank. *Sustainability Assessment of Rural Water Service Delivery Models: Findings of a Multi-Country Review*; World Bank: Washington, DC, USA, 2017; Volume 8. Available online: https://openknowledge.worldbank.org/bitstream/handle/10986/27988/W17055.pdf?sequence=4&isAllowed=y (accessed on 3 December 2017).
3. World Health Organization (WHO). *Safely Managed Drinking Water—Thematic Report on Drinking Water 2017*; World Health Organization: Geneva, Switzerland, 2017; ISBN 9789241565424.
4. World Health Organization (WHO). *Guidelines for Drinking-Water Quality*, 4th ed.; World Health Organization: Geneva, Switzerland, 2011; ISBN 9789241548151.
5. World Health Organization (WHO). *Guidelines for Drinking-Water Quality*, 3rd ed.; World Health Organization: Geneva, Switzerland, 2004; ISBN 9241546387.
6. Fewtrell, L.; Bartram, J. *Water Quality: Guidelines, Standards and Health: Assessment of Risk and Risk Management for Water-Related Infectious Diseases*; World Health Organization: Geneva, Switzerland, 2001; ISBN 924154533X.
7. World Health Organization (WHO). *Global Status Report on Water Safety Plans: A Review of Proactive Risk Assessment and Risk Management Practices to Ensure the Safety of Drinking-Water*; World Health Organization: Geneva, Switzerland, 2017; Licence: CC BY-NC-SA 3.0 IGO.

8. Gunnarsdottir, M.J.; Gardarsson, S.M.; Elliott, M.; Sigmundsdottir, G.; Bartram, J. Benefits of water safety plans: Microbiology, compliance, and public health. *Environ. Sci. Technol.* **2012**, *46*, 7782–7789. [CrossRef] [PubMed]
9. Summerill, C.; Pollard, S.J.T.; Smith, J.A.; Breach, B.; Williams, T. Securing executive buy-in for preventative risk management—Lessons from water safety plans. *Water Sci. Technol. Water Supply* **2011**, *11*, 682–691. [CrossRef]
10. Kot, M.; Castleden, H.; Graham, A.; Gagnon, G.A. The human dimension of water safety plans: A critical review of literature and information gaps. *Environ. Rev.* **2015**, *23*, 24–29. [CrossRef]
11. Bartram, J.; Corrales, L.; Davison, A.; Deere, D.; Drury, D.; Gordon, B.; Howard, G.; Rinehold, A.; Stevens, M. *Water Safety Plan Manual: Step-By-Step Risk Management for Drinking-Water Suppliers*; World Health Organization: Geneva, Switzerland, 2009; ISBN 9789241562638.
12. Global Water Partnership (GWP). *Integrated Water Resources Management*; Global Water Partnership (GWP) Technical Advisory Committee: Stockholm, Sweden, 2000; Background Paper No.4; ISSN 1403-5324.
13. WSP Training Package. Available online: http://www.who.int/water_sanitation_health/publications/wsp_training_package/en/ (accessed on 12 December 2017).
14. Omar, Y.Y.; Parker, A.; Smith, J.A.; Pollard, S.J.T. Risk management for drinking water safety in low and middle income countries—Cultural influences on water safety plan (WSP) implementation in urban water utilities. *Sci. Total Environ.* **2017**, *576*, 895–906. [CrossRef] [PubMed]
15. Parker, A.; Summerill, C. Water safety plan implementation in East Africa: Motivations and barriers. *Waterlines* **2013**, *32*, 113–124. [CrossRef]
16. Rotgans, J.I.; Schmidt, H.G. Situational interest and academic achievement in the active-learning classroom. *Learn. Instr.* **2011**, *21*, 58–67. [CrossRef]
17. Boekaerts, M.; Pintrich, P.R.; Zeidner, M. Self-Regulation: Theory, Research, and Applications. In *Handbook of Self-Regulation*; Academic Press: Orlando, FL, USA, 2000; ISBN 9780121098902.
18. Niemi, H. Active learning—A cultural change needed in teacher education and schools. *Teach. Teach. Educ.* **2002**, *18*, 763–780. [CrossRef]
19. Biggs, J.B.; Tang, C. *Teaching for Quality Learning at University*; Open University Press, McGraw-Hill Education: Maidenhead, UK, 2007.
20. Oblinger, D. The next generation of educational engagement. *J. Interact. Media Educ.* **2004**, *8*, 1–18. [CrossRef]
21. Jonassen, D.H.; Rohrer-Murphy, L. Activity theory as a framework for designing constructivist learning environment. *Educ. Technol. Res. Dev.* **1999**, *47*, 61–79. [CrossRef]
22. The Manager's Core Work in the New Economy. Available online: http://www.newmango.com/01iftf/henschel.html (accessed on 12 December 2017).
23. Dale, E. *Audiovisual Methods in Teaching*, 3rd ed.; The Dryden Press: New York, NY, USA, 1969.
24. Freeman, S.; Eddy, S.L.; McDonough, M.; Smith, M.K.; Okoroafor, N.; Jordt, H.; Wenderoth, M.P. Active learning increases student performance in science, engineering, and mathematics. *PNAS* **2014**, *111*, 8410–8415. [CrossRef] [PubMed]
25. Rusca, M.; Heun, J.; Schwartz, K. Water management simulation games and the construction of knowledge. *Hydrol. Earth Syst. Sci.* **2012**, *16*, 2749–2757. [CrossRef]
26. Ruben, B.D. Simulations, games, and experience-based learning: The quest for a new paradigm for teaching and learning. *Simul. Gaming* **1999**, *30*, 498–505. [CrossRef]
27. UNESCO. *Water, a Shared Responsibility*; The United Nations World Water Report 2, UNESCO and Berghahn Books; UN-HABITAT: Paris, France; New York, NY, USA, 2006; ISBN 9231040065.
28. Lautze, J. (Ed.) *Key Concepts in Water Resource Management: A Review and Critical Evaluation*; Routledge—Earthscan: New York, NY, USA, 2014; ISBN 9780415711739.
29. Castro, J. Water governance in the twenty-first century. *Ambient. Soc.* **2007**, *10*, 97–118. [CrossRef]
30. Lubell, M.; Robins, G.; Wang, P. Network structure and institutional complexity in an ecology of water management games. *Ecol. Soc.* **2014**, *19*, 23. [CrossRef]
31. Pahl-Wostl, C.; Holtz, G.; Kastens, B.; Knieper, C. Analyzing complex water governance regimes: The Management and Transition Framework. *Environ. Sci. Policy* **2010**, *13*, 571–581. [CrossRef]
32. Oskam, I.F. T-shaped engineers for interdisciplinary innovation: An attractive perspective for young people as well as a must for innovative organisations. In Proceedings of the 37th Annual Conference—Attracting students in Engineering, 1–4 July 2009.

33. Douven, W.; Mul, M.L.; Alvarez, B.F.; Son, L.H.; Bakker, N.; Radosevich, G.; van der Zaag, P. Enhancing capacities of riparian professionals to address and resolve transboundary issues in international river basins: Experiences from the Lower Mekong River Basin. *Hydrol. Earth Syst. Sci. Discuss.* **2012**, *9*, 3813–3849. [CrossRef]
34. Seibert, J.; Vis, M.J.P. Irrigania—A web-based game about sharing water resources. *Hydrol. Earth Syst. Sci.* **2012**, *16*, 2523–2530. [CrossRef]
35. Rusca, M.; Boakye-Ansah, A.S.; Loftus, A.; Ferrero, G.; van der Zaag, P. An interdisciplinary political ecology of drinking water quality. Exploring socio-ecological inequalities in Lilongwe's water supply network. *Geoforum* **2017**, *84*, 138–146. [CrossRef]
36. Zwarteveen, M.; Kemerink-Seyoum, J.S.; Kooy, M.; Evers, J.; Acevedo Guerrero, T.; Batubara, B.; Biza, A.; Boakye-Ansah, A.; Faber, S.; Cabrera Flamini, A.; et al. Engaging with the politics of water governance. *WIREs Water* **2017**, *4*, 1–9. [CrossRef]
37. Kolb, D.A. *Experiential Learning: Experience as the Source of Learning and Development*; Prentice Hall: Englewood Cliffs, NJ, USA, 1984.

water

MDPI

Article

Multi-Stakeholder Development of a Serious Game to Explore the Water-Energy-Food-Land-Climate Nexus: The SIM4NEXUS Approach

Janez Sušnik [1,*], Chengzi Chew [2], Xavier Domingo [3], Simone Mereu [4,5], Antonio Trabucco [4,5], Barry Evans [6], Lydia Vamvakeridou-Lyroudia [6], Dragan A. Savić [6], Chrysi Laspidou [7] and Floor Brouwer [8,*]

1. Integrated Water Systems and Governance Department, IHE Delft Institute for Water Education, PO Box 3015, 2601DA Delft, The Netherlands
2. DHI Group, Agern Allé 5, DK-2970 Hørsholm, Denmark; czc@dhigroup.com
3. Eurecat, Lleida Agri-food Science and Technology Park, ICT Building, Ground Floor, 2, 25003 Lleida, Spain; xavier.domingo@eurecat.org
4. Euro-Mediterranean Center on Climate Changes, IAFES Division, 07100 Sassari, Italy; si.mereu@gmail.com (S.M.); antonio.trabucco@cmcc.it (A.T.)
5. Department of Science for Nature and Environmental Resources (DipNeT), University of Sassari, 07100 Sassari, Italy
6. Centre for Water Systems, College of Engineering, Mathematics and Physical Sciences, University of Exeter, Exeter EX4 4QF, UK; b.evans@exeter.ac.uk (B.E.); l.s.vamvakeridou-lyroudia@exeter.ac.uk (L.V.-L.); d.savic@exeter.ac.uk (D.A.S.)
7. Civil Engineering Department, University of Thessaly, GR-38334 Volos, Greece; laspidou@uth.gr
8. Wageningen Economic Research, PO Box 29703, 2502LS The Hague, The Netherlands

* Correspondence: j.susnik@un-ihe.org (J.S.); floor.brouwer@wur.nl (F.B.)

Received: 15 December 2017; Accepted: 30 January 2018; Published: 1 February 2018

Abstract: Water, energy, food, land and climate form a tightly-connected nexus in which actions on one sector impact other sectors, creating feedbacks and unanticipated consequences. This is especially because at present, much scientific research and many policies are constrained to single discipline/sector silos that are often not interacting (e.g., water-related research/policy). However, experimenting with the interaction and determining how a change in one sector could impact another may require unreasonable time frames, be very difficult in practice and may be potentially dangerous, triggering any one of a number of unanticipated side-effects. Current modelling often neglects knowledge from practice. Therefore, a safe environment is required to test the potential cross-sectoral implications of policy decisions in one sector on other sectors. Serious games offer such an environment by creating realistic 'simulations', where long-term impacts of policies may be tested and rated. This paper describes how the ongoing (2016–2020) Horizon2020 project SIM4NEXUS will develop serious games investigating potential plausible cross-nexus implications and synergies due to policy interventions for 12 multi-scale case studies ranging from regional to global. What sets these games apart is that stakeholders and partners are involved in all aspects of the modelling definition and process, from case study conceptualisation, quantitative model development including the implementation and validation of each serious game. Learning from playing a serious game is justified by adopting a proof-of-concept for a specific regional case study in Sardinia (Italy). The value of multi-stakeholder involvement is demonstrated, and critical lessons learned for serious game development in general are presented.

Keywords: nexus; participatory modelling; serious game; system dynamics; water-food-land-energy-climate

1. Introduction

Water, energy, food, land and climate exist in a 'hyperconnected' state [1] and are bound together in an extraordinarily complex system in which each sector cannot be considered in isolation, as it has impacts on at least one of the other system sectors. The Bonn Nexus Conference in 2011 [2] focussed particularly on the water-energy-food (WEF) 'nexus'. Since then, the 'classical' WEF nexus has evolved to include land use, the environment, climate change, waste and/or the economy ([1,3–9] and www.flores.unu.edu/en/research/nexus). Others have used alternative terms to define a complex system connected at the global scale and in which actions of one sector can have significant impacts on other sectors (e.g., 'teleconnections', especially in the climate system; [10]). Global research teams, politicians and multinational corporations are showing increased interest in the nexus and its potential implications for business [1,3,8,11–13]. The nexus has global relevance and is extremely complex. However, global issues often require local actions [13]. Full understanding of this nexus and characterization of its internal feedbacks is required in order to be able to make informed, meaningful (policy) decisions [14]. Herein lies a current gap in understanding, as many previous studies either: (1) consider only one or perhaps two nexus sectors, neglecting impacts on the others [7,15]; or (2) consider more sectors, but only for very local, specific case studies that lack wider applicability [16,17]. Efforts are required, therefore, to consider many nexus sectors together at a range of scales from regional to international in order to develop a more general understanding of nexus behaviour, and the potential impact of climate, socio-economic and policy changes on this behaviour.

To enable full understanding of complex relations between natural resources and to promote the nexus concept as a means to enhance wider nexus-compliant practices require capacity building and education programs to engage current and future stakeholders. Learning is essential. However, the process of capacity building and education is complex, and there is a push to find innovative ways to make these processes more efficient, effective and scalable. A major opportunity to accelerate progress is through the utilization of opportunities provided by advances in and dissemination of information and communications technology (ICT).

One upside of ICT is the easy access to information and data, which is in turn related to one of its main downsides, namely the overabundance of information, not all of which may be useful, necessary or accurate. Consider the number of hits a search engine returns on a simple search, or the amount of online advertisements a typical Internet user may be exposed to during the course of day. According to [18,19], we are exposed to these sources at such a high rate, that it affects the way our brains process information. Today, we are immediately drawn to information mechanisms, for example, notifications from devices (mobile phones, tablets or computers) that 'pop-up', interact with us briefly and give us a very condensed information stream. This brings about additional challenges in learning and capacity building: besides having to make it more efficient and scalable, we also need to keep up with communication trends and technology to be able to attract the attention and interests of our target audience(s).

A serious gaming approach combines a learning objective with a fun activity in an attempt to increase the potential for learning uptake [20,21]. Perhaps one of the best-known applications of an electronic-based serious "game" is that of flight simulators used in pilot training. One of the main reasons for the success of this kind of application is that pilots need a very realistic learning environment, but also one that has minimal risk and allows them to make non-catastrophic mistakes from which they can learn. This learning-by-playing approach is commonly termed "meaningful play" [22], and a well-designed serious game environment provides a feedback mechanism that allows the player to reflect on his or her actions and adopt different approaches or strategies. The internalisation of actions and reactions stimulates learning, often resulting in an increase in self-learning and knowledge retention. Serious games, or games used for purposes other than mere entertainment [21], are often used for education, decision making and for public policy making. Since [23] introduced the notion of serious games, various applications have been developed, including, but not limited to, educational [24], science [25], training, psychology [26], military [27], government, corporate, healthcare [28] and water management domains [29]. However, no comprehensive serious game has been developed to deal with the water-energy-food-land-climate nexus, which represents a clear gap in the literature.

With respect to the above discussion, there is a clear need to: (1) improve scientific understanding of the water-food-land-energy-climate nexus in a holistic way and applicable at a range of scales; and (2) to develop serious games addressing the (global) water-energy-food-land-climate nexus, or parts thereof, at multiple scales from regional to international, and for a reasonable policy-relevant time dimension (decades at least). Such serious games should be grounded in robust science and analysis, using the latest data and state-of-the-art modelling tools and techniques. They should also be appropriately 'packaged' and targeted to certain users, namely those in decision-making or decision-influencing positions. These games should allow those users to explore nexus issues at many scales, allowing exploration of the trade-offs and synergies between sectors. They need to be coherent and understandable and convey clear policy-relevant messages and information based on the latest scientific understanding. The main purpose of these serious games should be to enable stakeholders to understand and learn about the medium and long-term implications of nexus-related policies applied to each case study. Using gaming (not necessarily as an ICT tool) has long been established as means for understanding policies [20], leading to acceptance, mitigating conflicts and compromise. Moreover, Medema et al. [30] discuss how serious games offers stakeholders an opportunity to voice their perspectives, and Gurung et al. [31] show how useful serious gameplay can be in the development of trust and empathy among stakeholders, as well as an improved understanding of systems and complex issues at play. A similar conclusion is reached by [32], who asked farmers in conflict over water allocations to play a simulation game in which the farmers swapped roles for one round of the game. When the roles of the farmers are swapped from upstream to downstream villages and vice versa, it was very effective at sharing different perspectives. This paper introduces a serious gaming approach, based on these concepts, aiming at creating an understanding of nexus issues related to food, energy, water, climate and land, across various stakeholders and the long-term implications of different policies that may be implemented at the regional, national or international level.

The work presented in this paper refers to an ongoing (2016–2020) EU Horizon2020 project "Sustainable Integrated Management FOR the NEXUS of water-land-food-energy-climate for a resource-efficient Europe (SIM4NEXUS; www.sim4nexus.eu)". SIM4NEXUS aims at developing serious games for each of the project's case studies, using an innovative methodological approach, described in this paper. The serious games in SIM4ENXUS will be operable at scales ranging from regional to national, to continental, to global, as well as at different time horizons: short-, medium and long-term to 2050. A specific project case study (Sardinia) and its serious game are presented, which was selected from the start of the project as a "fast-track" proof-of-concept exercise. The term "fast-track" refers to the time frame (as fast as possible) during which the serious game was developed, as well as to some relatively simplified approach in the modelling; we are interested in a methodological innovative high level approach, rather than detailed simulation of the region. On the other hand, the paper is meticulous in detailing the nexus approach and all the steps leading from "raw" data to the development of a serious game for a specific region; this same approach is being followed for all case studies (still an ongoing exercise until 2020). Furthermore, it has a wider potential for further applications and, to the best of our knowledge, does not exist in the literature.

While previous efforts have considered nexus elements either in complete isolation or only in 'partial integration' (i.e., considering only a few nexus elements at one time), SIM4NEXUS aims to address all the water-food-land-energy-climate elements together (Figure 1) and to account for the possible impact on these elements in response to climate and relevant policy changes. SIM4NEXUS aims to address both research and knowledge gaps regarding the global nexus at multiple scales and communicating critical results to policy-makers via a state-of-the-art serious game built on robust science. Considering Figure 1, the image on the far left reflects the 'traditional' approach with sectoral modelling (e.g., from models such as CAPRI (Common Agricultural Policy Regionalised

Impact Modelling System) or E3ME (energy-environment-economy[E3] macro-economic model)). The middle image reflects the most current advanced approaches (e.g., impact assessment tools, integrated assessment modelling, integrated water resources management). There is partial integration, e.g., linking the model MAGNET (Modular Applied GeNeral Equilibrium Tool) with IMAGE, used to assess cross-sectoral impacts of climate, social and/or policy changes, but complete nexus understanding is still missing. SIM4NEXUS integrates thematic models within a system dynamics modelling framework to integrate the knowledge and data from thematic models and local experts. SIM4NEXUS adds the learning from playing the serious game and testing nexus-compliant policies. In addressing nexus-wide interconnections, system-wide trade-offs and synergies can be sought, with benefits for multi-sectoral policy development and planning [14].

The objectives of this paper are two-fold. Firstly, the general process of serious game development in SIM4NEXUS that is common for all 12 case studies is presented. In this way, the consistent mode of game development in SIM4NEXUS, running from case study nexus qualitative conceptualization and problem mapping, right through to serious game development and implementation, is presented. Herein lies the first novelty: inclusion of local-level stakeholders from the very beginning, with each case study led by a local expert. The second objective is to highlight the central novelty of SIM4NEXUS. To the best of our knowledge, no other serious game has been developed that covers the range of nexus sectors (water, energy, food, land and climate), the range of spatial scales (regional to global) and that also includes comprehensive case study-relevant policy analysis to help frame and guide model and game development. In this sense, many of the gaps alluded to above regarding nexus modelling and serious gaming are being addressed in SIM4NEXUS. It is pointed out that SIM4NEXUS is ongoing (started June 2016, ending May 2020). Therefore, this paper does not present completed or final models or serious games, but focusses instead on the process and on intermediary results from a pilot fast-track study in the project. Lessons learned from the fast-track are presented. These lessons are useful not just in SIM4NEXUS, but for other serious game development exercises.

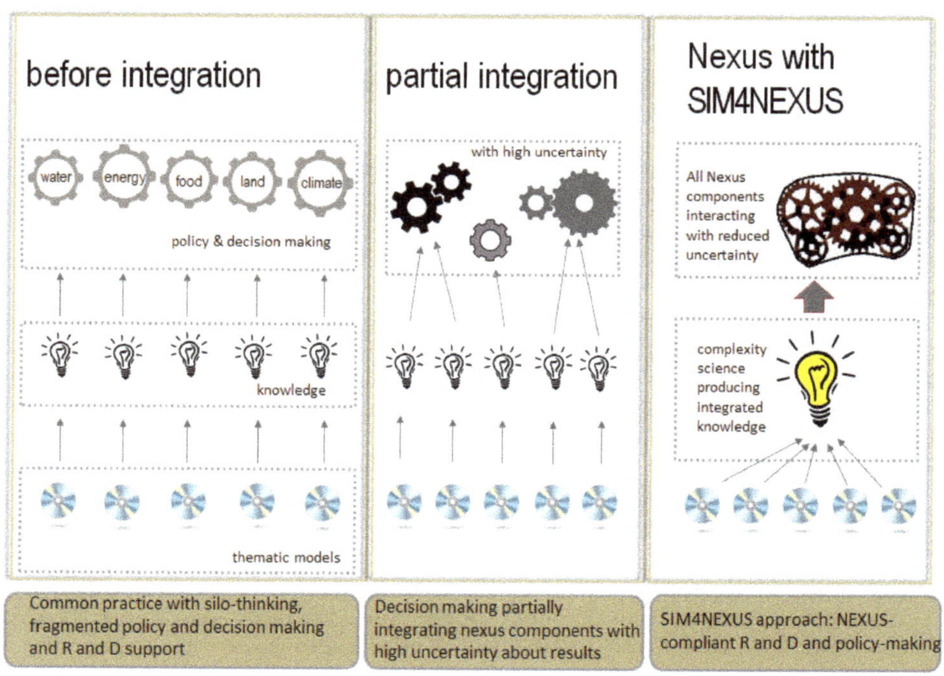

Figure 1. The SIM4NEXUS integration concept.

2. Serious Game Development: Process Overview

SIM4NEXUS will develop serious games for 12 case studies, ranging in scale from regional (e.g., southwest England, Sardinia) to global. For each, the general process of model development leading to serious game design and implementation will follow a similar path, only changing specific details particular to each case study. There are five main steps in the overall general model and serious game development process comprising: Step 1. Case study nexus system description, framing and conceptualisation; Step 2. Thematic data identification and collection; Step 3. System dynamics integration modelling and conversion to R script (www.r-project.org); Step 4. Communication of model output to the Knowledge Elicitation Engine; Step 5. Design and development of the serious game front-end. These five steps are elaborated in the following sections (Sections 2.1–2.4). For each step, the general process/methodology to be used in all SIM4NEXUS case studies is described, followed by the application to the pilot fast-track case study of Sardinia.

2.1. Case Study System Conceptualisation and Data Identification (Steps 1 and 2)

This step is concerned with framing the key nexus issue(s), within the remit of SIM4NEXUS, to be explored in each case study. Together with project case study leaders and diverse stakeholder groups, the main case study nexus issue(s) is identified in stakeholder workshops. For example, in Sardinia, the central issue is the long-term resilience of reservoir water supply to changes in climate and agricultural factors, while in the Netherlands, it is centred on biomass generation and a switch to a low-carbon economy. From the central theme, interactions with other nexus sectors are identified, including important feedback processes. A dedicated Work Package in SIM4NEXUS on nexus policy analysis is critical in offering input to this process, as it provides a first idea of the most pertinent nexus-relevant policies, goals and instruments for each case study. The central theme/issue identified by case study leaders and stakeholders can be cross-checked with the policy analysis to begin to identify likely policy scenarios/goals to be modelled and implemented in the serious game so as to have relevance for local players. Therefore, even at this early stage, stakeholder involvement is critical, and input is taken from case study-level policy analysis.

The end product is a conceptual (qualitative) diagram that defines the central nexus issue and that elaborates the key interactions with other nexus sectors. Key policies to potentially be played in the final game are already identified, but subject to changes. Data requirements for quantitative modelling are identified at this stage, and data are gathered.

In this paper, the general modelling process is elaborated in relation to a 'fast-track' proof-of-concept carried out in SIM4NEXUS using Sardinia as the pilot case study (Figure 2). The fast-track approach was adopted to test each step in the development process, identifying and addressing any major difficulties and issues before rolling out the process in detail for all 12 SIM4NEXUS case studies. Sardinia was chosen as the fast-track because knowledge of system dynamics modelling to assess resilience of reservoirs to meet multiple water demands (i.e., agriculture, domestic/tourism, hydroelectric) under climate change was already available from the EC FP7 project Water Availability and Security in Southern EuRope and the Mediterranean; 'WASSERMed' (see www.cmcc.it/projects/wassermed-water-availability-and-security-in-southern-europe-and-the-mediterranean for a general project description and [33] for results). Therefore, the team was not starting from a 'blank slate', allowing a facilitated process to integrate further nexus sectors and feedbacks in the quantitative modelling.

To articulate the conceptual framing of the Sardinia case, interactive workshops with local experts and stakeholders, including academics, public authorities, decision makers and unions, were carried out to define the key nexus sectors to consider, identify sector drivers, relevant key policies and, crucially, how sectors and policies interact. At the end of the process, the model was expanded in terms of: (i) nexus sectors, which now include energy, land and food, that were not included in the WASSERMed version; (ii) spatial scope, from district level to integrating sectorial interactions for the whole of the Sardinian region; and (iii) increasing the detailed representation of nexus sectors in the model, including the policies that affect them. Figure 3 shows the conceptual system diagram developed for the Sardinia fast-track, on which further quantitative model development was based.

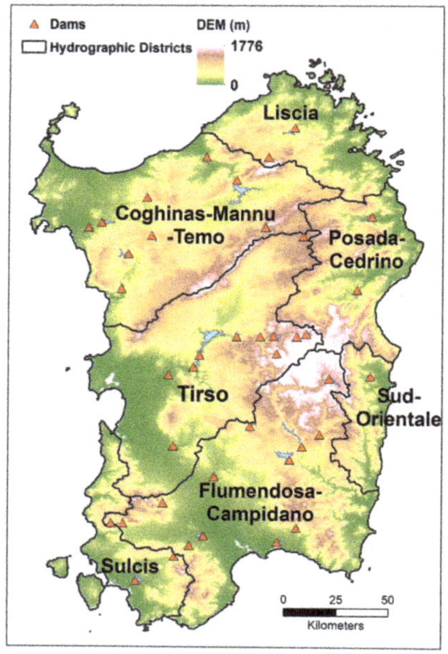

Figure 2. Map of Sardinia, showing the seven hydrological districts and the locations of the reservoirs studied.

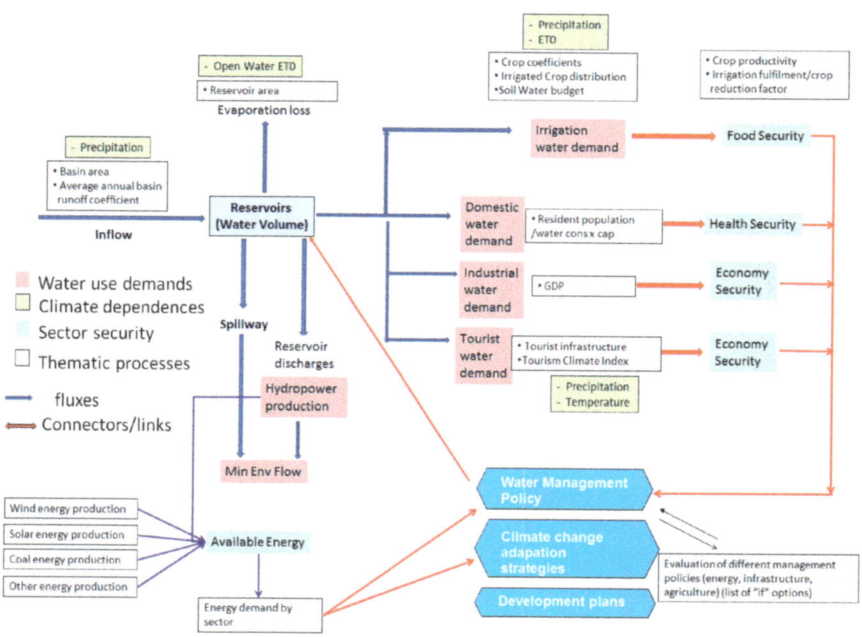

Figure 3. Conceptual diagram describing the major nexus components relevant for the Sardinia case study. ETO is reference evapotranspiration.

Based on the above conceptualisation, it was possible to identify the relevant 'thematic models' from within SIM4NEXUS from which data would be required. Here, data from CAPRI (a global agricultural and production model; [34]), the Global Trade Analysis Project (GTAP) database (www.gtap.agecon.purdue.edu/), E3ME (a global economic and energy model; [35]), downscaled climate data from HadGEM2-ES from the Inter-Sectoral Impact Model Intercomparison Project (ISIMIP) [36], as well as locally-relevant data (e.g., for reservoir operating rules and environmental flow regulations) were acquired. Data for 2010 (the baseline) and 2030 under RCP Climate Scenarios 4.5 and 8.5 [37] and Shared Socioeconomic Pathway 2 (SSP2) were gathered and used for quantitative model development. The conceptual framework will be further elaborated and improved during the SIM4NEXUS project, although the present framework already gives a reasonably accurate representation of the nexus in Sardinia. For full model development in SIM4NEXUS case studies, projections to 2050 will be simulated.

2.2. System Quantification: System Dynamics Modelling (Step 3)

Based on the conceptual framework (Figure 3), the qualitative descriptions of the case studies will be 'translated' into quantitative system dynamics nexus models built in STELLA Professional (www.iseesystems.com). STELLA is a system dynamics modelling (SDM; [38]) software that allows for the exploration of complex systems. STELLA, like most SDM software, uses the concepts of stocks, flows and convertors (see [38] for a description of these elements) to create models of the system under study. SDM is ideal for modelling complex systems governed by feedback, delay and cross-disciplinary problems. STELLA can integrate data from many disciplines. SDM is ideally suited to studying the trajectory of system parameters under change and for dealing with uncertainty. SDM is proven in modelling complex systems in a diverse range of disciplines and for communicating with non-expert stakeholders (e.g., [39–42]).

For the Sardinia case study, the main focus of the SDM was the representation of the reservoir water balance for the island, accounting predominantly for water supply and for water demand related to agricultural, energy and domestic/tourist consumption. Energy generation and consumption were also important along with the mode of generation and sector of consumption, as was modelling the change in crop types (i.e., land use and food production changes) and the crop water requirements associated with potential crop and cropped area changes, as well as in response to change in the local climate. While water is the central focus, this model is not only concerned with Sardinian hydrology and is not a hydrological model, but considers other nexus sectors including energy, climate, food and land use.

On the water supply side, the model (Figure 4) accounts for inflows to the reservoirs based on precipitation partitioning to runoff over the catchment area upstream of reservoirs. For the purposes of the fast-track, water supply for the 40 main reservoirs and multiple demands were aggregated at the island level. However, final case study results will aim at a more articulated disaggregation within seven hydrological districts in Sardinia (see Section 4).

For water demand, the model considers: (1) open-water evaporation from reservoir surfaces; (2) discharges for hydroelectric generation; (3) spillways in times of overflow; (4) irrigation requirements; (5) industrial demand; (6) domestic and tourist water requirements; and (7) environmental flows (i.e., the minimum amount of water needed to preserve ecological functions and values in watercourses). With irrigated agriculture being the largest water consumer, this sector was modelled in more detail. The crop water requirements per unit-area, and the area planted, were taken into consideration for 13 major crops in Sardinia as a function of current and changing climatic conditions.

Energy production is modelled from sources including oil and gas, solar, wind and hydropower, while energy demand comes from the agricultural, domestic, industrial and service sectors. Touristic fluxes and relative water demands are modelled based on a Touristic Climate Index [43] and socio-economic scenarios. Climate change will have an impact on evaporation rates, crop water requirements, precipitation recharge to reservoirs, but also touristic fluxes. Data from thematic models provide projected changes of irrigated area by crop (CAPRI), energy production and demand by sector (E3ME) and socio-economic factors (GTAP). All other data are from local Sardinian sources.

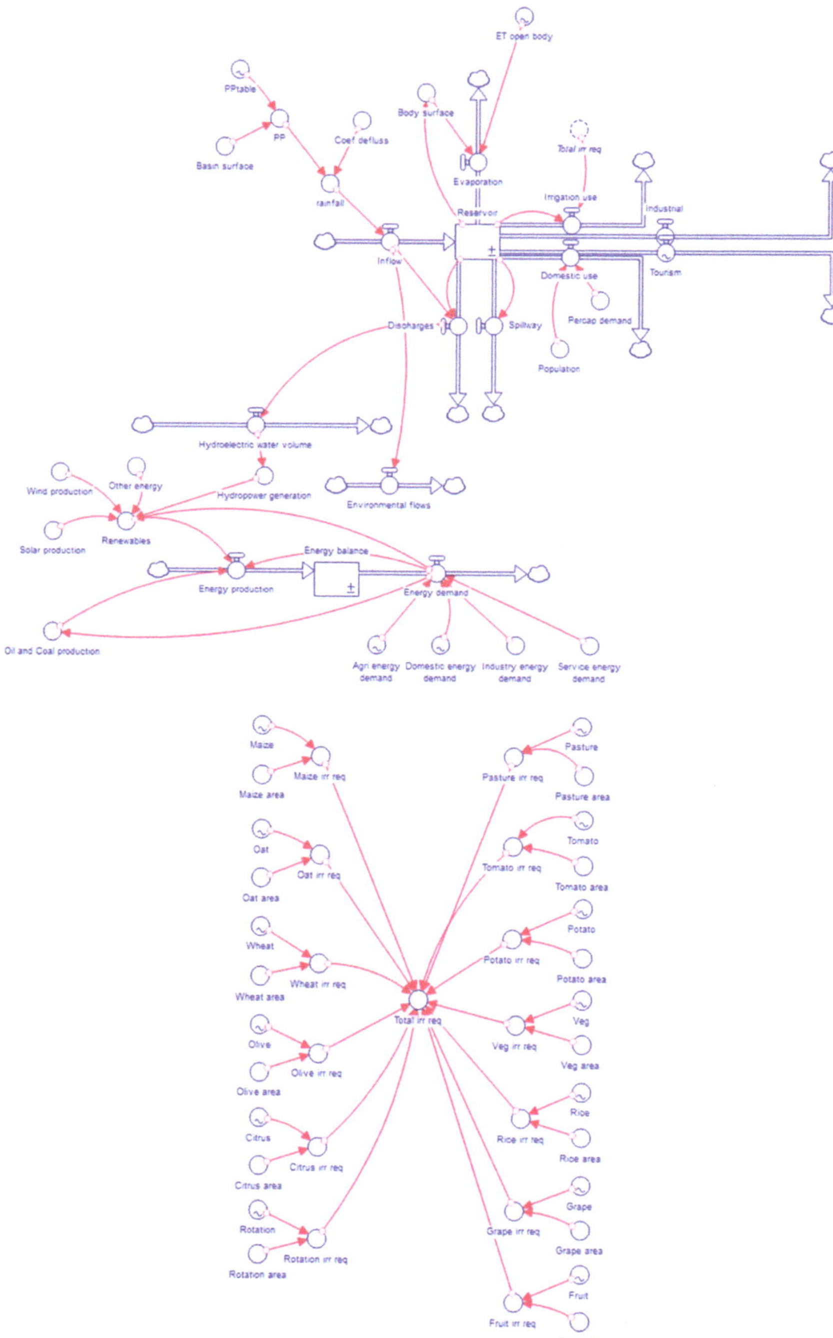

Figure 4. The developed system dynamics model for the Sardinia fast-track.

The model for Sardinia runs simulations for 2010 and 2030 for both the RCP4.5 and 8.5 scenarios. The model has a total of 73 variables accounting for each nexus sector and the interlinkages between the sectors. The modelling time-step is monthly, with all data scaled to this resolution. The model for Sardinia runs with five different climate models, accounting for inter-model variability. The variables of concern for the fast-track were chosen by the case study partner in SIM4NEXUS in consultation with local stakeholders.

Once the model structure was deemed representative (in consultation with local stakeholders), appropriate data were inputted into the model and the model outputs (in terms of the general trends of key variables such as the reservoir water balance over the year; see Section 3) were discussed with local case study experts to verify if the model was yielding sensible results. Once an agreed SDM representation was defined, the next step was to convert the model into an R script.

The primary reasons for the conversion of SIM4NEXUS STELLA models into an R script are that R is a freely-available open-source platform that can be distributed to all stakeholders, and the R script can be integrated easily with the Knowledge Elicitation Engine (KEE; Step 4) and the serious game (SG; Step 5). This conversion will therefore greatly speed up gameplay and will allow for much easier communication between the model, the KEE and the serious game.

The general conversion of the STELLA models to R is facilitated by the STELLA functionality allowing model equations and parameters to be exported to a text format (.txt; known as the 'equation file'). To achieve the conversion process, a Python script was created that reads the equation file and generates the R script and corresponding lookup files required for the R script to run. The script allows for policy changes to be implemented in the game front-end. In addition to the main R script, a number of accompanying files are produced (Table 1).

Table 1. Data outputs from STELLA conversion to R.

File	Purpose
R Function File	Contains details relating to time-series lookup tables used by the main script along with mathematical functions
Time-Series Files	Input data within the STELLA model are stored as time-series CSV files that are referenced within the R function file
Coefficient Files	A coefficient file is generated for each time-series variable to allow for policy implementation at a monthly scale, e.g., specific crop area can be reduced by 20% from June to August by switching the coefficient variable from 1.0 to 0.8 for those months.
Dynamic Variable Equations	Values within the SDM that change over time and/or are dependent on conditions of other dynamic variables are stored as a reference so that these lines can be substituted by variations of that line to allow for policy changes to be implemented

In order to implement and analyse the effects of policy changes (e.g., changes in energy mix or changes in cropping patterns) with the SDM file, variables within either the coefficient files or one of the dynamic variable equations (Table 1) can be altered to reflect new policy rules or conditions. To highlight the implementation of a policy change for the Sardinia case study within the R script, a working example looking at environmental flows is summarised in Table 2, which shows the present policy implemented in the SDM for Sardinia to foster ecosystem services and biodiversity downstream from the reservoir. Within the R code, the cases depicted in Table 2 are represented in Equation (1).

$$\begin{gathered}\text{Environmental\_flows} = \text{IFELSE (Reservoir/2000} > 0.20\text{, Inflow*0.1,} \\ \text{IFELSE (Reservoir/2000} > 0.1\ \&\ \text{Reservoir/2000} \leq 0.20\text{, Inflow*0.05, 0))} * \\ \text{Environmental\_flows\_Policy\_Coeff}\end{gathered} \tag{1}$$

where 'IFELSE' is the R representation for the IF-ELSE logic, 'Reservoir' is the volume of water stored in Sardinian reservoirs at a given time-step (in million cubic meters; Mm^3), 'Inflow' is the inflow volume to reservoirs in that time-step (Mm^3 $month^{-1}$) and 'Environmental_flows_Policy_Coeff' is a coefficient that can be changed to reflect different policy conditions. The number '2000' represents the cumulative maximum reservoir storage capacity.

Table 2. Policy rules to foster ecosystem services and the biodiversity of ecosystems downstream of Sardinian reservoirs.

Case	Rule
1	If water in the reservoir is above 20% of the reservoir capacity, then 10% of the monthly basin run-off is allowed to be environmental flow.
2	If water in the reservoir is below 20% of the reservoir capacity, then 5% of the monthly basin run-off is allowed to be environmental flow.
3	If water in the reservoir is below 10% of the reservoir capacity, then 0% of the monthly basin run-off is allowed to be environmental flow.

Once the policy changes are implemented within the R script, the script can be run, and resulting outputs from the R systems model can be exported as CSV or JSON files and referenced by the Knowledge Elicitation Engine (Step 4).

2.3. *Knowledge Elicitation Engine (Step 4)*

The Knowledge Elicitation Engine (KEE) in SIM4NEXUS focuses on integrating knowledge and strategies at different spatial and temporal scales, with resource-efficient land use, agricultural productivity improvements, sustainable water management and low carbon energy transition. The KEE enables the analysis of interactions within existing regulatory frameworks and barriers to implementation (Steps 1 and 2). The KEE provides the serious game with the system-wide impact of each action implemented under a case study or specific scenario, considering the interactions modelled in the SDM (Step 3). Moreover, the KEE permits a top-down learning approach based on serious game front-end user decisions (Step 5) and a bottom-up approach based on classical machine learning methodologies applied to data. Top-down learning focuses on expert validated decisions made by users playing the game and how they affect the nexus system being played, giving feedback explaining the consequences of the decisions they have made. The bottom up learning focuses on discovering new knowledge in the form of unknown relations between systems, basing its learning on the data generated by the SDM from different geographic and temporal scale points of view. All learning and relevant knowledge for the nexus are stored in a semantic repository, which is supported by a cross-domain ontology. This simplifies information exchange between different nexus components and fosters better understanding of nexus interactions from a holistic point of view.

The KEE comprises three main items: (i) the game decision support system (GDSS), which focuses on recommending serious game user actions to be taken considering the current status of the game (polices applied, indicators, etc.), the objectives of the scenario and the learning goals to be achieved; (ii) the inference engine, which generates new knowledge based on machine learning algorithms applied to previous user decisions playing under different scenarios; and (iii) an agent-based module, which acts as artificial intelligence-based players when required by the specific case study. Thus, the KEE integrates knowledge coming from Steps 1 and 2, nexus complexity science from Step 3 and the serious game front-end (Step 5).

The KEE modules rely on the SIM4NEXUS database and metadata ontology (semantic repository), which contains general information and knowledge about the nexus, specific information regarding each case study, historical data about user actions while playing the game and all necessary pre-calculated information (e.g., climate information such as precipitation projections) to accelerate real-time computational needs and to feed the serious game front-end to represent the different elements in the scenario. The semantic repository is focused on support of the fast-track described in this paper, with the resulting simplified schema of the current ontology shown in Figure 5. Historical data about user actions are used to create an artificial intelligence able to simulate human players in the game, when needed. Finally, pre-calculated information is used to assure appropriate response time while playing the game, as some calculations require a long time and high computer power to be executed.

The ontology (Figure 5) represents the SIM4NEXUS knowledge and scope. Currently, the top concept is the 'Session', which represents a game session in the serious game front-end. Linked to the 'Session' there is a 'User', representing the player, and a 'Case Study' (at the moment this is only Sardinia), which in turn is related to a specific 'Model' (currently the SDM for the Sardinia case). In order to represent the nexus state through the game, the 'Session' has a list of 'StateEvolution', which represents the changes from one state ('State') to the new state ('State'), after the user decides to apply certain polices ('Policy'). The 'State' is defined by a collection of 'NexusComponent' (for instance, 'Climate and environment' or 'Water'), and these components have specific related variables ('Parameter').

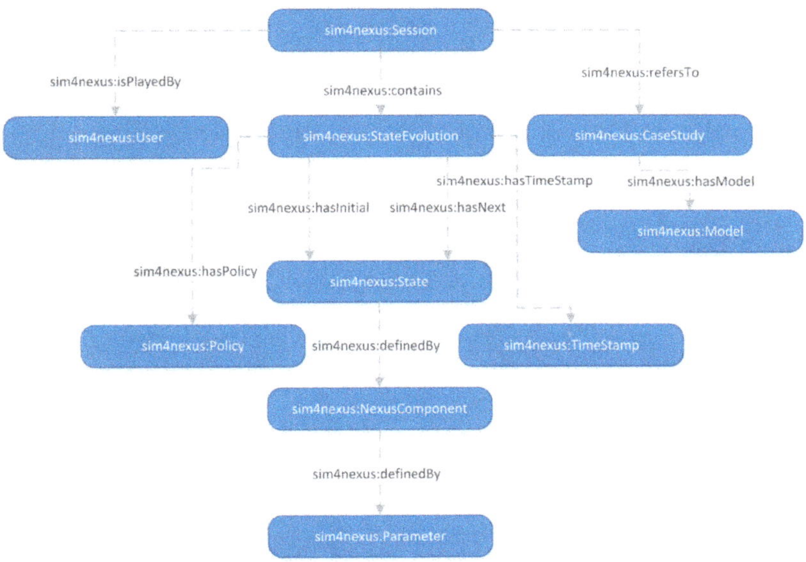

Figure 5. Simplified ontology schema for the Sardinia fast-track. See the text for details.

The ontology is defined using the Web Ontology Language (OWL), a semantic web language designed to represent rich and complex knowledge about things, and relations between things. Some existing ontologies, related to the nexus, have been included in the SIM4NEXUS context:

(1) WatERP ontology [44], developed under the EC FP7 project 'WatERP' (see HTTP://WWW.WATERP-FP7.ORG/ for a general project description), which reflects the water manager's expertise for managing water supply and demand. The novelty of the WatERP ontology lies in including interactions with natural processes as a mechanism to understanding how to affect changes in water resources management so as to achieve the objective of matching supply with demand. These interactions could range from infrastructures to management decisions.

(2) The WEFNexus ontology [45], which concerns water, energy and food derived from the European Directives: Article 2 of EU Directive 98/83/EC that defines the water intended for human consumption; Article 2 of EU Directive 2003/30/EC that defines bio-fuels; and Article 2 of EU Regulation 178/2002/EC that defines food.

For the Sardinia fast-track specifically, the KEE focuses on: (i) storing information about the case study in the Semantic Repository (e.g., climate data, CAPRI data, E3ME data, local reservoir operating rules); (ii) storing historical data about user actions for future analysis; and (iii) providing a standardised communication interface between the serious game front-end and the SDM simulations. The KEE needs to run SDM models repeatedly for each case study during a game session. At the moment, the computational requirement of the Sardinia model allows for real-time simulation, but this computational load will increase as models become more sophisticated. To enable this, the R script is called when necessary from a Python script, which is included and published in the SIM4NEXUS web services layer. This layer establishes the communication mechanisms among the different modules. The deployed web service follows the OpenGIS® Web Processing Service (WPS) Interface Standard [46] for standardizing the inputs and outputs for geospatial processing services. The standard defines how an execution of a process has to be requested by a client and how the process output is handled. It defines an interface that facilitates the publishing of geospatial processes and clients' discovery of and binding to those process.

Using this web service, the serious game front-end is able to send the current status of the scenario, define the policies to be applied and define the time-step. The web service runs the SDM to calculate the evolution of the scenario under that context and sends it back to the serious game front-end. Finally, that information is presented to the user, and new policies to be applied are offered.

2.4. Serious Game Front-End (Step 5)

The SIM4NEXUS development of the serious game front-end for each case study uses the AquaRepublica serious game developed by DHI [47] as a starting concept, but further develops that game for the specific requirements of each case study in SIM4NEXUS. The goal of the game is to bridge the gap between science and policy stakeholders by translating complex modelling results into an interactive virtual world. The serious game will be used in stakeholder workshops to explain and provide learning opportunities about the potential impact of different policies on the nexus and how these policies impact a case study through a "learning by playing" approach (Figure 6).

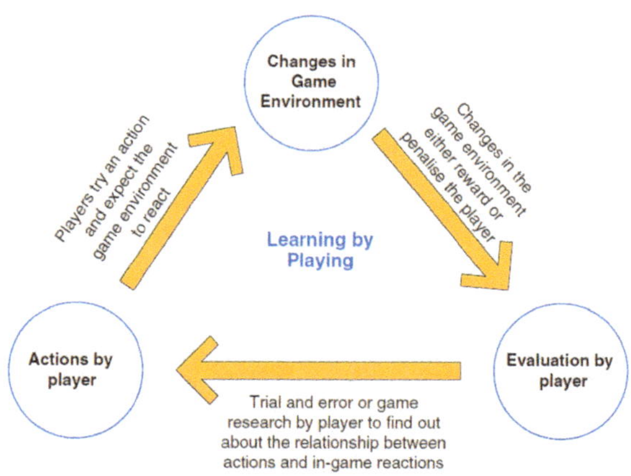

Figure 6. Learning by doing flowchart.

Based on these concepts, the serious game of SIM4NEXUS attempts to create the same understanding of nexus issues across stakeholders in food, energy, water, climate and land. The setup of the serious game is as follows:

As a player, you represent decision makers in the various sectors in a particular area: food, energy, water, climate and land. Your aim is to achieve the targets (objectives) set out by the national or international bodies by changing or adapting new policies in your area and to implement practices to comply with local policy objectives. To succeed in the game, you must learn how to achieve these targets by mixing and matching various cross-sectoral policies without compromising the existing 'goal attainment' of the other sectors.

The main content in the serious game is provided both through the interfaces and the logics that the game contains, as well as through the KEE (Step 4) that provides information and further knowledge to stress the immersions of the game user for each case study. The identification and formalization of the content to be imparted is important for the game logics requirements definition, as the logics behind the game must guide the user through these contents in order to impart the knowledge to be conveyed. This identification is required for the construction of the KEE, as well as the way both the game and the KEE architecture will communicate the knowledge and information. Therefore, Steps 4 and 5 are critically linked. This content is divided into three main parts:

1. Core experience: What are the players experiencing as they play the game?

The core experience is to play the role of policy makers in food, energy, water, climate and land management. The player will typically start off with separate "silo-thinking" approaches towards decision making and policy implementation. Over the course of playing the game, they will be encouraged to change towards a more integrated nexus-compliant policy implementation approach and decision making.

2. Base mechanics: What does the player do?

The player will have a target at the start of each turn of the game and he/she will have to implement policies to try to achieve the target. The turn ends when the player has decided on the policies to be implemented to achieve the targets and clicked on a "next turn" button. The game will compute the policies, and an analysis of the decisions will be displayed in the following turn, with a new target to achieve for the turn. The targets are envisioned to be displayed in a step-by-step manner to the player. This will help guide the player on what to do during the game play.

3. Penalties and Reward (P&R) system: What behaviour within the game is encouraged or discouraged?

Silo-thinking in decision making and policy implementation within the game is discouraged, whereas integrated nexus-wide decision making is encouraged. For every target in each turn, the player is encouraged to look at policies in all sectors and consider them to achieve a target. The P&R system will be provided in three parts:

(i) Key indicators across all nexus components. These key indicators will be defined with case study partners to ensure the game is suitably targeted for the local context. Local stakeholder meetings will be used to elucidate these indicators.
(ii) Events within the game. Events are news happening "on the ground", which add a societal and cultural aspect to the game. These events will be narrated in the same tone as the shared socioeconomic pathways (SSPs; [48]) and will be triggered based on the decisions the player made in the game. There will be uncertainties in event triggers to add more realism in the game (e.g., the occurrence of extreme events such as economic crisis or disaster events). There will be three categories of events: (a) informational events, which are neutral; (b) negative events, which will penalize the player by deducting points; and (c) positive events, which will reward the player with bonus points.
(iii) Score. There will be a score for the player. The score will indicate how successful the player is at applying nexus-compliant decision making to achieve the targets in the game. Every progression in the time-step of the game will add to the score to encourage the player to continue, with every policy implemented and events occurring adding further points to this score. The score is computed as follows:

$$S = 1 - \sum_i w_i \frac{(\max\{0,\ m_i(g_i - x_i)\})^2}{(g_i - x_{0i})^2}$$

where:

- the score S is based on a weighted sum of squared, normalized, differences between indicators x_i and targets g_i;
- the x_{0i} stands for the starting value of the indicator at the beginning of the game;
- for indicators that must be maximized $m_i = 1$, and for indicators that must be minimized $m_i = -1$;
- the sum of the weights $\sum_i w_i = 1$;
- $S = 1$ is the maximal score to be attained (full compliance with all targets).

Such a score can be computed over all indicators and per policy area, making clear on which areas to focus in order to improve the general score. Thus, it can serve as a basis to advise users and explain opportunities to improve their performance in nexus management.

In SIM4NEXUS, nexus-compliance refers to the degree to which policy choices made by the player tend to lead towards/away from policy objectives for that case study, as elucidated by both the detailed policy analysis work in SIM4NEXUS and by relevant nexus policies/objectives as indicated by stakeholders during the case study formulation and framing phase (Steps 1 and 2). The detailed nexus policy mapping will help define the scoring system for each case study. This may lead to some not necessarily 'bad' decisions (e.g., climate-neutral energy generation) perhaps not being scored as highly as expected if they do not fit strongly within the policy framework defined for a case study. However, through the policy work and stakeholder discussions, such instances should be minimized. Emphasis is placed on minimizing detrimental cross-sector impacts from a given policy choice.

The feedback between spatial scales is facilitated by the KEE. The KEE collects data about players' policy choices and the resulting consequences for the nexus components. This mechanism will be used to feed new targets resulting from games at the global or continental scale, into games at the national or regional scale. For instance, some policy may result from playing a game on the European level. This policy may entail new targets set at national and regional levels. By playing games with these new targets with national and regional policy makers, data can be collected about their reactions and the consequences for the nexus components. The KEE will then inform the policy makers at the European level (Figure 7).

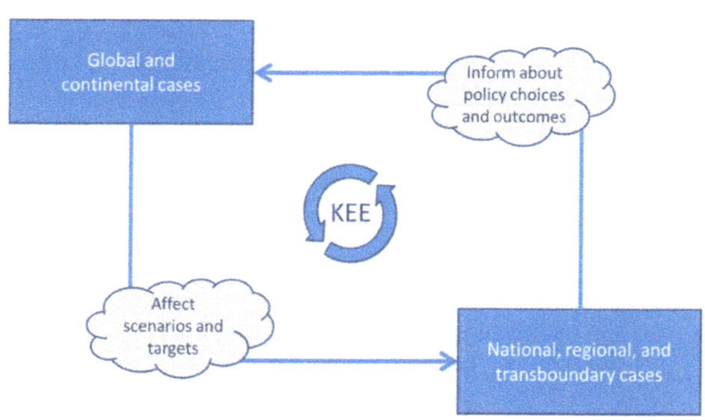

Figure 7. Feedback across spatial scales facilitated by the Knowledge Elicitation Engine (KEE).

The game can be played in different modes. It can be used by a single player, controlling all policy options. It can be used for playing the game in sessions led by a trainer or group facilitator, where participants play roles of policy makers in particular nexus domains. Single- and multi-player games can both be played. Lastly, it can be played with artificial agents. In the latter case, users take the roles of particular policy makers, while other roles are fulfilled by artificial agents, based on data collected by the KEE.

- Single-player controlling all policy options

The user has opened a new game session and initialized a predefined scenario. The user does not select a role as policy maker in a particular nexus domain. In the main gameplay screen, the user is presented with policy options enabled for all nexus domains. While playing subsequent turns, the user can adjust policies in all domains, attempting to achieve a balanced set of targets across the nexus. This setting for playing the game is particularly suitable for education and training to offer insight into relations across the entire nexus.

- Multiple players, each taking policy makers' roles on particular nexus domains

Playing games with groups where participants take different roles requires a group facilitator to set up the game. In the simplest case, the facilitator asks the players for their policy decisions and enters these into a game set up as in the single-player case described above.

As a future extension, a more advanced multi-player setting is envisaged, where the facilitator starts a game and assigns the players roles as policy makers in a particular nexus domain. The game logic is running on a central server, but players are presented with an individual gameplay screen in which all policy options and indicators are visible, but only the policy options related to the selected role can be changed. Turns are synchronised on the central server, and the next time-step is not taken until all participants have made their policy choices. Individual players are able to try various policy options available to them and check the consequences, but only have to commit to a policy once they are happy with it. At this point, the policy option and its consequences are revealed to other players. This stops exploratory and sub-optimal policies affecting other players too often, causing a disruption to the game play.

- Playing the game with artificial agents

As another future extension, it is envisaged to have artificial agents participating in the game. During the games described above, the KEE collects data on the policy choices the players make and the resulting outcomes in terms of indicators. Since users have supplied information on their background and expertise when registering for the game, the KEE can learn the behaviours of policy makers in particular domains. To this end, several machine learning techniques will be tested, including, but not limited to: decision trees, nearest neighbour and support vector machines. Combined with other knowledge, the collected data and generated knowledge and models can be used to design artificial agents, playing the roles of policy makers. The results attained by human players can be used to configure agents on a scale ranging from a focus on one particular nexus domain to full awareness of all domains. In this setting, users can select a role as policy maker in one of the nexus domains and then play the game as in the multiplayer setting. Moreover, game designers and academics are able to understand the usual behaviour of specific stakeholders, helping them tailor the contents of the game sessions to better fit their needs and to fulfil the learning goals.

3. Results from the Sardinia Fast-Track

Here, results from the fast-track SDM for Sardinia are shown. Since the Sardinia island-wide reservoir system was represented for simplicity as an aggregated system (i.e., geographically lumped; see the conceptual framework in Figure 3 and the model schema in Figure 4), an overestimated efficiency of the system to capture and manage water was introduced, thereby overestimating resilience to changes in water supply and demand (in Mm^3 $month^{-1}$). However, the model did accurately capture the *pattern* of reservoir filling and emptying through the year (Figure 8a), and it captured the fact that energy supply and demand (in GWh $month^{-1}$) are balanced over a year (Figure 8b).

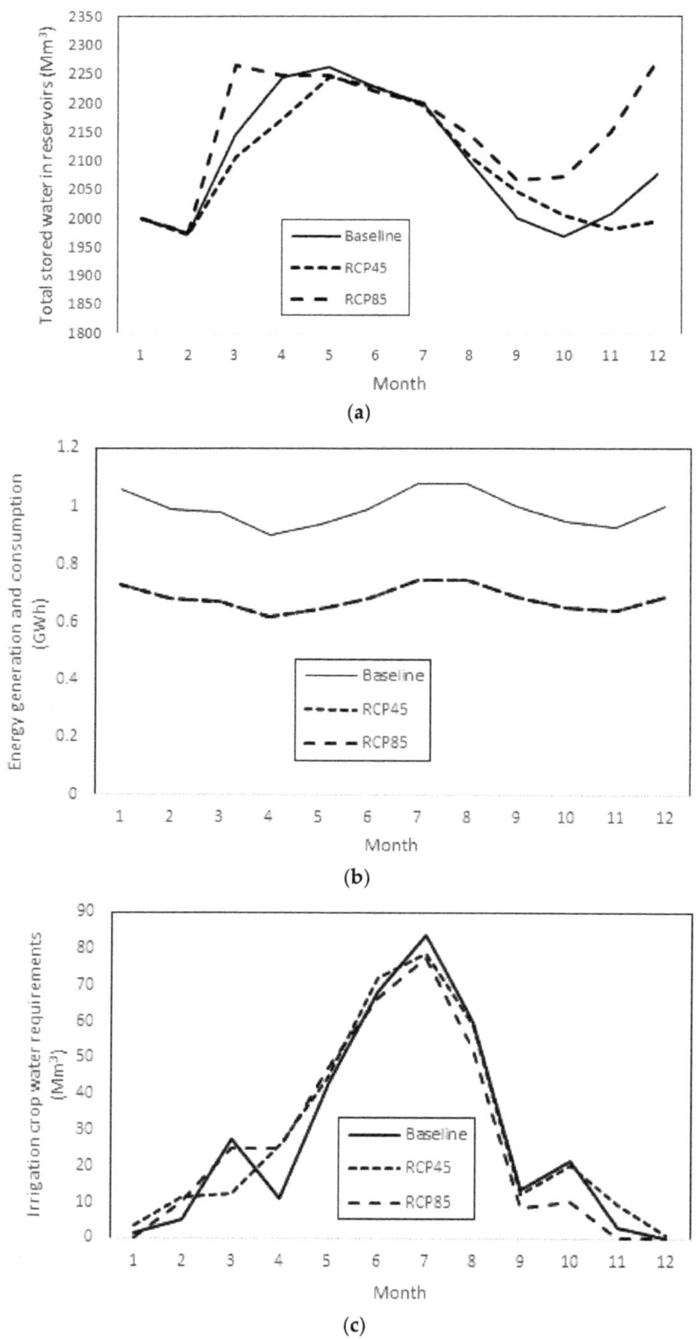

Figure 8. (**a**) Showing the results of the Sardinia fast-track simulation for the baseline (2010) and for 2030 under RCP4.5 and 8.5. Results show the total water stored in all Sardinian reservoirs. (**b**) Energy generation and consumption for the baseline and both RCP scenarios. Energy generation equals consumption in each month. The predicted generation/consumption in the RCPs are the same. (**c**) Irrigation crop water requirements.

An interesting feature regarding energy generation is the predicted reduction in 2030. This is largely due to a reduction in fossil-based energy generation, which is not fully replaced by greener energy sources. Demand is still expected to be met, but increasing pressure from the tourist sector, especially in the summer months, could strain the system. It could be expected that the replacement with a greener energy system may take longer than planned if system strain becomes too large.

Figure 8c shows the modelled baseline and future irrigation crop water requirements. The irrigation requirements are similar in terms of the summer peak in all three simulations; however, the annual totals are slightly higher in the RCPs due to the increase in requirement in the spring months. It is expected that these requirements will increase further by 2050, increasing pressure on the water system and on the energy system for increasing pumping.

As an additional element for the Sardinia fast-track, a working prototype front-end for the serious game has been developed (Figure 9). The prototype includes a photo-realistic map of the study area, together with a toggle display of key nexus indicator values. In Figure 9, example indicators for the water sector are displayed. A timeline slider is shown at the top illustrating how far through the game the player is. Future extensions will include boxes/options to adjust various key nexus and policy parameters and to observe the nexus-wide impacts of these changes. It is noted that Figure 9 is a prototype. It will be further developed and refined throughout SIM4NEXUS. It is shown here as an illustrative example of what the final serious game front-ends may look like.

Figure 9. Screenshot from the developed prototype serious game front-end for the Sardinia fast-track case study.

4. Lessons Learned and Future Developments in SIM4NEXUS

The general process of serious game development in SIM4NEXUS has been presented, going from initial case study nexus framing and conceptualization, through to serious game development. Two key findings and general conclusions are found:

(1) Collaborative and constructive stakeholder involvement from project inception is critical in the development of meaningful models and games that are more likely to have impact. Suitable policy analysis is an important aspect in appropriate case study framing and model/game design and;

(2) Appropriate spatial scale selection for the case study and nexus sectors is important in producing reliable model outputs. Deciding on the spatial scale and potential disaggregation of a case study is intimately linked to Lesson 1 above.

These lessons were learned from direct feedback from stakeholders regarding model outputs to case study leaders and model/game developers, from case study leaders themselves and from academics within the project through development iterations. At present, feedback is based on the model outputs: the game is not yet developed due to the stage within the project, although discussions on game expectations are ongoing.

It is shown through the Sardinia fast-track (and reinforced as more case studies get further into the development process) how stakeholder interaction, contribution and verification of model veracity are critical from the very beginning, the value of which can be overlooked in serious game development. Case study-level policy analysis is central in guiding case study development, another element that can easily be neglected, but something that is centrally built into SIM4NEXUS. Critical and constructive stakeholder involvement, along with relevant and detailed policy analysis to aid case study framing and gameplay, are key conclusions that can be adopted in many serious game development exercises, but they must be well thought through and implemented from early stages.

The second key conclusion, linked to the first, is deciding on appropriate levels of model/game spatial disaggregation. Discovering the appropriate spatial scale might depend on a number of iterations, but will depend on the feedback from local expert stakeholders. It was only during the verification process for the Sardinia fast-track that it was found that while the timing of reservoir filling and emptying processes was correct, the modelled volume of water entering the system was vastly overestimated. This would have had significant implications for any subsequent analysis and potential policy recommendations. As a result, for the final SIM4NEXUS Sardinia model, the island will be disaggregated into a number of hydrological basins to better reflect spatial variability. Cross-checking with stakeholders will remain an integral element. As seen from the Sardinia fast-track results, it is important to articulate correctly the spatial scale of any model for the case study under consideration and for the nexus sectors being modelled. In future developments, Sardinia will be split into seven hydrological basins to better capture hydrological dynamics across the island. Aggregation to the island scale will take place a posteriori.

During the Sardinia fast-track, a number of different groups and individuals were involved at different stages in the process. Without this cooperation, the process could easily stall or fail, or a poor/unrepresentative model could be developed, with consequences for the robustness and veracity of the messages being communicated by the game. It could also mean limited uptake and therefore limited impact of the game and the results. It is important to invest time and resources in bringing together individuals with different experience, integrating their knowledge and focusing discussions on interactions between different sectors. As anticipated, difficulties emerged during stakeholder interactions with respect to expanding the experience of individuals beyond their specific sectors (i.e., breaking 'silo-thinking'). Only by including stakeholders in cooperative forums could these difficulties start to be addressed and compromises found. It is expected that similarly diverse groups will contribute to the full development of all 12 SIM4NEXUS models and serious games. Indeed, other case studies (Andalucía, Greece, The Netherlands, Latvia and the southwest U.K.) are all currently at various stages of development within the project. In each case, a wide range of relevant, but diverse stakeholders is actively involved to determine: (i) the case study nexus sectors(s) of key importance based on locally-relevant policy issues; (ii) critical interactions between nexus sectors; (iii) data availability; (iv) the main (policy-relevant) questions that the game(s) should try to address; (v) what the game might look like, how it might be played and the added value for participants; and (vi) how best to market and 'sell' the game in order to reach as wide and diverse an audience as possible. These lessons are generally applicable for other serious game development projects.

The lesson on correct spatial representation is important because better information and knowledge will be generated for policy and decision making, with results more accurately expressing on-the-ground situations. By sticking with a lumped approach, model results would not have been representative, and therefore, confidence in the results and the implications for nexus sectors discovered during game play would not be taken seriously. This finding has already had important implications for model development in other cases (e.g., Andalucía, Greece, southwest England), where accounting for local level dynamics will be critical for modelling success and by extension for providing the end-user of the serious game(s) with robust policy-relevant messages. By taking the pilot case seriously and rigorously assessing the outputs, important implications were highlighted. By intense and ongoing cross-project communication, these important lessons were noted and are being implemented by other project case studies. Within-project learning also follows as a critical lesson. As with the first key lesson, this lesson of tailoring spatial detail appropriate to the study under consideration is generally applicable, with the appropriate scale perhaps only becoming apparent after trial-and-error and the analysis of initial results.

The findings resulting from the SIM4NEXUS fast-track are more generally applicable. This is not to say that implementing these lessons is easy, but where possible, time and effort should be appropriately invested in these activities to ensure robust and policy-relevant modelling activities and serious game development (including the aims, objectives and game-play criteria). These lessons are being implemented within SIM4NEXUS and could prove useful for similar coupled modelling-serious gaming exercises.

Acknowledgments: The work described in this paper has been conducted within the project SIM4NEXUS. This project has received funding from the European Union's Horizon 2020 research and innovation programme under Grant Agreement No. 689150 SIM4NEXUS. This paper and the content included in it do not represent the opinion of the European Union, and the European Union is not responsible for any use that might be made of its content. We also thank two anonymous reviewers whose comments contributed to significantly improving the quality of this paper.

Author Contributions: All authors cconceived and designed the simulations and serious game development concept. J.S., B.E., S.M. and A.T. performed the model simulations. J.S., A.T. and S.M. analyzed the data; C.C. and X.D. contributed analysis tools (the Knowledge Elicitation Engine, and its relation to the SDM and serious game); all authors wrote the paper.

Conflicts of Interest: The authors declare no conflict of interest.

References

1. World Economic Forum. *The Global Risks Report 2016*, 11th ed.; World Economic Forum: Cologny, Switzerland, 2016; p. 103. Available online: http://wef.ch/risks2016 (accessed on 18 January 2016).
2. Hoff, H. *Understanding the Nexus*; Background paper for the Bonn2011 Conference: The Water, Energy and Food Security Nexus; Stockholm Environment Institute: Stockholm, Sweden, 2011; p. 52. Available online: https://www.sei-international.org/publications?pid=1977 (accessed on 17 June 2017).
3. WWF (World Wide Fund for Nature) & SABMiller. *The Water-Food-Energy Nexus*; Insights into Resilient Development; WWF (World Wide Fund for Nature): Gland, Switzerland; SABMiller: London, UK, 2014; p. 20. Available online: http://assets.wwf.org.uk/downloads/sab03_01_sab_wwf_project_nexus_final.pdf (accessed on 3 August 2014).
4. Sušnik, J. Economic metrics to estimate current and future resource use, with a focus on water withdrawals. *Sustain. Prod. Consum.* **2015**, *2*, 109–127. [CrossRef]
5. WWAP (World Water Assessment Programme). *The United Nations World Water Development Report 2015: Water for a Sustainable World*; UNESCO: Paris, France, 2015; p. 139. Available online: http://www.unesco.org/new/en/natural-sciences/environment/water/wwap/wwdr/2015-water-for-a-sustainable-world/ (accessed on 2 February 2017).
6. Fasel, M.; Brethaut, C.; Rouholahnejad, E.; Lacayo-Emery, M.A.; Lehmann, A. Blue water scarcity in the Black Sea catchment: Identifying key actors in the water-ecosystem-energy-food nexus. *Environ. Sci. Policy* **2016**, *66*, 140–150. [CrossRef]
7. Feng, M.; Liu, P.; Li, Z.; Zhang, J.; Liu, D.; Xiong, L. Modelling the nexus across water supply, power generation and environment systems using the system dynamics approach: Hehuang Region, China. *J. Hydrol.* **2016**. [CrossRef]

8. World Bank. *High and Dry: Climate Change, Water and the Economy*; World Bank: Washington, DC, USA, 2016; p. 69.
9. Sušnik, J.; van der Zaag, P. Correlation and causation between the UN Human Development Index and national and personal wealth and resource exploitation. *Econ. Res. (Ekonomska Istraživanja)* **2017**, *30*, 1705–1723. [CrossRef]
10. Najibi, N.; Devineni, N.; Lu, M. Hydroclimatic drivers and atmospheric teleconnections of long duration floods: An application to large reservoirs in the Missouri River Basin. *Adv. Water Res.* **2017**, *100*, 153–167. [CrossRef]
11. IMechE (Institute of Mechanical Engineers). *Global Food: Waste Not, Want Not*; IMechE: London, UK, 2013; p. 35.
12. EEA (European Environment Agency). *European Environment—State and Outlook 2015: Assessment of Global Megatrends*; European Environment Agency: Copenhagen, Denmark, 2015; p. 140. [CrossRef]
13. OECD. *The Land-Water-Energy Nexus: Biophysical and Economic Consequences*; OECD Publishing: Paris, France, 2017. [CrossRef]
14. Brouwer, F.; Avgerinopoulos, G.; Fazekas, D.; Laspidou, C.; Mercure, J.-F.; Pollitt, H.; Ramos, E.P.; Howells, M. Energy modelling and the Nexus concept. *Energy Strateg. Rev.* **2017**. [CrossRef]
15. Simonovic, S.P. World water dynamics: Global modelling of water resources. *J. Environ. Manag.* **2002**, *66*, 249–267. [CrossRef]
16. Hussein, W.A.; Memon, F.A.; Savić, D.A. An integrated model to evaluate water-energy-food nexus at a household scale. *Environ. Model. Softw.* **2017**, *93*, 366–380. [CrossRef]
17. Valek, A.M.; Sušnik, J.; Grafakos, S. Quantification of the urban water-energy nexus in México City, México, with an assessment of water-system related carbon emissions. *Sci. Total Environ.* **2017**, *590*, 258–268. [CrossRef] [PubMed]
18. Small, G.; Vorgan, G. Meet your iBrain. *Sci. Am. Mind* **2008**, *19*, 42–49. [CrossRef]
19. Gee, J.P. What video games have to teach us about learning and literacy. In *Theoretical and Practical Computer Applications in Entertainment*; ACM: New York, NY, USA, 2003; Volume 1, p. 20. [CrossRef]
20. Mayer, I.S. The gaming of policy and the politics of gaming: A review. *Simul. Gaming* **2009**, *40*, 825–862. [CrossRef]
21. Savić, D.A.; Morley, M.S.; Khoury, M. Serious Gaming for Water Systems Planning and Management. *Water* **2016**, *8*, 456. [CrossRef]
22. Salen, K.; Zimmerman, E. *Rules of Play: Game Design Fundamentals*; The MIT Press: Cambridge, MA, USA; London, UK, 2004; p. 668.
23. Abt, C.C. *Serious Games*; Viking: New York, NY, USA, 1970.
24. Van der Graaf, J.; Segers, E.; Verhoeven, L. Discovering the laws of physics with a serious game in kindergarten. *Comput. Educ.* **2016**, *101*, 168–178. [CrossRef]
25. Gugerell, K.; Zuidema, C. Gaming for the energy transition: Experimenting and learning in co-designing a serious game prototype. *J. Clean. Prod.* **2017**, *169*, 105–116. [CrossRef]
26. Kang, J.; Liu, M.; Qu, W. Using gameplay data to examine learning behaviour patterns in a serious game. *Comput. Hum. Behav.* **2017**, *72*, 757–770. [CrossRef]
27. Planchon, J.; Vacher, V.; Comblet, J.; Rabatel, E.; Darses, F.; Mignon, A.; Pasquier, P. Serious game training improves performance in combat life-saving situations. *Injury* **2017**, *49*, 86–92. [CrossRef] [PubMed]
28. Tan, A.J.Q.; Lee, C.C.S.; Lin, P.Y.; Cooper, S.; Lau, L.S.T.; Chua, W.L.; Liaw, S.Y. Designing and evaluating the effectiveness of a serious game for safe administration of blood transfusion: A randomized controlled trial. *Nurse Educ. Today* **2017**, *55*, 38–44. [CrossRef] [PubMed]
29. Van der Wal, M.M.; de Kraker, J.; Kroeze, C.; Kirschner, P.A.; Valkering, P. Can computer models be used for social learning? A serious game in water management. *Environ. Model. Softw.* **2016**, *75*, 119–132. [CrossRef]
30. Medema, W.; Furber, A.; Adamowski, J.; Zhou, Q.; Mayer, I. Exploring the Potential Impact of Serious Games on Social Learning and Stakeholder Collaborations for Transboundary Watershed Management of the St. Lawrence River Basin. *Water* **2016**, *8*, 175. [CrossRef]
31. Gurung, T.R.; Bousquet, F.; Trebuil, G. Companion modelling, conflict resolution, and institution building: Sharing irrigation water in the Lingmuteychu watershed, Bhutan. *Ecol. Soc.* **2006**, *11*, 36. [CrossRef]
32. Hagen, E. New approaches in the Potomac river basin and beyond—Pioneering work in the development of shared vision planning. In *Converging Waters: Integrating Collaborative Modeling with Participatory Processes to Make Water Resources Decisions*; Bourget, E., Ed.; Institute for Water Resources, U.S. Army Corps of Engineers: Alexandria, VA, USA, 2011; pp. 35–58.

33. Mereu, S.; Sušnik, J.; Trabucco, A.; Daccache, A.; Vamvakeridou-Lyroudia, L.; Renoldi, S.; Virdis, A.; Savić, D.A.; Assimacopoulos, D. Operational resilience of reservoirs to climate change, agricultural demand, and tourism: A case study from Sardinia. *Sci. Total Environ.* **2015**, *543*, 1028–1038. [CrossRef] [PubMed]
34. Britz, W.; Witzke, H.P. *CAPRI Model Documentation*; Institute for Food and Resource Economics University of Bonn: Bonn, Germany, 2014.
35. Cambridge Econometrics. *E3ME Technical Manual Version 6*; Cambridge Econometrics: Cambridge, UK, 2014; p. 136. Available online: www.camecon.com/how/e3me-model/ (accessed on 12 November 2017).
36. Hempel, S.; Frieler, K.; Warszawski, L.; Schewe, J.; Piontek, F. A trend preserving bias correction—the ISI-MIP approach. *Earth Syst. Dyn.* **2013**, *4*, 219–236. [CrossRef]
37. Van Vuuren, D.P.; Edmonds, J.; Kainuma, M.; Riahi, K.; Thomson, A.; Hibbard, K.; Hurtt, G.C.; Kram, T.; Krey, V.; Lamarque, J.F.; et al. The representative concentration pathways: An overview. *Clim. Chang.* **2011**, *109*, 5–31. [CrossRef]
38. Ford, A. Modeling the Environment. In *An Introduction to System Dynamics Modeling of Environmental Systems*; Island Press: Washington, DC, USA, 1999; p. 401.
39. Rehan, R.; Knight, M.A.; Haas, C.T.; Unger, A.J.A. Application of system dynamics for developing financially self-sustaining management policies for water and wastewater systems. *Water Res.* **2011**, *45*, 4737–4750. [CrossRef] [PubMed]
40. Sahin, O.; Siems, R.S.; Stewart, R.A.; Porter, M.G. Paradigm shift to enhanced water supply planning through augmented grids, scarcity pricing and adaptive factory water: A system dynamics approach. *Environ. Model. Softw.* **2014**, *75*, 348–361. [CrossRef]
41. Sušnik, J.; Vamvakeridou-Lyroudia, L.S.; Savić, D.A.; Kapelan, Z. Integrated modelling of a coupled water-agricultural system using system dynamics. *J. Water Clim. Chang.* **2013**, *4*, 209–231. [CrossRef]
42. Tidwell, V.C.; Passell, H.D.; Conrad, S.H.; Thomas, R.P. System dynamics modeling for community-based water planning: Application to the Middle Rio Grande. *Aquat. Sci.* **2004**, *66*, 357–372. [CrossRef]
43. Mieczkowski, Z. The Tourism Climatic Index: A Method of Evaluating World Climates for Tourism. *Can. Geogr.* **1985**, *29*, 220–233. [CrossRef]
44. Anzaldi, G. *Generic Ontology for Water Supply Distribution Chain*; Eurecat Technology Center: Barcelona, Spain, 2015. Available online: http://waterp-fp7.eu/Downloads/deliverables/D1.3_Generic_Ontology_for_water_supply_distribution_chain_v1.3.pdf (accessed on 15 December 2017).
45. WEFNexus Ontology. Available online: https://github.com/mizanur3/WEFNexus (accessed on 15 December 2017).
46. OGC (Open Geospatial Consortium). *OGC® WPS 2.0 Interface Standard*; Open Geospatial Consortium: Wayland, MA, USA, 2015. Available online: http://www.opengeospatial.org/standards/wps (accessed on 15 December 2017).
47. Chew, C.; Lloyd, G.J.; Knudsen, E. Capacity building in water with serious games. In *Subconscious Learning via Games and Social Media*; Sourina, O., Wortley, D., Kim, S., Eds.; Springer: Singapore, 2015; pp. 27–43.
48. O'Neill, B.C.; Kriegler, E.; Ebi, K.L.; Kemp-Benedict, E.; Riahi, K.; Rothman, D.S.; van Ruijven, B.J.; van Vuuren, D.P.; Birkmann, J.; Kok, K.; et al. The roads ahead: Narratives for shared socioeconomic pathways describing world futures in the 21st century. *Glob. Environ. Chang.* **2015**, *42*, 169–180. [CrossRef]

water

MDPI

Article

Models, Simulations and Games for Water Management: A Comparative Q-Method Study in The Netherlands and China

Qiqi Zhou [1,*] and Igor Stefan Mayer [2]

[1] TIAS School for Business and Society, Tilburg University, Warandelaan 2, 5037 AB Tilburg, The Netherlands
[2] Academy for Digital Entertainment, NHTV Breda University of Applied Sciences, Monseigneur Hopmansstraat 1, 4817 JT Breda, The Netherlands; i.s.mayer@hotmail.com
* Correspondence: qiqi.zhou@tias.edu

Received: 4 October 2017; Accepted: 13 December 2017; Published: 24 December 2017

Abstract: How do policy analysts perceive the various roles that Models, Simulations and Games (MSG) have, or can have in Integrated Water Resources Management (IWRM)? Fifty-five policy analysts in water management in The Netherlands and China were interviewed, following the procedure of the Q-method. Comparative analysis of the combined quantitative and qualitative data show that: (1) The debate on the role of MSG for IWRM is structured around five frames in The Netherlands and three frames in China. (2) The frames in The Netherlands and China are significantly different. (3) In China, there is a predominant frame that perceives MSG for IWRM as data driven simulation technology for rationalization of water management, which is less significant in The Netherlands. (4) The reverse is true with regard to MSG for stakeholder interaction, learning and integrated assessment, which are significant frames in The Netherlands, but not in China. The conclusion is that frame differences can easily confuse professional and academic debate about MSG for water management; within the same institutional and cultural context, but even more so in Netherlands–China co-operation projects. Frames are also relevant when designing, using or evaluating innovative methods for integrated water resources management.

Keywords: simulations; serious games; Q-method; integrated water resources management; policy analysis

1. Introduction

How do policy analysts perceive the various roles that Models, Simulations and Games (MSG) have, or can have for public policy making in general, and for Integrated Water Resources Management (IWRM) in particular? What do policy analysts in water management expect of MSG, if anything at all? These questions are relevant in light of a growing advocacy to develop and apply advanced models, simulations and (serious) games, in particular, for IWRM [1–5].

The professional and academic debate on the use and usefulness of MSG for policy-making and management in general has quite a tradition [6,7]; however, it is also rife with strong and diverging opinions about what MSG can or should deliver. Some expect nothing less than a digital revolution in the way we make, or will make policy decisions; a Futurium where scientific evidence and public participation are fully integrated [8]. In the near or more distant future, we may well be managing natural systems, such as oceans and river basins, through advanced and highly interactive and participatory management and control systems, built upon big data, geoinformation systems, artificial intelligence, scientific models and simulations combined with advanced game-technology, and Virtual and Augmented Reality (VR and AR, respectively) [9–12]. This is also the rationale behind

"Understanding Game-based Approaches for Improving Sustainable Water Governance: The Potential of Serious Games to Solve Water Problems", the theme of this Special Issue.

Others claim that such technological innovations are rather trivial, because they are not well aligned with the true and underlying reality of politics and management; the blunt exercise of power to protect particular interests [13]; and economic growth and the accumulation of capital. Analytical methods in the policy process, including MSG, may be seen as irrelevant [14], or as an intricate form of manipulation [15,16]. Conversely, they may be seen as a last resort to rescue evidence-based policy making from the growing influence of a "post-truth" belief system [17]. From that perspective, the advocacy for things such as serious games, gamification and virtual reality can be seen as the response of policy analysis to a growing social need to get "engaged and entertained", also in public policy making.

Other voices start from the normative end. The urgency and complexity of contemporary problems—climate change and the rapid degradation of natural systems in particular—necessitates a radical transformation of the way we make policy decisions [18,19]. This has led to a strong call for a post-normal science [20]. Holistic interventions should drive for transformative change at the level of values and fundamental beliefs of stakeholders [21,22]. Advocates of such view commonly expect a lot of social learning with or without the support of interactive methods (serious games), immersive technologies (VR/AR, Virtual Reality/ Augmented Reality) and big data [2,23–25].

Amidst such diverging voices, empirical evidence about the use and usefulness of MSG for transformative policy making does not really take us much further, unfortunately. There is a considerable and fast-growing list of research publications where MSG have been used for IWRM (e.g., [3,10,12,26–39]). There are a few observations that can be taken from this literature. Most of the empirical research is done in an educational context, often with university students as players. Professionals in such game sessions are usually experts who work at the boundary of science and policy; academics, consultants, and strategy officers. Game sessions are often part of a general conference, or a dedicated seminar or workshop with reflective and social purposes. Most of the empirical studies can be classified as case-based, tool-oriented design and validation studies. Learning effects are mainly considered at the individual level, or player-group level [3,37]. Some MSG have been used effectively however as intervention method with local stakeholders at a community level, for instance with local farmers or fishermen in rural areas of developing countries. See for a few examples: [33,36,40,41].

The most significant observation however is that research on how MSG actually contributes to, or even influences, policy making and management is scarce, perhaps because it requires an evaluation type of research that is quite difficult to set up. It would need to build on a comparative analysis of a rich and varied set of cases where such innovative approaches have been used for policy making, over a longer period (for a few exceptions, see [5,6]). Furthermore, how do we observe and value the different functions of policy analysis against one another; their instrumental, communicative, enlightenment and empowerment functions [42,43]?

The objective of this paper therefore is to take a step back by reconstructing the different fundamental and coherent views, called frames, that policy-analysts and policy makers have about the use and usefulness of MSG for IWRM. To do so systematically and profoundly, we use the Q-method, explained in greater detail in the Materials and Methods Section. There are three key concepts that provide the basis for the operationalization of statement in the Q-study: (1) Policy Analysis (PA); (2) Integrated Water Resource Management (IWRM); and (3) Models, Simulations and (Serious) Games (MSG).

Policy analysis is defined by Dunn [44] as "an intellectual and practical activity aimed at creating, critically assessing, and communicating knowledge of and in the policy-making process". Policy analysis finds itself at the heart of the Science–Policy Interface (SPI) [13,45–48]. This is the boundary sphere were scientific and other knowledge oriented institutions interact closely with political institutions and public agencies. Research has shown that the different rationalities in the world of science and the world of politics lead to many kinds of tensions, such as the risk that results of studies tend to come too late, and after the political facts. These boundary tensions need to be managed and mediated roughly by accommodating policymaking to science (speaking truth to power) or, vice versa, by accommodating science to policymaking (sciences that serve power) [49–51]. Models, Simulations and Games (MSG) are among the more sophisticated methods in the policy analysis' tool box, and as such subject to many of the tensions at the SPI [45,48,52]. The boundary tensions with regard to the use and usefulness of MSG for policy making are the focus of our research.

Integrated Water Resources Management is the locus—the domain application area—of our research. IWRM has been defined by the Global Water Partnership (GWP) [53] as "a process which promotes the coordinated development and management of water, land and related resources, in order to maximize the resultant economic and social welfare in an equitable manner without compromising the sustainability of vital ecosystems". In this research, we will consider two water management regimes, in The Netherlands and China, which may be very different in the way they view the science–policy interface in IWRM.

According to some authors, IWRM is closely connected to the transformation of a "prediction and control" regime to an "adaptive, learning" type of regime [54]. Similar to major regions in China, The Netherlands is a river delta, although of a different size than for instance the Pearl River delta. Most of The Netherlands lies below sea level and a substantial proportion of land has been won from the sea through land reclamation and the building of dikes. Since the early 1990s, however, there is an ongoing paradigm shift from engineering-driven solutions (more, higher and stronger levees) to solutions where nature, technology and human aspects are in balance. Since the "Room for the River" policy in the early 1990s, IWRM is well established in The Netherlands. Due to global environmental changes, flooding is becoming a major problem in other parts of the world too, where there is much less expertise in water management. China is one of the countries to which The Netherlands actively disseminates and exports the principles and methods of IWRM, along with many of the tools to realize it. With the increasing occurrence and scale of flooding, the central government of the Peoples Republic of China (PRC) is reforming from a highly fragmented and engineering driven water management administration, towards an IWRM approach. The gives an excellent opportunity to compare the two countries. We will therefore reconstruct the different frames of policy analysts and policy makers in the two countries with respect to IWRM on aspects such as governance, sectoral and functional integration—and examine if and how these frames influence their perceptions on the use and usefulness of MSG. We intentionally do not narrow down the understanding of IWRM—s.a. into urban/rural or water quality/quantity—as this would need to come to the surface in the interviews.

Models, Simulations and (Serious) Games is increasingly being used as a general reference term for a varied cluster of related methodologies and techniques for understanding and managing complexity in a context of decision-making. This includes systems dynamics, agent-based modelling, Geographical Information Systems (GIS) and gaming-simulation. More recent innovations in this area are (serious) gaming, Virtual Reality (VR) and Augmented Reality (AR). There is considerable overlap and cross fertilization among these methods.

The term model mainly refers to the manner in which a part of reality—called the "reference system"—is captured and represented, in a mental, physical, digital or virtual form. Simulation mainly refers to the manner in which the dynamic behavior of the reference system is represented in such a way that it enables human experimentation and manipulation over time. When one or more human actors can give input into the simulation—the user can play with certain input parameters—it is not uncommon to call it an interactive simulation. When multiple users interact with the simulation as well as with each other, it is commonly called a social simulation. When there is also a defined set of rules for human–computer or human–human interaction, in the form of play rules, player roles, player goals and a narrative, it becomes a simulation-game or a serious game. The term simulation-game or serious game generally refers to the societal utility of games, the game technology, as well as the game principles and the game formats. This utility of games is always mediated through a form of learning or change, at the individual, organizational or system level. The almost infinite number of ways how external purposes, applications domains, game technologies and game mechanics, can be combined creates an enormous variety in serious games. This also explains why they are also called under different names, such as policy exercises or gamification [55]. In this study, we will consider all kinds of genres of serious games—including interactive models, simulations and VR/AR—designed and used for IWRM, be it learning, research, validation, intervention and more.

2. Materials and Methods

2.1. Q-Method

A comparative Q-method study in The Netherlands and China was designed and executed to understand how policy analysts—e.g., the experts, policy makers, managers, etc. who operate at the science–policy interface—frame the use and usefulness of MSG for IWRM. The Q-method is particularly suited to uncover the underlying structure of human subjectivity around a certain topic [56]. It combines qualitative and quantitative data collected in personal interviews, and works as follows:

- Collection of several statements (the Q-set, usually between 30 and 40) that are representative for diverging views on a certain topic.
- Identification of respondents (P-set, usually between 25 and 40) who are representative for the diverging viewpoints.
- Personal interviews in which each respondent ranks the collection of statements (Q-sort) on a scale (Q-scale, −3 to +3, strongly disagree–strongly agree), thus laying out a quasi-normal distribution of the cards, as in Figure 1, while giving detailed explanations and arguments to the interviewer.
- Statistical analysis (an "inverted factor analysis") of the Q-sorts (called the Q-data) and coding of the qualitative information.
- Reconstruction of a limited set of frames based on the factor analysis but interpreted and labelled with the qualitative data.

2.2. Q-Set

The set of statements—called the Q-sort—was collected by scanning the literature in the widest and most diverse coverage possible, as well as interviewing a few experts. The initial Q-sample consisted of sixty-five statements divided over the three main issues IWRM, Policy Analysis and MSG. The initial statements were improved and reduced to forty-three statements in several Q-trials with peer experts. The statements were then translated from English to Dutch and Chinese. The quality of the translation was verified in a few more trials with bilingual colleagues, with English or Chinese as a native language. The final set of Q-statements was put on laminated cards, as well as into the online Flash Q system for online use, mainly in China. To make the interviews in China shorter and more practical, it later proved necessary to reduce the number of statements to thirty-four and adapt the Q-scale accordingly. Although this affected comparability, the Q method is positioned as a flexible method that relies heavily on the explanations that the respondents give for their positioning of the Q-statements.

2.3. P-Set

The respondents in The Netherlands and China were senior professionals, mostly male, with quite a lot of work experience in the field. Both in China and in The Netherlands, we selected respondents from all relevant fields and points of view: water policy makers, water managers, water stakeholders, water researchers and consultants, MSG developers and users. In The Netherlands, we approached candidates for an interview through organizations such as the ministry, the water boards, provincial and local governments, universities, consultant and engineering companies. The long list of candidates went up to around fifty in The Netherlands, of which thirty-three were interviewed with a fully completed Q-sort.

Interviewees in China were selected through a similar procedure, making use of our professional contacts in cities of Dalian, Nanjing and Guangzhou. We contacted officials in the local municipality, water management sectors, infrastructure design institutes, water research institutes and universities. Finally, we had useful interviews with twenty-two persons from different institutions who also completed the Q-sort. All interviews were done by the first author, born and raised in China, educated in universities in The Netherlands and proficient in the Dutch and English language.

The whole interview would take between one and three hours, with the card sorting consuming about an hour. In The Netherlands, all but five respondents used the card version. In China, we only used the digital Flash Q-software version (not the cards) during live, personal interviews. At the end of the interview, respondents were asked to answer some questions related to their professional affiliation, their expertise, and knowledge about MSG and additional things if there were any. We coded all additional information and inserted the data into one database. All interviews were fully transcribed. After some introduction of interviewer, objective and technique, the respondents were asked to read (aloud) the statements on the cards and subsequently rank the cards in a quasi-normal distribution (scale from −3 to +3). Figure 1 gives an impression of a quasi-normal distribution as in this research in paper-based form.

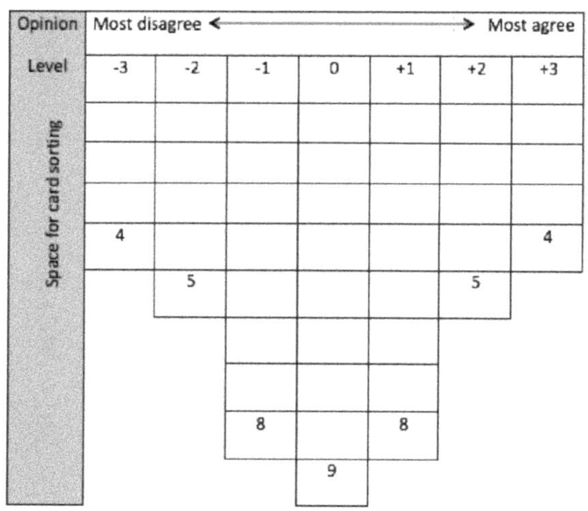

Figure 1. Card sorting example.

2.4. Analysis

The results of the Q sorts were analyzed with factor analysis, supported with the software PQ Method. With the factor analysis (Varimax rotation) respondents with the same answer patterns were grouped together. The factor analysis extracts factors on two layers, a frame about IWRM and a frame about MSG, that matched together. The frames suggested by the statistical analysis were combined with coded background data (position, work field, experience with MSG, etc.), general questions (familiarity with and trust in MSG) as well as quotations from the interviews.

3. Results

3.1. The Netherlands

Twenty-three out of thirty-three respondents in The Netherlands had quite some experience with the use of MSG for policymaking. Three respondents, all in frame NL1 (see below) had "no trust at all" in MSGs and one had a "little doubt". The respondents could reflect on the conditional usefulness of MSG, such as "when it is based on scientific data", "when it is used with the right participants", or "yes, but it is costly and time consuming". From the analysis, five frames are distinguished:

- Frame NL1: Bureaucratic alignment
- Frame NL2: Stakeholder interaction
- Frame NL3: Learning
- Frame NL4: Uncertainty
- Frame NL5: Science vs. emotions.

3.1.1. Frame NL1: Bureaucratic Alignment

Frame NL1 is represented by eleven respondents, three respondents belonging to the world of science, and eight to the world of politics. Table 1 shows the distinguished statements about IWRM in this frame. Distinguished statements combine extreme answers (−3; +3) with strong factor loadings.

Table 1. Distinguished statements about IWRM in frame NL1.

Number	Statement	Ranking
16	Reinforcing levees (dikes) etc., is insufficient to keep The Netherlands safe from flooding in the 21st century.	+3
18	Socioeconomic developments in flood-prone areas should be mitigated through spatial planning and construction regulation.	+3
11	A strong degree of integration of water management and spatial planning at different administrative and spatial levels is crucial for water management.	+3
5	Uncertainty in water management is deepened by a lack of integration among social, political, technological, ecological, economic (etc.) knowledge.	−3
7	Water managers should set more 'social learning' activities on their agenda.	−3
8	A centralized form of governance, with sufficient authority and decision power at the national level, is crucial for water management.	0

The statements give a rather clear picture of how the respondents frame the issue of IWRM, namely as predominantly a matter of integrating engineering works and infrastructures with spatial planning, regulation and levels of governance. Water management at the different administrative levels of governance should be integrated. At the other extreme, the integration of stakeholders' values and things like collaborative learning and stakeholder participation are most strongly rejected. We therefore label this frame "bureaucratic alignment", because it is mainly the administrators and experts who integrate (or should integrate) their expert knowledge and authority. One interviewee who scored high on frame NL1 expressed quite honestly why he rejects learning and stakeholder interaction:

> "At the national level, the spatial solution has been chosen as the necessary solution to flood management in the future. However, until now the top-down decision is not fully accepted and cooperated with, at regional and local levels. [...] To deal with these issues you need to have a clear understanding of the power and interest of the stakeholders. [...] When people in a group are against the spatial solution you brought, it is hard to talk to and convince them. However, if you come in the evening to communicate individually, the chance of negotiation will increase. They will ask you what your offer is." (Interviewee No. 16).

In other words, agreement with stakeholders can better be reached through power, by making "package deals" or by "compensation". Top-down decisions simply need to be sold locally. There is little need for or influence of learning, since stakeholders already know what they do, and do not want. Now, the question should be asked what the role of integrated approaches and MSG methods is. Probably not a lot, except for making package deals and getting things done, preferably behind closed doors. This is reflected in the respondents' framing of MSG in IWRM. Five respondents had no experience of MSG; one had, but does not trust it; and a few others seemed politely skeptical. Table 2 shows the distinguished statements about MSG presented in this frame.

Table 2. Distinguished statements about MSG in frame NL1.

Number	Statement	Ranking
23	The key function of 'policy analysis' is to support the stakeholders' learning process.	−3
31	Policy simulation does not need to be computerized. 'Low-tech' gaming based on human behavior is also a scientifically proven method for water policy analysis.	−2
43	Playing together in a simulation game increases the stakeholders' willingness to cooperate in the real world.	−2
36	Visualization (e.g., by pictures, animations or 3D graphics) significantly increases the users' understanding of models and simulations.	+3
39	Computer simulations can accommodate poorly with conflicting values and interests of stakeholders in water management and water policy.	−3

In line with the bureaucratic tendency in this frame and the rejection of things such as learning and stakeholder participation, low-tech games and human play are regarded as not very useful. From the interviews, it becomes clear that in frame NL1 the bureaucratic process can best be supported with simulation and visualization technology. Think of models that can calculate the consequences of plans, or visualization tools to integrate plans in 3D. Such analyses are powerful for bureaucratic integration and when negotiating with stakeholders. The social learning value of MSG, however, is deemed irrelevant and impossible in this frame. The reason for this is nicely illustrated in the following quotations:

> "Simulating the richness of social values is impossible because a lot of social values, individual values, are not possible to involve in the model. […] A spatial solution needs people's property. They have different reasons to refuse to give you their property. […] How can you explore those through gaming? […] In such an environment the most important issues of integration are network cooperation and visualization technology. Visualization increases the policymakers' understanding of technical analysis and therefore contributes to cooperation." (Interviewee No. 16).

> "Policymakers look for excuses to not to learn from the game. Gaming is not the thing to change the behaviours of individuals. However, it can be used strategically to show the community the need to make the long-term decision and stimulate the discussion." (Interviewee No. 20).

3.1.2. Frame NL2: Stakeholder Interaction

Nine respondents loaded in frame NL2. We coded four respondents as belonging to the world of science (researchers, consultants) and five as belonging to the world of politics (public policymakers, water managers). All had experience with MSG, and trusted it to be useful for policy making. We label this frame "stakeholder interaction". Table 3 shows the distinguished statements about IWRM.

Table 3. Distinguished statements about IWRM in frame NL2.

Number	Statement	Ranking
22	The key solution to the consequences of climate change lies in active public involvement and stakeholder participation. Societal interaction will provide the most significant contribution to water management and policymaking in the near future.	+3
2	The key problems in water management today are more socio-political than technological-infrastructural in nature.	+2
9	A network type of governance, with interaction between interdependent stakeholders, is crucial for water management.	+2
11	A strong degree of integration of water management and spatial planning at different administrative and spatial levels is crucial for water management.	+2
18	Socioeconomic developments in flood-prone areas should be mitigated through spatial planning and construction regulations.	0
3	The increasing complexity of society leads to a problematic compartmentalization and fragmentation in water management.	−3
8	A centralized form of governance, with sufficient authority and decision power at the national level, is crucial for water management.	−3

This frame relies more strongly than frame NL1 on a network view of society and policymaking, whereby many stakeholders are and should be involved in water management. Water management problems are more socio-political than technological–infrastructural in nature; solutions should therefore come from stakeholder interaction and public involvement. Governance should accommodate the network character of society. The centralization of governance should be strongly rejected (difference from frame NL1). Power, authority and competences are unequally distributed in the interaction between water authorities and stakeholders and other authorities: the region of a water board can be as big as ten municipalities or a province. This can cause a lack of trust and frustrate the cooperation between water boards and municipalities. This needs some time and effort to change, to build enough trust to cooperate. Table 4 shows the distinguished statements about MSG in this frame.

Table 4. Distinguished statements about MSG in frame NL2.

Number	Statement	Ranking
23	The key function of policy analysis is to support the stakeholders' learning process.	−2
31	Policy simulation does not need to be computerized. Low-tech gaming based on human behaviour is also a scientifically proven method for water policy analysis.	+3
28	The outcomes of computer simulation are generally more authoritative (trustworthy) for water policymakers and water managers than the outcomes of a simulation game with real stakeholders.	−3
42	Simulation gaming can effectively facilitate and support the interaction among stakeholders from different governance sectors.	+3
43	Playing' together in a simulation game increases the stakeholders' willingness to cooperate in the real world.	+3
37	Computer simulations for water management and water policymaking should be easy to use and understand by non-expert users.	−3

As in frame NL1, the respondents in this frame disagree that the function of policy analysis is to support a learning process. However, in contrast to frame NL1, playing together in a game is seen to have effects on stakeholders' interaction in the real world. According to frame NL2, strategic interaction in a serious game is a social intervention to further stakeholder interaction, and not a learning process (difference from frame NL3). Two quotations illustrate how respondents in this frame distinguish serious games from computer simulation, that is, they are not comparable.

3.1.3. Frame NL3: Learning

Only three respondents loaded in frame NL3, and we coded all of them as water manager. In contrast to the other frames, these three respondents put significant emphasis on learning, that is, increasing stakeholders' understanding of the complexity of water management. As in frame NL1, but in contrast to frame NL2, the three respondents in this frame do not consider problems in water management to be mainly socio-political in nature. From the statements, it also appears that respondents are rather technology-oriented (contrast with frames NL1 and NL2), and rejecting of stakeholder interaction and network governance. Table 5 shows the most distinguished statements about IWRM in this frame.

Table 5. Distinguished statements about IWRM in frame NL3.

Number	Statement	Ranking
23	The key function of policy analysis is to support the stakeholders' learning process.	+3
2	The key problems in water management today are more socio-political than technological–infrastructural in nature.	−2
22	The key solution to the consequences of climate change lies in active public involvement and stakeholder participation. Societal interaction will provide the most significant contribution to water management and policymaking in the near future.	−3
16	Reinforcing levees (dikes) etc., is insufficient to keep The Netherlands safe from flooding in the 21st century.	−2
9	A network type of governance, with interaction between interdependent stakeholders, is crucial for water management.	−1
11	A strong degree of integration of water management and spatial planning at different administrative and spatial levels is crucial for water management.	−1
29	Rational thinking should always be combined with human emotions in policy analysis for integrated water management.	−2

The following quotation illustrates the frame:

> "The water management problem is very dependent on the local conditions. Generating the solution needs a lot of local knowledge. Water governance should be more decentralized. In The Netherlands, the surroundings of the local area is really different, both the socio-political issues and the characteristics of the water problem. The solutions must satisfy the needs of local development. For example, agriculture in greenhouses is a typical economic activity in the area around [city name]. The policy strategy needs to involve the calculation of the cost and impact of policies on this activity, which is not necessary in the other areas. Central government cannot generate solutions but only make political choices. The practitioners in the local sectors are the experts to get the job done." (Interviewee No. 1).

The focus on learning in a complex technological setting, rather than stakeholder interaction, seems to be why the respondents consider the combination of computer simulation and stakeholder participation important. SG is valued as a good method for innovative learning in water management. It can be effective to analyze the future, to test policy options in a safe environment, etc. Table 6 shows the most distinguished statements about MSG in this frame.

Table 6. Distinguished statements about MSG in frame NL3.

Number	Statement	Ranking
20	Methods that combine computer simulation with stakeholder participation are supportive of water management.	+2
30	A simulation game with real stakeholders as players is generally more effective to foresee and analyse what can happen in the near future than a computer simulation.	+3
24	Most computer models are not flexible enough to deal with complex water problems. Models that can be quickly developed and changed to fit the circumstances are needed.	+3
27	Gaming simulation with real stakeholders as players is a better strategy for the innovative process than using computer simulations in integrated water management.	+2
30	A simulation game with real stakeholders as players is generally more effective to foresee and analyse what can happen in the near future than a computer simulation.	+3
34	Testing various policy options in a safe environment (such as simulation gaming with real stakeholders as players) is crucial to avoid serious consequences of water policymaking to the real world.	+2
43	Playing together in a simulation game increases the stakeholders' willingness to cooperate in the real world.	−3
37	Computer simulations for water management and water policymaking should be easy to use and understand by non-expert users.	−3

What is interesting is that the respondents strongly disagreed with the statement that gaming increases stakeholders' cooperation in the real world, whereas this statement was strongly valued in frame NL2. Upon closer inspection, however, this is not so strange. Respondents value MSG for their capacity to learn from it, and not as socio-political intervention, as in frame NL2. We believe that frame NL3 is present among a small number of experts in water management who have a focus on engineering but are open to social innovation. They believe in interactive simulations and games for learning, but reject the negotiated nonsense and the wheeling and dealing that commonly occur in interactive stakeholder processes:

> "We do not believe in the complex integrated model. You can, for example, combine the groundwater model with the surface water model. In such a model you get more parameters that can also be wrong. It will not give more certain results but create more doubt. For instance, models often give incorrect predictions of the water level rise. Simulation games should be used first to explore the possibilities. It can mean a lot at the start of a process to explore each other's views and understand the opportunities and constraints analysis. Computer simulation can be used after a game to analyse the best option." (Interviewee No. 33).

3.1.4. Frame NL4: Uncertainty

Six respondents loaded in frame NL4. We coded two respondents as belonging to the world of science (researchers) and four as belonging to the world of politics (public policymakers, water manager). Frames NL4 and NL2 have in common a strong preference for network governance, cooperation and integrated policymaking among administrative levels. Like frame NL1, frame NL4 does not agree that water management problems are more socio-political than technological-infrastructural in nature. Table 7 shows the most distinguished statements about IWRM in this frame.

Table 7. Distinguished statements about IWRM in frame NL4.

Number	Statement	Ranking
23	The key problems in water management today are more socio-political than technological–infrastructural in nature.	−1
4	There are significant uncertainties about the local and regional impacts of global climate change.	+3
17	There is a need to collaboratively find local solutions to water problems (flooding, draughts, pollution, etc.).	+2
9	A network type of governance, with interaction between interdependent stakeholders, is crucial for water management.	+3
11	A strong degree of integration of water management and spatial planning at different administrative and spatial levels is crucial for water management.	+3
12	A strong degree of cooperation among public water management authorities is crucial for water management.	+3

The main distinction between frame NL4 and the other frames lies in the emphasis put upon uncertainty and the interactions between the global and the local system. The respondents in frame NL4 markedly agree that the local and regional impacts of climate change are very uncertain and that this makes integration between planning scales and water management authorities necessary. This becomes clear in the interviews:

> "There are a lot of technical uncertainties and they are rarely communicated to policymakers. At the same time, decision makers don't like to take uncertainty into their policy. This brings the risk that we spend a lot of money on analysing the measures, which may not be as useful as we think. More effort should be made to increase the communication of uncertainty to decision makers. In this way, decisions can be made in a more robust and flexible way to deal with uncertain situations, instead of aiming to reach the number that indicates the coming water level, a goal that can be both unrealistic and risky." (Interviewee No. 15).

Respondents in this frame have a strong systems orientation towards policy analysis. They like to see the big picture and the longer term future. In contrast to frame NL3, they do not attach much value to learning from MSG, but they do seem to approach MSG as a kind of "integrated assessment". The respondents are very aware that politicians have a limited capacity to incorporate scientific information in policymaking, and that this is a problem. Enhanced cooperation and communication between the world of science and the world of politics is necessary. Table 8 shows the most distinguished statements about MSG in this frame.

Table 8. Distinguished statements about MSG in frame NL4.

Number	Statement	Ranking
6	A system approach is useful for water management only when it addresses the techno-physical and socio-political aspects in an integrated fashion.	+2
23	The key function of policy analysis is to support the stakeholders' learning process.	−3
27	Simulation gaming with real stakeholders as players is a better strategy for the innovative process than using computer simulations in integrated water management.	−2
30	A simulation game with real stakeholders as players is generally more effective to foresee and analyse what can happen in the near future than a computer simulation.	−2
39	Computer simulations can accommodate poorly with conflicting values and interests of stakeholders in water management and water policy.	−2
36	Visualization (e.g., by pictures, animations or 3D graphics) significantly increases the users' understanding of models and simulations.	+1

Quotations about MSG in frame NL4:

"Simulation should not be only technical, but also involve socio-political aspects. Computer simulation can shine a light on the conflicts. If you have a clear view on the social conflicts and values you should be able to put them in the computer simulation as well, in graphics or in other forms. But it has not been done very well yet. A lot of experience of technological development has been gained. However, social simulation is very hard because reading the exact interests of stakeholders is difficult. We don't know how far computer simulation can go, but we think technologies for such analysis have improved. But there is a lot of room to improve them further." (Interviewee No. 4).

"I do think it's useful to talk to each other and share information and ideas. But it's only good when you have a good start, already have the information and foundation. For example, the model can show which approach is more promising and do the analysis. In many cases, the information is available. You just need to study more to get it. However, the situation in The Netherlands is that in some areas they really talk too much. They have so many workshops to talk about things that are easier to study by water modelling and analysis. We think they should study more before doing the workshops, do more of the analysis." (Interviewee No. 6).

3.1.5. Frame NL5: Science vs. Emotions

Four respondents loaded in frame NL5. We coded three respondents as belonging to the world of science (researchers, consultants) and one as belonging to the world of politics (public policymaker). Table 9 shows the most distinguished statements about IWRM in this frame.

Table 9. Distinguished statements about IWRM in frame 5.

Number	Statement	Ranking
2	The key problems in water management today are more socio-political than technological–infrastructural in nature.	+3
12	A strong degree of cooperation among public water management authorities is crucial for water management.	+3
29	Rational thinking should always be combined with human emotions in policy analysis for integrated water management.	+2
16	Reinforcing levees (dikes) etc. is insufficient to keep The Netherlands safe from flooding in the 21st century.	−3

A somewhat cynical attitude towards the science–policy interface and the value of SG emerges in frame NL5. Based upon the interviews, it appears that some of the statements were answered with a kind of alternative interpretation (see quotation below). First, frame NL5 most strongly believes in "reinforcing levees" as the solution to keep The Netherlands from flooding. Secondly, and equally strongly, the frame agrees that problems in water management are more socio-political than techno-infrastructural. It becomes clear from the interviews that respondents in this frame believe that the technical solutions are available and that science has provided most of the answers, but that politicians and societal stakeholders do not listen: they should trust water experts to get the job done, but unfortunately emotions and irrationalities play too big a role. It seems that the respondents have come to accept it:

"Science and knowledge generation are not the problem in the current water decision-making process. In The Netherlands a lot of investigations have been made on scientific research for the long-term water management. The result is based on very good investigation and therefore does not need to be doubted. But on the other hand, the lack of communication of management sectors is the big problem in The Netherlands.

A lot of failures to make a decision on a development plan happened due to the lack of willingness to cooperate. It is very often that sectors make plans by themselves; there is not so much communication." (Interviewee No. 19).

"Science is no longer taken seriously enough in decision making. Emotion and power dominate the decision-making process. The politicians are not interested in rational evidence. The priority of interest and power determines what will happen. Scientific evidence can help, but it depends on the political situation. It can be easily denied if it does not match the political interest in the problem. We should move back to the situation that the socio-political power does not constrain the technical power". (Interviewee No. 11).

If this is the case, what is the role of MSG? In the interviews, some respondents indicated that it could be useful to find a balance between science and emotions. However, it is not clear whether MSG can help to find a balance. For the most part, the respondents seem to have mixed feelings about MSG. They disagree strongly that computer simulations are difficult for policy stakeholders to use and understand. They also highlight the importance of visualization to increase policymakers' understanding of models and simulation. Low-technology games are not scientifically valued (see Table 10). From the discussions, it seems that gaming is regarded as useful to reflect the human emotions (but not to change them) and it needs the support of technology to gain the trust of players if it tries to simulate the reality (see quotations below).

Table 10. Distinguished statements about MSG in frame NL5.

Number	Statement	Ranking
28	The outcomes of computer simulation are generally more authoritative (trustworthy) for water policymakers and water managers than the outcomes of a simulation game with real stakeholders.	−3
31	Policy simulation does not need to be computerized. Low-tech gaming based on human behaviour is also a scientifically proven method for water policy analysis.	−1
33	The process of decision making simulated by human players in a gaming environment is generally more useful for learning than for real policy analysis.	−2
36	Visualization (e.g., by pictures, animations or 3D graphics) significantly increases the users' understanding of models and simulations.	+3
38	Computer simulations in water management are generally difficult to use and understand by policy stakeholders.	−3

Quotations about MSG in frame NL5:

"For me, the concept of gaming means computer model based, role playing, rules and group activity. […] We never use a game in a real decision-making process. We use games in academic exercises. With the students the experience often shows the non-rational outcome, which is not what we expected. The decision always depends on the political and social power of some of the roles. We think the reason behind it is that people are selfish. If they are powerful enough they will push their selfish interest. In such a situation, gaming does not help at all. So we think gaming can help to make a quicker decision, but it does not help to make a better decision." (Interviewee No. 11).

"The dilemma is that gaming works with respondents who are willing to be involved and communicate. But if the respondents are already open and willing to interact, is the value of gaming still significant, considering the time and money consuming process to organize it?" (Interviewee No. 9).

3.1.6. Intermediate Conclusions

Through the analysis of data, we found some rationality between how the respondents view IWRM and the role that MSG can or should play. The familiarity, experience with and trust in MSG was generally high. Table A1 present an overview of the five frames. They are separated in their views on governance (hierarchy, power, and public participation?), IWRM (what is integrated? administration, sectors, disciplines, and stakeholders?), the role of science in policy making (science or evidence based, political power play, social and human factors, and role of emotions). These factors correspond with different views on the role of MSG: (1) strengthen the evidence and science base of decision-making; (2) strengthen the interaction and involvement of stakeholders; (3) for learning; (4) for integrated, long term assessment; and (5) for dealing with public emotions.

3.2. Frames in China

The twenty-two respondents in China were quite unfamiliar with MSG and therefore either curious, or skeptical about the use of MSG for IWRM. To facilitate the Q-method, we decided to show a few good international MSG examples before the interview. We identified three Chinese frames:

- Frame CH1: Doctrine of the mean
- Frame CH2: Modern and rational governors
- Frame CH3: The open-minded reformer

3.2.1. Frame CH1: The Doctrine of the Mean

This frame is shared by 12 respondents, the biggest group among the four frames. We coded all of them as belonging to the world of politics (provincial and local water managers). The attitude in this frame provides quite a good understanding of the values in the view of the doctrine of the mean: "Say as little as possible while knowing perfectly well what is wrong, to be worldly wise and play it safe". Table 11 shows the most distinguished statements about IWRM in this frame. The best form of governance is for central government to have a strong leading position, and more interaction and cooperation among different levels of government sectors. Methods to enhance cooperation are needed, but methods to analyze conflicts are not needed. The topics related to the science–policy interface, such as the impact of uncertainty and fragmentation, and matters like stakeholder participation and collaborative learning for policy analysis, are regarded as unimportant or are strongly rejected (see Table 11).

Table 11. Distinguished statements about IWRM in frame CH1.

Number	Statement	Ranking
2	The key problems in water management today are more socio-political than technological-infrastructural in nature.	+2
6	A centralized form of governance, with sufficient authority and decision power at the national level, is crucial for water management.	+2
7	A network type of governance, with interaction between interdependent stakeholders, is crucial for water management.	+1
8	There is a need for methods that can enhance the cooperation among different sectors and levels of governance in water management.	+2
9	There is a need for methods that can analyse the conflicts and cooperation among different regions in river basins.	0

Another interviewee in this frame admitted quite honestly that the interaction among government sectors is more of a network routine than for meaningful policy analysis (see quotation below).

> "Interactive, participatory policy analysis in China is still more of a 'lip service'." (Interviewee No. 11).

"The focus of water management and flood control in China is still on the development of infrastructure. [. . .] With such a focus, a centralized government is efficient. Enhancing the cooperation among sectors increases the efficiency of management. However, the advanced development in Western countries is dependent not on participation and social interaction, but on the standardization of rules. The standardization in China is still at a low level. This is the critical reason for the problems in water management." (Interviewee No. 7).

Table 12 presents the distinguished statements about MSG. It is quite obvious that in this frame MSG is neither trusted nor preferred. All the key statements about the values of MSG for policymaking, whether for providing real-life insights or for increasing cooperation, are strongly rejected. Low-tech games are certainly not a scientific method. An integrated computer model—data driven—is more trustworthy for analysis.

Table 12. Distinguished statements about MSG in frame CH1.

Number	Statement	Ranking
19	By letting stakeholders play their own role (interests, behaviour, etc.) in a gaming environment, we can simulate real problems and solutions in water management and derive valuable insights for water policymaking.	−2
25	Policy simulation does not need to be computerized. Low-tech gaming based on human behaviour is also a scientifically proven method for water policy analysis.	−2
34	Playing together in a simulation game increases the stakeholders' willingness to cooperate in the real world.	−2
33	Simulation gaming can effectively facilitate and support the interaction among stakeholders from different governance sectors.	−2
22	The outcomes of computer simulation are generally more authoritative (trustworthy) for water policymakers and water managers than the outcomes of a simulation game with real stakeholders.	+1

MSG is regarded as too Western or too early for the complex Chinese politics. This is especially the situation in the regional and local areas. The culture gap can also explain the distrust of MSG, as expressed by two of the interviewees (see quotation below).

Quotations about MSG in frame CH1

"The situation of 'treatment after pollution' is not avoidable in the developing process, which the developed countries also experienced. [. . .] However, whether the Western method of such participatory role playing game is also useful for the 'Chinese solution' is still too early to see." (Interviewee No. 18).

"The power relation and strategic game in China's policy environment is deeply embedded in its routine. Chinese politicians follow 'the doctrine of the mean' to be able to survive in the environment, which makes it impossible for them to articulate their needs and interests, and express their emotions. The Chinese political game contains many uncertain and un-parameterized variables to design a game for." (Interviewee No. 22).

3.2.2. Frame CH2: Modern and Rational Governors

Five respondents loaded in frame CH3; two of them are coded as belonging to the world of science (researchers), while the other two belong to the world of politics (regional and local policymakers). In contrast to the other frames, the five respondents in this frame gave the impression that they appreciate a network type of governance. We label this frame the "modern and rational governors": they modestly agree on the importance of socio-political problems in water management, which reflects their modern outlook. They are also rational governors, because they rely on science and strongly disagree with the contribution of public participation to solving the problem and consequences of climate change. They reject the suggestion that policy analysis should involve human and social aspects. Table 13 shows the most distinguished statements about IWRM in this frame.

Table 13. Distinguished statements about IWRM in frame CH3.

Number	Statement	Ranking
6	A centralized form of governance, with sufficient authority and decision power at the national level, is crucial for water management.	−2
2	The key problems in water management today are more socio-political than technological-infrastructural in nature.	+1
7	A network type of governance, with interaction between interdependent stakeholders, is crucial for water management.	+1
8	There is a need for methods that can enhance the cooperation among different sectors and levels of governance in water management.	+2
16	The key solution to the consequences of climate change lies in active public involvement and stakeholder participation. Societal interaction will provide the most significant contribution to water management and policymaking in the near future.	−2
23	Rational thinking should always be combined with human emotions in policy analysis for integrated water management.	−2

Quotations below gives a respondent's thought on the urgent needs to develop a rational, science-based cooperative government.

> "Rational, science-based governance is urgently needed to improve the efficiency of management. However, the bottom-up type of social participation is not a suitable method due to the very complex social situation in China. It will lead to a big crisis and loss of control when too much emotion is allowed in the policy analysis process. A good governmental regulation system based on rational priorities is the proper way to achieve better water management." (Interviewee No. 4; water manager).

Frame CH2 also agrees on such things as the need for social simulation, the limitation of using computer modelling alone, the importance of visualization and using MSG for learning rather than for policy analysis (see Table 14).

Table 14. Distinguished statements about MSG in frame CH2.

Number	Statement	Ranking
15	There is a need for socio-political simulations that provide valuable insights into the multi-actor complexity of water management.	+2
18	Most computer models are not flexible enough to deal with complex water problems. Models that can be quickly developed and changed to fit the circumstances are needed.	+2
25	Policy simulation does not need to be computerized. Low-tech gaming based on human behaviour is also a scientifically proven method for water policy analysis.	+1
30	Computer simulations for water management and water policymaking should be easy to use and understand by non-expert users.	+1
26	It is not enough to rely on computer simulation for the exploration of policy problems and the testing of policy options (even they have been developed on the basis of best-available scientific knowledge).	+1
29	Visualization (e.g., by pictures, animations or 3D graphics) significantly increases the users' understanding of models and simulations.	+1
27	The process of decision making simulated by human players in a gaming environment is generally more useful for learning than for real policy analysis.	+2

Quotations about MSG in frame CH2

> "The development and application of integrated models is quite advanced due to the large investment from the national government. However, developing socio-political

simulation is a different topic. In developed countries such as The Netherlands, they are interested because the development of infrastructures is completed. Water management can now focus more on the small-scale, 'soft' issues and use the more 'soft integrated' approaches such as gaming for less urgent issues in a long-term perspective. It is important to address the long-term planning in water management, but in China the more urgent issue is developing infrastructures, especially in the northwest area. Gaming will not be considered in these tasks. It is useful to learn new perspectives in policy analysis, but only after the fundamental structure has been completed." (Interviewee No. 8).

3.2.3. Frame CH3: The Open-Minded Reformers

Six respondents loaded in frame CH3; two of them also loaded on frame CH1. We coded two respondent as belonging to the world of science (researcher) and four as belonging to the world of politics (public policymaker). Table 15 shows the most distinguished statements about IWRM in this frame. The views of these open-minded reformers.

Table 15. Distinguished statements about IWRM in frame CH3.

Number	Statement	Ranking
4	Uncertainty in water management is deepened by a lack of integration among social, political, technological, ecological, economic, etc. knowledge.	−2
6	A centralized form of governance, with sufficient authority and decision power at the national level, is crucial for water management.	0
7	A network type of governance, with interaction between interdependent stakeholders, is crucial for water management.	+1
2	The key problems in water management today are more socio-political than technological-infrastructural in nature.	−1

This frame shows quite strong agreement with the use and usefulness of MSG for IWRM in China. Games are regarded a more social-innovative strategy for IWRM than computer simulation, and it can bring valuable insight for real-life policymaking. Besides the scientific value, MSG is also regarded useful for learning, increasing the real-life communication and willingness of collaboration among government sectors (see Table 16).

Table 16. Distinguished statements about MSG in frame CH3.

Number	Statement	Ranking
19	By letting stakeholders play their own role (interests, behaviour, etc.) in a gaming environment, we can simulate real problems and solutions in water management and derive valuable insights for water policymaking.	+2
21	Simulation gaming with real stakeholders as players is a better strategy for the innovative process than using computer simulations in integrated water management.	+2
22	The outcomes of computer simulations are generally more authoritative (trustworthy) for water policymakers and water managers than the outcomes of a simulation game with real stakeholders.	−2
25	Policy simulation does not need to be computerized. Low-tech gaming based on human behaviour is also a scientifically proven method for water policy analysis.	+1
32	Simulation gaming with real stakeholders as players integrates 'soft knowledge' from stakeholders with 'hard knowledge' from scientific research.	+2
34	Playing together in a simulation game increases the stakeholders' willingness to cooperate in the real world.	+2
33	Simulation gaming can effectively facilitate and support the interaction among stakeholders from different governance sectors.	+2
27	The process of decision making simulated by human players in a gaming environment is generally more useful for learning than for real policy analysis.	−2

Quotations about MSG in frame CH3

> "Developing methods and technology for socio-technical integration in China is only a matter of time. [...] In China there are already some demonstration projects going on at the national level, big institutions. However it is still quite new and needs more time to be introduced to the local level [...]. The technology in China has been developing quite rapidly in recent years. We now have a lot of advanced 3D visualization technology and integrated simulation models. They are used successfully in technological control and management in large-scale infrastructures. So far, however, there has not been much convincing evidence [...]" (Interviewee No. 5; senior researcher).

4. Conclusions and Discussion

The empirical studies in this manuscript provide insights into the use and usefulness of MSG from a macro level. Not from one case or one experiment, but from the generalized overviews at the higher level of institutions and political system, where knowledge, expertise, interests, power and stakes are interwoven at the science–policy interface.

Comparative analysis of the combined quantitative and qualitative data show that the debate on MSG for IWRM is structured around five frames in The Netherlands and three frames in China. Table A1 gives an overview of the frames in key words. The frames look at IWRM regimes—control vs. adaptive—through different lenses that are internally consistent in the way they perceive MSG at the science–policy interface. In and between the frames, there are significant differences. The words that are used—such as integration, stakeholder involvement, visualization, game, and evidence-based—might be the same, but they have very different meanings and connotations—even within the same country.

It is difficult to compare the frames in The Netherlands and China. Many respondents, also in China, agree that IWRM is a general trend in both countries and also globally. Water management is moving from its hard-core engineering domain to become an inherent part of integrated spatial development in order to achieve the ecology-based, sustainable, risk-involved management of water resource. However, the pathway to IWRM in both countries will be curved differently. There is also a similar "skeptic" tendency in both countries which has to do with the respondents' view on whether political reality can, and should, be influenced by science, evidence, participation and learning. In The Netherlands, there is a growing frame that perceives MSG for water management as "data driven simulation and visualization technology" especially in light of emerging trends in systems science and AI. A similar trend is noticeable in China, however there it also seems to have a function as political legitimation—the appearance of being modern, science-based. However, the use of MSG for understanding and managing complex systems, social learning or as persuasive technology, are significant frames in The Netherlands, but not as yet in China.

The findings and conclusions have scientific and practical relevance. First, the Q-method was found to be a valuable method for comparative, cross cultural research around topics that show a diversity of understandings, opinions and values. In countries such as China, officials are not very inclined to give interviews and speak openly. The Q method proved practical and effective, because it was novel, gave just enough structure and soon proved *fun* to do. The definition of a stable Q-set of statements that can be used in multiple contexts, requires pre-study validation efforts to avoid that the statement set needs intermediate revisions.

Second, many Dutch companies are now taking their advanced IWRM approaches and MSG methods across the globe, especially to Asia, and into China. Frame differences can easily confuse professional and academic debate about MSG for policy making and IWRM; within the same institutional and cultural context, but even more so in Netherlands–China water management co-operation projects. Frames are relevant when designing, using or evaluating innovative methods for IWRM. MSG are now increasingly advocated for social learning in a context of natural resource management. Learning to understand, design and work with multiple frames is a prerequisite for the relevance and impact of innovations in IWRM and MSG.

Acknowledgments: This paper is based on the Ph.D. thesis by Qiqi Zhou [57]. The research was sponsored by the Dutch–China Research Centre on Urban Systems & Environment (USE), TU Delft and the Next Generation Infrastructures program.

Author Contributions: The authors conceptualized the study design together. Qiqi Zhou conducted the interviews and data analysis, and authored the results and conclusion parts of the paper. Igor Stefan Mayer and Qiqi Zhou co-authored the Introduction, and Materials and Methods Sections. Igor Stefan Mayer final edited the paper.

Conflicts of Interest: The authors declare no conflict of interest.

Appendix A

Table A1. The frames in key words.

The Netherlands					
Frame	**Bureaucratic Alignment**	**Stakeholder Interaction**	**Learning**	**Uncertainty**	**Science vs. Emotions**
Policy-science interface	Power, authority, multi-level public governance.	Policy networks, Stakeholder interdependency.	Social-technical complexity; social dimension of technology.	Future and complex system orientation.	Engineering and science.
IWRM	Engineering solutions into (local, regional) planning; Alignment between administrative authorities and levels.	Stakeholder views, interests; interdependencies.	Social and technical dimensions of water management.	Local and global integrated future perspectives (climate change);	Science has to deal with irrationality, and emotions.
MSG	Minor relevance; Moderate trust in games; Only as supportive simulation and visualization technology.	Moderate trust and relevance of social games and simulations for better network interaction, trust, collaboration.	High trust and relevance of interactive, social-technical simulations and games for social learning.	Moderate trust and relevance of MSG as integrated assessment, analysis of possible futures.	Not much relevance of MSG; perhaps as a way to deal with public emotions.
Respondents (33)	33% (11 persons)	27% (9 persons)	9% (3 persons)	18% (6 persons)	12% (4 persons)
China					
Frame	**Doctrine of the Mean**	**Modern and Rational Governors**	**Open Minded Reformers**		
Policy-science interface	Centralization, hierarchy; policy analysis not very relevant.	Network governance; Rational, evidence-based decision-making.	Network type of governance; Science-based but open for social dimensions.		
IWRM	Better co-operation with regional and local authorities.	Sectoral and expert integration.	Integration of more perspectives, disciplines, possibly stakeholders.		
MSG	No trust and relevance in MSG. Support and legitimation of policy with data driven scientific computer models and simulations.	Low trust and relevance of MSG for different purposes. May be good for experts, not public or stakeholders.	Moderate trust and relevance of MSG. May be useful for broad array of uses, possibly social learning.		
Respondents (22)	54% (12 persons)	18% (4 respondents)	27% (6 persons)		

References

1. Savic, D.; Morley, M.; Khoury, M. Serious gaming for water systems planning and management. *Water* **2016**, *8*, 456. [CrossRef]
2. Medema, W.; Furber, A.; Adamowski, J.; Zhou, Q.; Mayer, I.S. Exploring the potential impact of serious games on social learning and stakeholder collaborations for transboundary watershed management of the St. Lawrence River Basin. *Water* **2016**, *8*, 175. [CrossRef]
3. Van der Wal, M.M.; de Kraker, J.; Kroeze, C.; Kirschner, P.A.; Valkering, P. Can computer models be used for social learning? A serious game in water management. *Environ. Model. Softw.* **2016**, *75*, 119–132. [CrossRef]

4. Wu, J.S.; Lee, J.J. Climate change games as tools for education and engagement. *Nat. Clim. Chang.* **2015**, *5*, 413–418. [CrossRef]
5. Rumore, D.; Schenk, T.; Susskind, L. Role-play simulations for climate change adaptation education and engagement. *Nat. Clim. Chang.* **2016**, *6*, 745–750. [CrossRef]
6. Meadows, D.H.; Robinson, J.M. The electronic oracle: Computer models and social decisions. *Syst. Dyn. Rev.* **2002**, *18*, 271–308. [CrossRef]
7. Mayer, I.S. The gaming of policy and the politics of gaming: A review. *Simul. Gaming* **2009**, *40*, 825–862. [CrossRef]
8. Accordino, F. The futurium—A foresight platform for evidence-based and participatory policymaking. *Philos. Technol.* **2013**, *26*, 321–332. [CrossRef]
9. White, D.D.; Wutich, A.Y.; Larson, K.L.; Lant, T. Water management decision makers' evaluations of uncertainty in a decision support system: The case of WaterSim in the Decision Theater. *J. Environ. Plan. Manag.* **2015**, *58*, 616–630. [CrossRef]
10. Edsall, R.M.; Larson, K.L. Effectiveness of a semi-immersive virtual environment in understanding human–environment interactions. *Cartogr. Geogr. Inf. Sci.* **2009**, *36*, 367–384. [CrossRef]
11. Diehl, A.; Delrieux, C. Applications of serious games in Geovisualization. In *Handbook of Research on Serious Games as Educational, Business and Research Tools*; Manuela Cruz-Cunha, M., Ed.; IGI Global: Hershey, PA, USA, 2012; pp. 25–46, ISBN 9781466601499.
12. Den Haan, R.-J.; Cortes Arevalo, V.J.; van der Voort, M.; Hulscher, S. Designing virtual river: A serious gaming environment to collaboratively explore management strategies in river and floodplain maintenance. In *Games and Learning Alliance*; Springer: Cham, Switzerland, 2016; pp. 24–34, ISBN 9783319501819.
13. Gluckman, P. The science–policy interface. *Science* **2016**, *353*, 969. [CrossRef] [PubMed]
14. Te Brömmelstroet, M.; Schrijnen, P.M. From planning support systems to mediated planning support: A structured dialogue to overcome the implementation Gap. *Environ. Plan. B Plan. Des.* **2010**, *37*, 3–20. [CrossRef]
15. Robinson, J.B. Of maps and territories: The use and abuse of socio-economic modelling in support of decisionmaking. *Technol. Forecast. Soc. Chang.* **1992**, *42*, 147–164. [CrossRef]
16. Brewer, G.D. *An Analyst's View of the Uses and Abuses of Modeling for Decisionmaking*; The Rand Corporation: Santa Monica, CA, USA, 1975.
17. Sismondo, S. Post-truth? *Soc. Stud. Sci.* **2017**, *47*, 3–6. [CrossRef] [PubMed]
18. Carrozza, C. Democratizing expertise and environmental governance: Different approaches to the politics of science and their relevance for policy analysis. *J. Environ. Policy Plan.* **2015**, *17*, 108–126. [CrossRef]
19. Saltelli, A.; Giampietro, M. What is wrong with evidence based policy, and how can it be improved? *Futures* **2017**, *91*, 62–71. [CrossRef]
20. Funtowicz, S.O.; Ravetz, J.R. Uncertainty, complexity and post-normal science. *Environ. Toxicol. Chem.* **1994**, *13*, 1881–1885. [CrossRef]
21. De Graaf, R.E.; Dahm, R.J.; Icke, J.; Goetgeluk, R.W.; Jansen, S.J.T.; van de Ven, F.H.M. Receptivity to transformative change in the Dutch urban water management sector. *Water Sci. Technol.* **2009**, *60*, 311. [CrossRef] [PubMed]
22. Ferguson, B.C.; Brown, R.R.; Deletic, A. Diagnosing transformative change in urban water systems: Theories and frameworks. *Glob. Environ. Chang.* **2013**, *23*, 264–280. [CrossRef]
23. Pahl-Wostl, C.; Mostert, E.; Tàbara, D. The growing importance of social learning in water resources management and sustainability science. *Ecol. Soc.* **2008**, *13*, 24. [CrossRef]
24. Medema, W.; Wals, A.; Adamowski, J. Multi-loop social learning for sustainable land and water governance: Towards a research agenda on the potential of virtual learning platforms. *NJAS-Wagening. J. Life Sci.* **2014**, *69*, 23–38. [CrossRef]
25. Wals, A.E.J.; Rodela, R. Social learning towards sustainability: Problematic, perspectives and promise. *NJAS-Wagening. J. Life Sci.* **2014**, *69*, 1–3. [CrossRef]
26. Stefanska, J.; Magnuszewski, P.; Sendzimir, J.; Romaniuk, P.; Taillieu, T.; Dubel, A.; Flachner, Z.; Balogh, P. A gaming exercise to explore problem-solving vs. relational activities for river floodplain management. *Environ. Policy Gov.* **2011**, *21*, 454–471. [CrossRef]

27. Schulze, J.; Martin, R.; Finger, A.; Henzen, C.; Lindner, M.; Pietzsch, K.; Werntze, A.; Zander, U.; Seppelt, R. Design, implementation and test of a serious online game for exploring complex relationships of sustainable land management and human well-being. *Environ. Model. Softw.* **2015**, *65*, 58–66. [CrossRef]
28. Ducrot, R.; van Paassen, A.; Barban, V.; Daré, W.; Gramaglia, C. Learning integrative negotiation to manage complex environmental issues: Example of a gaming approach in the peri-urban catchment of São Paulo, Brazil. *Reg. Environ. Chang.* **2015**, *15*, 67–78. [CrossRef]
29. Rusca, M.; Heun, J.; Schwartz, K. Water management simulation games and the construction of knowledge. *Hydrol. Earth Syst. Sci.* **2012**, *16*, 2749–2757. [CrossRef]
30. Chew, C.; Lloyd, G.J.; Knudsen, E. Capacity building in water with serious games. In *Gaming Media and Social Effects*; Sourina, O., Wortley, D., Kim, S., Eds.; Springer: Singapore, 2015; pp. 27–43, ISBN 978-981-287-407-8.
31. Hoekstra, A.Y. Computer-supported games and role plays in teaching water management. *Hydrol. Earth Syst. Sci.* **2012**, *16*, 2985–2994. [CrossRef]
32. Rijcken, T.; Stijnen, J.; Slootjes, N. "SimDelta"—Inquiry into an internet-based interactive model for water infrastructure development in The Netherlands. *Water* **2012**, *4*, 295–320. [CrossRef]
33. Hertzog, T.; Poussin, J.C.; Tangara, B.; Kouriba, I.; Jamin, J.Y. A role playing game to address future water management issues in a large irrigated system: Experience from Mali. *Agric. Water Manag.* **2014**, *137*, 1–14. [CrossRef]
34. Abrami, G.; Ferrand, N.; Morardet, S.; Murgue, C.; Popova, A.; De Fooij, H.; Farolfi, S.; Du Toit, D.; Aquae-Gaudi, W. Wat-A-Game, a toolkit for building role-playing games about integrated water management. In Proceedings of the International Congress on Environmental Modelling and Software (iEMSs), Leipzig, Germany, 1–5 July 2012.
35. Kos, Z.; Prenosilova, E. Simulation and gaming in water management in the Czech Republic. *Simul. Gaming* **1999**, *30*, 476–481. [CrossRef]
36. Le Bars, M.; Le Grusse, P.; Albouchi, L. AquaFej: A simulation game for planning water management—An experiment in central Tunisia. *Int. J. Sustain. Dev.* **2014**, *17*, 242. [CrossRef]
37. Haug, C.; Huitema, D.; Wenzler, I. Learning through games? Evaluating the learning effect of a policy exercise on European climate policy. *Technol. Forecast. Soc. Chang.* **2011**, *78*, 968–981. [CrossRef]
38. Craven, J.; Angarita, H.; Corzo Perez, G.A.; Vasquez, D. Development and testing of a river basin management simulation game for integrated management of the Magdalena-Cauca river basin. *Environ. Model. Softw.* **2017**, *90*, 78–88. [CrossRef]
39. Hummel, H.G.K.; van Houcke, J.; Nadolski, R.J.; van der Hiele, T.; Kurvers, H.; Löhr, A. Scripted collaboration in serious gaming for complex learning: Effects of multiple perspectives when acquiring water management skills. *Br. J. Educ. Technol.* **2011**, *42*, 1029–1041. [CrossRef]
40. Magombeyi, M.S.; Rollin, D.; Lankford, B. The river basin game as a tool for collective water management at community level in South Africa. *Phys. Chem. Earth* **2008**, *33*, 873–880. [CrossRef]
41. Barreteau, O.; Le Page, C.; Perez, P. Contribution of simulation and gaming to natural resource management issues: An introduction. *Simul. Gaming* **2007**, *38*, 185–194. [CrossRef]
42. Weiss, C.H.; Bucuvalas, M.J. Truth tests and utility tests: Decision-makers' frames of reference for social science research. *Am. Sociol. Rev.* **1980**, *45*, 302. [CrossRef]
43. Dunn, W.N. Assessing the impact of policy analysis: The functions of usable ignorance. *Knowl. Policy* **1991**, *4*, 36–55. [CrossRef]
44. Dunn, W.N. *Public Policy Analysis: An Introduction*; Prentice Hall: Upper Saddle River, NJ, USA, 1994; ISBN 9780137385508.
45. Scheer, D. Computer simulation at the science-policy interface: Assessing the policy relevance of Carbon Capture & Storage simulations. *Energy Procedia* **2011**, *4*, 5770–5777. [CrossRef]
46. Pallett, H.; Chilvers, J. Organizations in the making. *Prog. Hum. Geogr.* **2015**, *39*, 146–166. [CrossRef]
47. Katyaini, S.; Barua, A. Water policy at science-policy interface—Challenges and opportunities for India. *Water Policy* **2016**, *18*, 288–303. [CrossRef]
48. Pahl-Wostl, C.; Schlumpf, C.; Büssenschütt, M.; Schönborn, A.; Burse, J. Models at the interface between science and society: Impacts and options. *Integr. Assess.* **2000**, *1*, 267–280. [CrossRef]
49. Hoppe, R. Policy analysis, science and politics: From "speaking truth to power" to "making sense together". *Sci. Public Policy* **1999**, *26*, 201–210. [CrossRef]

50. Wildavsky, A. *Speaking Truth to Power: The Art and Craft of Policy Analysis*; Little Brown & Company: Boston, MA, USA, 1979.
51. Guston, D.H. *Between Politics and Science: Assuring the Integrity and Productivity of Reseach*; Cambridge University Press: New York, NY, USA, 2000; ISBN 0521653185.
52. Van Pelt, S.C.; Haasnoot, M.; Arts, B.; Ludwig, F.; Swart, R.; Biesbroek, R. Communicating climate (change) uncertainties: Simulation games as boundary objects. *Environ. Sci. Policy* **2015**, *45*, 41–52. [CrossRef]
53. Global Water Partnership (GWP); Technical Advisory Commitee (TAC). *Integrated Water Resources Management*; Global Water Partnership: Stockholm, Sweden, 2000.
54. Pahl-Wostl, C. Transitions towards adaptive management of water facing climate and global change. *Water Resour. Manag.* **2006**, *21*, 49–62. [CrossRef]
55. Mayer, I.; Warmelink, H.; Zhou, Q. A frame-reflective discourse analysis of serious games. *Br. J. Educ. Technol.* **2016**, *47*, 342–357. [CrossRef]
56. Ramlo, S. Mixed method lessons learned from 80 years of Q methodology. *J. Mixed Methods Res.* **2016**, *10*, 28–45. [CrossRef]
57. Zhou, Q. The Princess in the Castle: Challenging Serious Game Play for Integrated Policy Analysis and Planning. Ph.D. Thesis, Delft University of Technology, Delft, The Netherlands, 2014.

water

MDPI

Review

Towards a Framework for Designing and Assessing Game-Based Approaches for Sustainable Water Governance

Alice H. Aubert [1,*], Wietske Medema [2] and Arjen E. J. Wals [3]

1 Eawag, Swiss Federal Institute of Aquatic Science and Technology, Überlandstrasse 133, CH-8600 Dübendorf, Switzerland
2 Department of Bioresource Engineering, McGill University, 21 111 Lakeshore, Ste Anne de Bellevue, QC H9X3V9, Canada; wietske.medema@mcgill.ca
3 Department of Social Sciences, Education and Learning Sciences, Wageningen University, 6706 KN Wageningen, The Netherlands; arjen.wals@wur.nl
* Correspondence: alice.aubert@eawag.ch; Tel.: +41(0)58-765-5688

Received: 1 February 2019; Accepted: 19 April 2019; Published: 25 April 2019

Abstract: Most of the literature on serious games and gamification calls for a shift from evaluating practices to using theories to assess them. While the former is necessary to justify using game-based approaches, the latter enables understanding "why" game-based approaches are beneficial (or not). Based on earlier review papers and the papers in this special issue of Water entitled "Understanding game-based approaches for improving sustainable water governance: the potential of serious games to solve water problems", we show that game-based approaches in a water governance context are relatively diverse. In particular, the expected aims, targeted audience, and spatial and temporal scales are factors that differentiate game-based approaches. These factors also strongly influence the design of game-based approaches and the research developed to assess them. We developed a framework to guide and reflect on the design and assessment of game-based approaches, and we suggest opportunities for future research. In particular, we highlight the lack of game-based approaches that can support "society-driven" sustainable water governance.

Keywords: gamification; serious games; water governance; stakeholder participation; sustainability

---

1. Game-Based Approaches for Water Governance Lack Systematic Assessments

Using game-based approaches in a water governance context is not new. For instance, the role-playing games "Water on the West Bank" [1] and "the river Wadu" [2] date to 1988 and 1989, respectively, while the simulation game "FishBanks" dates to 1993 [3]. At the moment, interest in game-based approaches in a water governance context is growing. Illustrative of this interest is the recent general assembly of the European Geosciences Union that featured many serious games on water [4]. In this section, we first define what we mean by "game-based approaches". We then define the dimensions and variables that characterize them based on a review of the literature on water governance, decision analysis, and game design (both in gray (e.g., [5]) and peer-reviewed e.g., [6–9] literature). We then address the current state of assessment of game-based approaches.

1.1. Characterizing Game-Based Approaches for Governance

A wide range of games for water governance currently exists, such as serious games, role-playing games, simulation games, applied games, and educational games. To them are added the inter-related phenomena of "gamification" and "gamified applications". Given the various definitions and interpretations of "games" and "gaming", it is useful to provide a more detailed outline of the emerging landscape of game-based approaches.

A dictionary definition of a "game" is an activity solely aimed at leisure or fun, defined by rules, and unsolicited (intrinsically motivating) [10]. Games often offer a chance to win or lose. They also use various amounts of physical and/or mental skills, and luck [10]. The term "serious game" refers to games developed for a purpose beyond entertainment alone [6]. Serious games address societal challenges (e.g., ecological, social, economic, environmental, or a combination thereof) [6,11–15]. Serious games tend to combine elements of entertainment (e.g., fun, suspense, mystery, inquiry) with elements of learning (e.g., developing knowledge, insights, skills). Gamification or gamified applications, on the other hand, are applications or services that include some game elements as motivational affordances [16] to motivate desired psychological processes [11,16,17] to prompt particular desired behaviors [18]. The following elements are usually considered: Points, achievements/badges, leaderboards, clear goals, challenge, progression, feedback, status, levels, rewards, roles, story/theme [16,19]. Designing gamified application and services implies the intention that the game elements create engagement, as a game does [17,18,20].

The boundaries between "serious games" and "gamification" are inevitably somewhat blurry. For instance, a stakeholder workshop can be "gamified" by introducing sessions dedicated to playing a serious game focusing on particular desired learning outcomes. The serious game thus represents the game element designed to help achieve the workshop's desired outcomes. The workshop remains a solicited activity, with introduction and debriefing sessions that are not part of the serious game. In addition, some serious games used during stakeholder workshops are not fully fledged, in the sense that they are not stand-alone or fully self-engaging. In this paper, we use the term "game-based approaches" to reflect the blurred boundary between the two terms and to recognize the multiple interpretations of them, all having some merit, that can be found in the literature. To provide more clarity, we now focus on the dimensions and variables that characterize game-based approaches for governance.

Based on recent and established literature on games and game-based approaches, we defined these dimensions and several associated variables (Table 1). The three dimensions are (a) game design and technical aspects, (b) the diversity of people who participate and how game-based processes are facilitated, and (c) the intended purpose and expected outcomes of game-based approaches. Some variables could be associated with more than one dimension. The framework we propose in Section 2 influenced the present classification. An example of an associated variable is the degree of complexity, which refers to the extent to which the game design (i.e., data, models, and game interface) resembles real-world complexity [6,9,21,22]. Other variables are levity and transference of learning. Levity is the extent to which a game is designed to draw targeted participants into playing it and to collaborate with each other [23–25]. Transference of learning refers to the extent to which a game enables participants to capture the emerging complexity of game play, to make sense of individual and isolated experiences, and to transfer what they learn in the game to the real-world context in which they need to act [25,26].

Table 1 summarizes the key variables characterizing game-based approaches. This overview of variables is not meant to be exhaustive or definitive, but rather to indicate the wide range of possibilities that currently exist.

Table 1. Variables of the dimensions of game-based approaches.

Variable	Description	Example References
	GAME DESIGN & TECHNICAL ASPECTS	
Immersive experience	Extent to which a game makes players feel that they are an intrinsic part of the game world through visualization and interactive stories	Dede 2009 [27], Burke 2014 [23], Zhou 2014 [25]
Levity	Extent to which a game draws target players into playing it and interacting with each other through an effective balance of seriousness and playfulness	Lankford and Watson 2007 [21], Burke 2014 [23], Hertzog et al. 2014 [24], Zhou 2014 [25]
Complexity	Extent to which the game design (i.e., data, models, game interface) resembles real-world complexity	Lankford and Watson 2007 [21], Zhou 2014 [25], Wesselow and Kleemann 2018 [26], Aubert et al. 2018 [6]
Game/motivational affordances	Extent to which game/motivational affordances are included (e.g., points, leaderboards, achievements/badges, levels, theme, clear goals, rewards, progress, challenge)	Burke 2014 [23], Hamari et al. 2014 [16], Seaborn and Fels 2015 [19], Aubert et al. 2018 [6]
Action- consequence feedback	The evaluation of actions taken to assess their effectiveness and to determine future action	Mendler de Suarez et al. 2012 [28], Plass et al. 2015 [29], Aubert et al. 2018 [6]
Flow experience	Amount of deep absorption that the game facilitates for players (i.e., symbiosis between challenges and the skills needed to meet them, skills neither outmatched nor under-used)	Csikszentmihalyi 1990 [30], 1997, Nakamura and Csikszentmihalyi 2002 [31], Sweetser and Wyeth 2005 [32]
	PEOPLE & PROCESSES	
Representation	Extent of diversity/heterogeneity of the viewpoints and interests of those who play the game	Simon and Etienne 2010 [33], Barreteau et al. 2014 [34], Hertzog et al. 2014 [24], Medema et al. 2017 [35], Wesselow and Kleemann 2018 [26]
Level playing field	The relative absence of hierarchy and power inequities among the participants	Zhou 2014 [25], Ubbels and Verhallen 2000 [36]
Open-endedness	The space for emergence where participants accept that goals cannot be pre-determined in advance	Zhou 2014 [25], Ubbels and Verhallen 2000 [36]
Internalization of motivation	Extent of players' engagement to continue the game until the challenge is met	Rigby 2014 [17], Zhou 2014 [25]
Reflection/transformative learning	Amount of individual/collective reflection of players' game experiences that is facilitated during the game	Lim et al. 2006 [37], Jean et al. 2018 [38]
Emotional involvement	Extent to which game play affects players' bodily responses to, emotions about, and motivation for the outcome of the game	Zhou 2014 [25]
	PURPOSE & OUTCOMES	
Shared understanding of facts	Extent to which the game enhances players' understanding of the ecological system in which they operate	Medema et al. 2017 [35], Jean et al. 2018 [38]
Shared understanding of values	Extent to which the game enhances players' understanding of different perspectives/interests in the socio-ecological system in which they operate	Medema et al. 2017 [35], Jean et al. 2018 [38]
Commitment to solution finding	Extent to which the game enhances intrinsic motivation of players to find a solution to real-world issues	Csikszentmihalyi 1990 [30], 1997, Nakamura and Csikszentmihalyi 2002 [31], Hamari et al. 2014 [16], Aubert et al. 2018 [6]
Social capital	Extent to which the shared game-experience can form a collective memory, create a sense of togetherness/trust, and strengthen social ties	Lankford and Watson 2007 [21], Zhou 2014 [25]
Sense of ownership/self-organization	Extent to which the game creates a sense of individual and collective ownership in players through the game experience and outcomes	Zhou 2014 [25], Jean et al. 2018 [38]
Transference of learning	Extent to which a game enables players to capture the emerging complexity of game play, help make sense of individual and isolated experiences, and fill the transfer gap between the game and the world outside	Zhou 2014 [25], Wesselow and Kleemann 2018 [26]

1.2. Assessment of Game-Based Approaches

The literature (see Section 2) reveals that game-based approaches in water governance tend to focus mostly on their *potential* to advance Multi-Criteria Decision Analysis [6], watershed governance [7,8], and water systems planning and management [9]. A common conclusion in papers studying game-based approaches is that research should go beyond suggesting potentials, and, instead, systematically assess how players experience game-based approaches and the degree to which they realize intended and unintended outcomes. In this section, we review the assessment of game-based approaches for water governance.

A review of game-based approaches for water management and planning [9] reported varying degrees of assessment of serious games among several studies. It noted that some studies did not clearly indicate how players had played and experienced games, or what the game outcomes were. Other studies, however, did describe game sessions that were successfully completed, with clear evidence from participants who provided feedback using, for instance, pre- and post-game questionnaires. A more recent review [6] indicated that many games developed and played in the water sector have not been adequately studied. Another review [39], focusing on the use of serious games to address sustainability issues by increased social learning, agreed: Few systematic assessments of game-based approaches exist.

Likewise, Soekarjo and Oostendorp [40] conclude that none of the 15 studies they assessed used control treatments, or if they did, they provided no information about them. In the closely related field of climate change, assessment of game-based approaches also varies widely [41]. Reviews report either a lack of proper assessment (e.g., small sample sizes, lack of control treatments) or equivocal assessment results (e.g., supporting the benefits of game-based approaches only under specific conditions or for specific user groups).

A recent review of game literature [18] calls for better scientific design in the emerging field of gamification science. In particular, it calls for more reliable constructs, internal validity studies (i.e., control treatments to manipulate study variables, systematic experimental design), and external validity studies to replicate and cross-validate the findings (e.g., using a heterogeneous sample or multiple samples). The general insight is that systematic scientific assessment of game-based approaches is essential to determine whether they fulfill the purposes and objectives for which they were developed. Systematic assessment would also contribute to the development of gamification theories, which hardly exist at present.

Hence, even though the "did it work?" question needs to be asked and researched, this question is insufficient in itself [7]. Following the trend in economics and operations research [42], research on game-based approaches should also contribute to a deeper understanding of people's actual behavior and help unravel the reasons for, or processes leading to, that behavior. Along these lines, Seaborn et al. [19] distinguish gamification-in-action (i.e., did it work?) from gamification-in-theory (i.e., why/how was the observed outcome reached?). Developing a deeper and more detailed understanding of how and why certain game-based approaches and processes lead to specific outcomes is receiving growing attention e.g., [18,43,44]. This will likely contribute to further development of gamification theories. This will also inform frameworks that not only will help understand how gamification works but also help improve game design, which will ultimately lead to (more) effective serious games. In short, there is a need for systematic assessment of game-based approaches to (a) properly assess their benefits, (b) develop theories, and in turn (c) design effective approaches.

In the next section, we develop a framework to reflect on, in a structured way, the diversity of existing game-based approaches used for sustainable water governance. Using this framework should help researchers to systematically assess (and design) game-based approaches. Thus, the framework aims to help address the limits identified above. We then present examples, structured according to this framework, using the papers in this special issue (SI) of Water entitled "Understanding game-based approaches for improving sustainable water governance: the potential of serious games to solve water problems" [45]. We conclude by discussing key findings of the assessment of game-based approaches for water governance as well as windows of opportunity for further research.

2. Framework to Map the Diversity of Game-based Approaches

2.1. Methodology

Our literature review focused on water governance and/or decision-analysis, and game design. We searched the Web of Science for recent (2015–2018) reviews at the intersection of these topics and consulted some of the references that these reviews cited. We supplemented this thematic literature review with a review of recent literature that focuses on the interface between society and the environment (e.g., the Open Traceable Accountable Policy [46] or the related Record of Engagement [47]). Identifying similarities within this extensive body of literature, we developed a framework that helps inform and structure the design and assessment of game-based approaches.

Much of research in this vein provides good-practice guidelines for participation in policy development. A recent theory of participation [48] proposes moving from Arnstein's ladder of participation [49] to a "wheel of participation", defined as having two major dimensions: (a) Hierarchy (e.g., bottom-up participation vs. top-down instruction) and (b) interaction (e.g., one-way communication vs. two-way deliberation or co-engaged co-production). The theory's authors suggest that the type of participation itself does not predict its success. Instead, the success of any participatory process will depend on (a) the context (i.e., previous experiences and failures or successes with participation), (b) the design (i.e., representation of the parties, transparency and structure), (c) the power (i.e., management of power relations), and (d) the scalar fit (i.e., spatial and temporal scales) [48].

This wheel of participation resembles a heuristic (i.e., a tool to promote discussion) developed in the context of education for sustainability by Jickling and Wals [50], who distinguish processes by what we consider to be the same dimensions. They refer to the "bottom-up/top-down" dimension as the degree of authority vs. participation, and the "one-way/two-way" dimension as the extent to which outcomes are fixed (i.e., predetermined to a transmissive end) vs. open (i.e., emergent, jointly established to a socio-constructivist end) (Figure 1). They refer to the former as more instrumental and the latter as more emancipatory. In other terms, one can define "society-driven" processes as processes in which multiple perspectives are represented (implicitly involving many stakeholder groups and citizens) and that are co-designed.

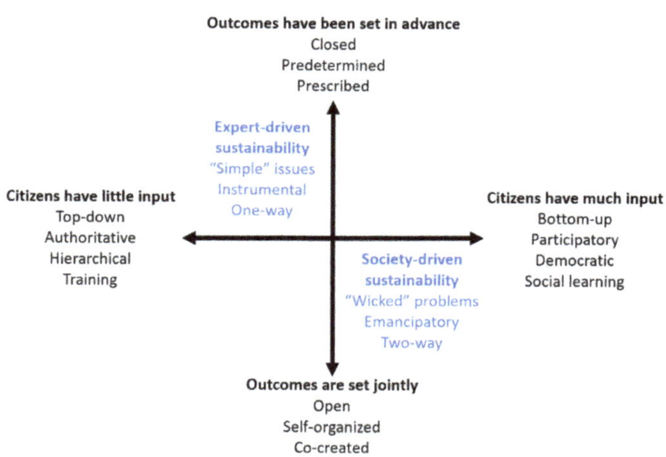

Figure 1. Overview of outcome (vertical) and participation (horizontal) dimensions of game-based approaches (based on Jickling and Wals [50]).

This literature from various disciplines shared similarities, which helped us identify four important aspects for classifying and assessing game-based approaches (Figure 2):

- "What?", i.e., the topic of the game-based approach (here, water governance)
- "Why?", i.e., the purpose of the approach
- "Who?", i.e., the stakeholders participating in the approach
- "When and where?", i.e., the temporal and spatial contexts of the process

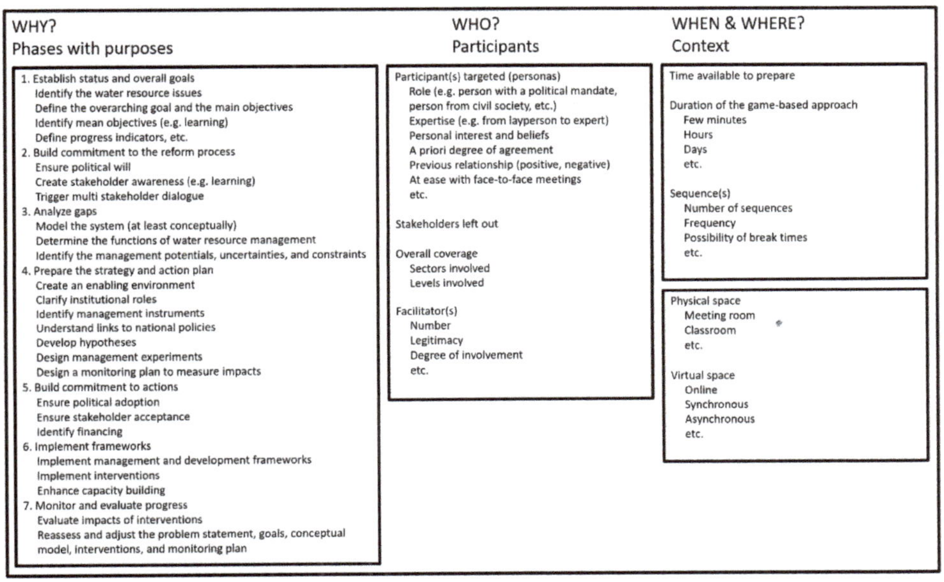

Figure 2. Defining the "why, who, when and where" of game-based approaches.

These aspects are in line with current game and gamification literature [12,20]. They also match the essential questions used to develop theories in the social sciences [51]. We argue that they inform (a) the three dimensions of game-based approaches and their characteristics (Table 1) and (b) the assessment of game-based approaches. The remainder of this section presents each aspect individually and ends by integrating them into a framework.

2.2. *Why Are Game-based Approaches Used for Water Governance?*

Water governance can be defined as "the set of rules, practices, and processes through which decisions for the management of water resources and services are taken and implemented, and decision-makers are held accountable" [52]. The Organization for Economic Co-operation and Development further specifies that water governance engages not only governments but also other stakeholders. Based on two popular management concepts for water resource planning and management—Integrated Water Resource Management and Adaptive Management [53,54]—water governance can be roughly captured by the iteration of seven phases, each with defined purposes (Figure 2, left side).

Many of these water governance purposes aim for some kind of learning (as defined in Reference [55]): From discovering, explaining, clarifying, changing (e.g., views) to creating (e.g., new alternatives). These types of learning can result from specific actions that lead directly to clear and immediate effects ("single-loop" learning) or from reflecting on the assumptions underlying these specific actions ("double-loop" learning) [56]. Sometimes, because of single- and double-loop learning, values and norms underpinning assumptions and actions are challenged, triggering "triple-loop" learning [57]. To some extent, water governance always aims to enable multi-loop social learning [58].

While learning is a common purpose of game-based approaches, engagement of participants is equally important. In their own review, Den Haan and Woort [39] report at least six other recent reviews about learning with and within serious games stemming from the educational sciences. Such educational serious games on societal challenges are also important for water governance because they help build awareness (phase 2 of water governance). On the other hand, Aubert et al. [6] suggest that game-based approaches are used mainly to create engagement in learning and other related spin-off activities. Creating or enhancing engagement is a goal for the entire spectrum of game-based approaches. The game elements included are usually affordances to make playing the game engaging and worthwhile [10,59]. Aubert et al. discuss how the engagement promoted by game-based approaches can benefit each step of a structured decision-making process ([6], Figure 1). Engagement can also be a means to achieve other outcomes, such as enhancing commitment to solution finding, creating social capital, and creating a sense of ownership of the decision (Table 1, *purpose & outcomes*).

2.3. Who is Involved in Game-Based Approaches Used for Water Governance?

By definition, a game is played by individuals or one or more groups of players. In the classification of Djaouti et al. [12], player characteristics are contained by the "scope" of the game. Recent reviews of game-based approaches for water governance [6,9] summarize several characteristics. For instance, game-based approaches can be single-player or multi-player. Some game-based approaches target a specific age group, but game-based approaches on water can be found for all age groups [6]. Players can vary from laypeople to experts. These two player characteristics (i.e., age and expertise) strongly influence the choice of the game-based approach. Most often, the targeted players come from a specific geographic area (e.g., due to the language used in the game, contextualization of the game to increase engagement). It is not necessarily exclusive, however; players from other regions can also use such regionalized game-based approaches. Some reviews e.g., [6,9,41] also highlight the involvement of a facilitator (sometimes assisted by other people). Flood et al. [41] point out that the experience of the facilitator and her/his knowledge on the topic strongly influences the success of the process.

Depending on specific water-governance phases and related objectives (Figure 2, left side), the following stakeholders can be targeted: Experts with a political mandate, politicians (with varying degrees of knowledge on the topic), technical experts (e.g., engineers, scientists), citizens enrolled in opinion groups (with some knowledge about the topic), and citizens as affected laypeople/end-users. In a multi-scale or multi-sector context, stakeholders can come from different sectors (e.g., agriculture, water supply). In multi-level contexts, stakeholders can come from different levels of decision-making or governance (e.g., local, regional, national) [60]. From a water governance perspective, the extent of representation (i.e., who participates and who does not) is important. In addition, consideration of previous relationships between participants can be critical, especially if there have been conflicts [48].

Considering this "who" question, several foci for studying the assessment of game-based approaches appear. One can study the game itself, independent of players (e.g., whether it is playable, how it works). One can study the individual, such as whether and how s/he reacts to the affordance of the game-based approach, or whether there are consequences for her/him (e.g., learning, engagement). One can focus on group dynamics, since interactions between individuals in a group often become complex and can lead to issues of power, conflict management, group thinking, etc. Depending on the purpose of the game and the stakeholder groups involved, a fourth focus would be the societal level, or the impact in the real world [61] that transcends the game-based approach itself.

2.4. What Are the Spatial and Temporal Contexts of the Game-Based Approach Used for Water Governance?

Games are characterized by spatial and temporal contexts, which, together, Huizinga calls the "magic circle" (i.e., the space and time of play) [62]. Within this magic circle created by the game, specific rules prevail. Game-based approaches last from a few minutes to several days [6]. Some game-based approaches enable players to take breaks, or they are developed as a sequential process, defined by the number of sequences (e.g., different scenarios), their frequency (i.e., time between two sequences), and whether there is a difficulty gradient (i.e., levels within a game). Regarding the spatial aspect, two types of spaces are reported [6]: Physical (e.g., meeting room, classroom) or virtual (e.g., online). The latter creates an additional option for the temporality of multi-player approaches: Contributions can be asynchronous.

In a water governance context, one-time meetings occur according to the phases previously described (Figure 2). Some of their outcomes should last long after the meeting. Water governance is sequential and iterative. In vertical governance, some of the stakeholders involved may be detached from the local water issue. In other cases, administrative boundaries do not match watershed boundaries [8]. These aspects could complicate physical meetings. Some patterns of scalar fit between space and time can be observed: a mediation process would most likely involve stakeholders within a local network, over a short period [48]. In contrast, deliberative democratic processes most likely involve stakeholders at a large scale (e.g., national) over a long period [48]. A final temporal aspect to consider is whether deadlines are set, which influences the time available to prepare meetings.

2.5. Framework to Define the Most Appropriate Game-Based Approach and Assess It

We argue that combining answers to the questions "why", "who", and "where and when" will help to (a) design the most effective game-based approach to reach the targeted outcomes and (b) assess the game-based approach, adequately monitoring the achievement of the targeted outcomes (Figure 3). In particular, answers to the "why" question guide the choice of the theoretical background to use to design and assess the game-based approach. Answers to the "who" question influence the focus of study: The game itself, the individuals involved, the groups involved, or the organization/society. Combining the answers to the "why" and "who" questions fully determines the focus. Combining answers to the "when and where" and "why" questions determines the research design (i.e., the methodological plan to collect assessment data). In certain spatial and temporal contexts, qualitative inquiry may be more appropriate than quantitative assessment.

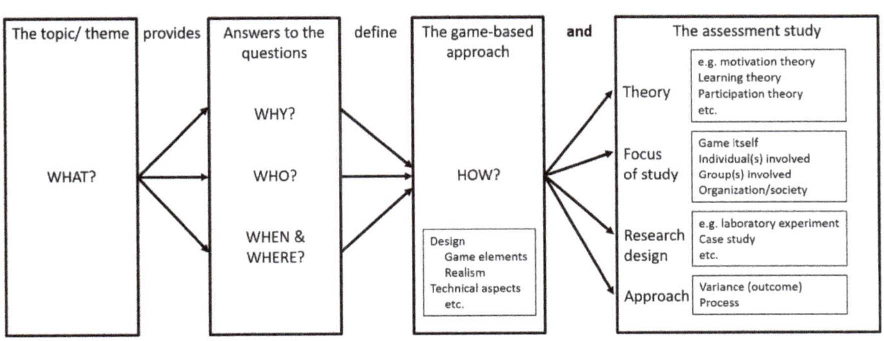

Figure 3. Framework defining the diversity of game-based approaches.

Finally, design and assessment can focus on the outcome (i.e., variance approach) and/or on understanding the process (i.e., process approach), as is defined in the assessment of decision-making [63]. Combining both approaches, researchers assess not only whether the expected outcomes have been achieved (i.e., variance approach) but also how and why they were achieved (if at all) (i.e., process approach). These two complementary approaches involve different scientific methods, and thus imply a different research design and/or focus.

As mentioned, this framework encompasses the three dimensions of game-based approaches and their variables (Table 1). In particular, the "why" question addresses the purpose and outcomes, the "who" question partly addresses the people and processes, and the "where and when" questions address the context and strongly influence the design and technical choices ("how" question). In the following section, we use this framework to review the game-based approaches featured in the SI. Based on the combinations of "why, who, when and where" in them, we identify similarities and differences in the use and assessment of their game-based approaches.

3. Results

3.1. The Context

3.1.1. Reasons for Using Game-Based Approaches and Topics Addressed

Predominantly, the game-based approaches for water governance described in this SI aim to (a) facilitate understanding of the overall complexity of the real world; (b) foster stakeholder collaboration, cross-sectoral integration, and/or multi-stakeholder dialogue; (c) experiment with multiple scenarios in a safe trial environment; (d) facilitate capacity building; and (e) overall, enable single- to triple-loop learning [64–72]. These aims correspond mostly to phase 2 of the water governance process (i.e., create stakeholder awareness and trigger multi-stakeholder dialogue), as well as phase 4 (i.e., creating an enabling environment) (Figure 2). Rodela et al. [70] call these game-based approaches "learning interventions". In addition, Galván-Pérez et al. [73] reviewed 20 game-based approaches ("educational videogames") that aimed to educate players in ecology, to raise awareness. Although not explicitly mentioned, the connection to water governance is that the younger generation will govern in the future. Presumably, the game authors aimed to prepare an enabling environment for sustainable future water governance.

The game-based approaches described in the SI papers focused on different water topics: Water supply safety plans at the watershed and household scales [64], water supply in peri-urban areas [66], local to transnational maritime spatial planning [65,67], flood mitigation in urban-rural watersheds [68], water resource management at the watershed scale [69], mangrove shrimp farming [70], the water-energy-food-land-climate nexus [71], how water managers perceive "models and interactive interfaces (serious games)" [72], and a review of games on the water cycle, water and aquatic ecosystem management, and the human right to water [73].

3.1.2. Stakeholders Involved

The game-based approaches described in the SI papers targeted diverse stakeholders. Overall, each of the studies included the key stakeholders concerned with the issue, from citizens (as end-users) directly impacted (e.g., [66,68]) to those involved in the decision-making process (i.e., stakeholders with a political mandate, such as water authorities) (e.g., [69,71,72]). Thus, the degree of expertise of those involved varied from being a layperson (e.g., [66,68]) to being an expert in the field (e.g., [65,72]). In the case of role-playing games, some of the participants sometimes played the role of another stakeholder not involved in the game (yet being represented in it) (e.g., playing the role of farmers [64,69]).

This diversity of stakeholders encompassed different levels of governance (e.g., from local coastal organizations to national and international organizations in charge of maritime spatial planning [65,67]). Stakeholder diversity also involved different sectors (e.g., fisheries, energy, trade [65,67]). Sometimes, a game-based approach tried but failed to include participants from all relevant sectors (e.g., private companies and land development sectors did not join [66]).

All of these game-based approaches required at least one facilitator and often more than one. In addition to guiding the game-based process, facilitators sometimes played a role in the game itself (e.g., the Game Overall Director (G.O.D.) [67]). In most cases (e.g., [68–70]), however, facilitators played no active role in the game but rather helped participants understand the rules and the technology used. The targeted players described in the review of Galván-Pérez et al. [73] differed significantly from an adult audience: They were teenagers (around 12). Most of the games reviewed in this paper were single-player games, and none of the games involved facilitators.

3.1.3. When and Where Game-Based Approaches Were Used

The game-based sessions described in the SI papers occurred in recent years, all over the world, from Bangladesh [66] and Vietnam [70] to Europe (e.g., [69,71]). Zhou and Mayer [72] reported on experts' perceptions of game-based approaches in China and the Netherlands. Jean et al. [67] focused primarily on a Canadian context, although they also included findings from Europe. Game-based approaches reviewed by Galván-Pérez et al. [73] were developed by European, American, and Australian organizations, as well as international committees.

Regardless of differences in the general setting, the spatial and temporal contexts of the game-based sessions were similar. Each of the game-based approaches discussed required interactive sessions (i.e., workshops that included specific game elements) (discussed in Section 3.2). These workshops required participants to meet in person; thus, they occurred in a physical space, usually a conference room or classroom. Sometimes additional arrangements were required to organize the room according to the required and available paraphernalia.

The workshops were also defined in time. They lasted from a few hours (e.g., [68]), to an entire day (e.g., [66]), to a few days (e.g., [67]). In one case, participants would have been willing to participate for even longer [66]. Workshops were divided into sessions. Most workshops started with a welcoming and game-introduction session. They then continued with the game experience itself, the longest session, which could be divided into a sequence of several rounds. The number of rounds depended on the total length of the workshop. Finally, the workshop ended with a debriefing session. Between the sessions and rounds, participants could usually take breaks.

The spatial and temporal contexts of the games reviewed by Galván-Pérez et al. [73] ranged from virtual space (e.g., online) to the classroom, with combinations being possible (e.g., players in a computer room playing an online game). The game-based approaches usually lasted from a few minutes to several hours. These game sessions were part of the school curriculum.

3.2. *Design of Game-Based Approaches*

Given their similar aims, targeted stakeholders, and settings, the game-based approaches described in the SI papers were somewhat alike. Primarily, they were simulation games: Based on models representing the real world, the stakeholders could "experience" consequences of their choices under different scenarios. Motivation to explore the scenarios came from a challenge given to workshop participants or to students in a course. The simulation game was sometimes combined with role-playing (e.g., [69]) or not (e.g., [68]). Simulation games described in the SI papers were multi-player games in which participants mainly had to perform various water governance steps in a safe trial environment to achieve the challenge at hand (e.g., [65–67,70,71]). The games created environments in which mistakes could be made without consequences besides the ability to learn from them, to reduce the chance of repeating them in the real world. In most cases, the games had well-defined rules. In a few cases, participants needed only to discover answers to guiding questions (e.g., [68]), and the rules of the game consisted of answering those questions.

We also observed some differences among the games. The technology varied from a board game (e.g., [66]), to a hybrid game (using a mix of physical and virtual elements) (e.g., [65]), to a virtual game (e.g., [68]). Partly due to the choice of technology, the degree of complexity represented varied. Several of the game-based approaches described in the SI papers used simplified models of stakeholder interactions that fit with the local regional context (e.g., [66]), but others were more complex 3D GIS simulations (e.g., [68]). Simplified representations of the complex real world were less realistic than complex representations of the real world. As the representation of the world became simpler, however, generality increased (e.g., representing no specific location [69]). The opposite was found as well: Complex, and thus more accurate, models were associated with specific geographic contexts (e.g., [71]).

In turn, these differences led to varying degrees of play vs. seriousness. In the papers from this SI, we observed simulation games at the serious end of the continuum (e.g., [68]), while others were designed to offer a more playful experience (e.g., [70]).

The educational video games reviewed by Gálvan-Pérez et al. [73] were single-player computer games. In addition to simulation games, there were also adventure games, platform or question games, and tile-based games. They differed in the game elements that they included, for instance, in adventure games, the narrative was salient. The level of complexity fit the audience: the complexity was often simplified, and thus the realism somewhat low. The play dimension tended to be as important as the seriousness.

3.3. How Game-Based Approaches Were Assessed

The papers in the SI report various ways of assessing game-based approaches. As mentioned (Section 3.1.1), the aim of most game-based approaches was related to some extent to a learning outcome. Consequently, we found assessments based on Kolb's experiential learning cycle [64], the knowledge co-creation cycle [67], the relational theory of multi-party collaboration processes ([69], though the authors did not refer to a specific seminal theory), and multi-level social learning (e.g., [67,69,70]). These various theories stress that learning is usually an iterative process, occurring when acquiring new knowledge, which in turn may create, change, or strengthen the belief of the individual. The last three theories listed suggest that knowledge is acquired through dialogical exchanges among individuals in a social context.

In some cases, theories were mentioned in reference to the design and conceptualization phase of the game. Among them, we found the constructivist learning theory [65], the Socratic method [68], and game theory [66]. The first two connect to the learning theories highlighted in the previous paragraph, but in these cases, they guided design of the game-based approach. The latter theorizes the behaviors of agents, in particular when they share a common and limited resource. Finally, Marini et al. [74] propose using Schwartz's socio-psychological theory of basic human values to design game-based approaches in which participants could move beyond their self-interest to more transcendental values. Some papers did not refer to any theories (e.g., [71]). Galván-Pérez et al. [73] assessed game-based approaches by developing an integrated quality indicator based on an ISO standard applied to video games and inspired by the Social Discourse of Video Games Analysis Model.

We observed different foci of assessment. Most studies had two to three foci, including the game itself (e.g., playability [70], an integrated quality indicator [73]), the individual (e.g., individual learning [68]), and/or the group involved (e.g., interaction analyses [67]). Only one paper [68] reported an outcome of a game session with the focus on society. Nonetheless, others (e.g., [66]) also highlighted the importance of follow-up studies to assess the outcomes of game-based approaches for society.

We observed a variety of research designs. Some assessments were based on case studies involving the targeted stakeholders (e.g., [69,71]). Others were based on single to multiple experimental game sessions (e.g., [64,65,67]). In these cases, the participants were not necessarily the targeted audience but a sub-sample of it (e.g., students). Finally, two studies stood apart from the previous assessments. The study of Zhou and Mayer [72] consisted of expert interviews based on the comparative Q-method, independent of a specific game session. Comparing the Netherlands and China, it assessed how experts perceived this common type of game-based approach, including a cultural dimension. The review of Galván-Pérez et al. [73] included the building of expert consensus on the rating of 20 games using Delphi expert consultation. The assessment intended to explain the process of increasing educational potential using narrative and gameplay.

In the papers of this SI, we encountered both the variance (i.e., outcome) and process approaches. Most of the papers, however, focused on the variance approach (e.g., did the individual learn? e.g., [68]). One paper that did focus on the process approach [67] monitored the game session while trying to explain the interactive processes among players during the session. We found a relatively rich diversity of assessment studies.

4. Discussion and Conclusion

4.1. Reflecting on the Diversity of Game-Based Approaches

4.1.1. Strong Points

As discussed by Barreteau ([75], and references therein), a dominant game-based approach for water governance is simulation role-playing, during which workshop participants can make decisions in a safe trial environment. This approach is justified by the enhanced potential to communicate about the complexity of the real world, to enhance collaboration and/or sharing of worldviews, and thus to promote multi-level learning. Another common game-based approach consists of games to engage in learning and create awareness about environmental issues. Thus, overall, game-based approaches for water governance aim to engage stakeholders in some learning activity about the complexity of water issues.

Of note, all papers in the SI include terms relating to stakeholder participation, such as "engagement", "collaboration", "participatory modelling", and "co-operation". This indicates that studying game-based approaches through the lens of a theory of participation might be promising [48]. Most game-based approaches described in the papers of SI appear to be located in the "expert-driven sustainability" corner (Figure 1, upper left). In this corner, there is strong consensus among experts with a policy and/or science background about what needs to happen or what is the right approach or solution to address the issue in a sustainable way. This is particularly true for educational games, which promote a "correct" solution. Most simulation games described in the papers of the SI, with or without role-playing, were designed by experts.

Despite the clearly dominant aim of using game-based approaches for a water governance, we observed finer variations in the characteristics of the game-based approaches identified (Table 1). Although the list of variables is not exhaustive, and no measurable attributes exist yet to characterize game-based approaches objectively, we used another heuristic to help design and assess game-based approaches: A radar chart (Figures 4 and 5). The radar chart shows key variables identified as continuums that range from low (center) to high (edge). Depending on the key challenge and main purpose, the starting situation for each of these variables is likely to differ, as are the expected changes in them resulting from using a game-based approach. Certain questions will need to be asked early in the process, such as (a) what needs to change (e.g., increasing knowledge, awareness, or understanding; optimizing current knowledge; engaging, co-creating or transitioning)? and (b) what is the nature of the problem (i.e., simple, complex, wicked)?

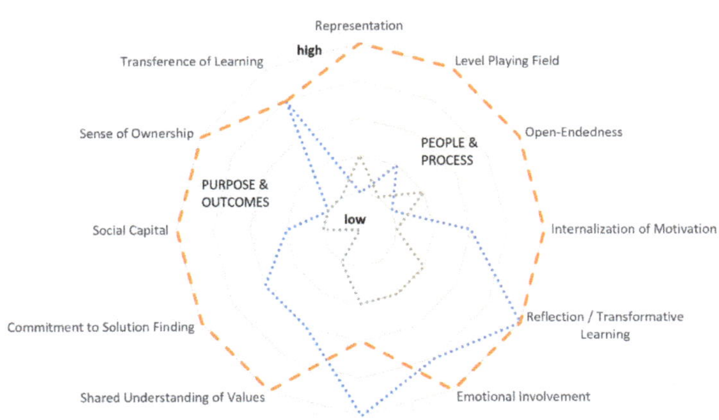

Figure 4. Heuristic representation of variables of purpose and outcomes (shaded green area) and people and processes (shaded orange area) of three hypothetical game-based approaches.

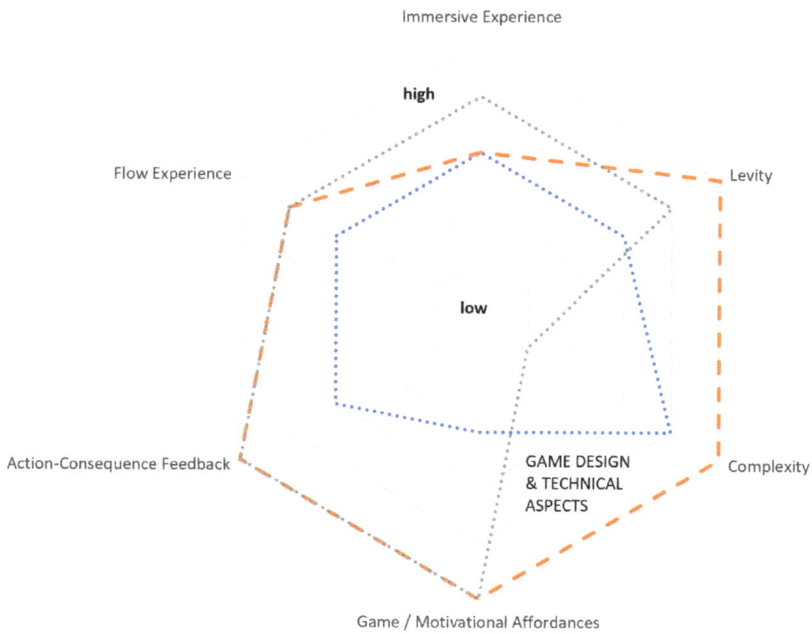

Figure 5. Heuristic representation of variables of game design and technical aspects for three hypothetical game-based approaches.

For instance, one can consider three hypothetical examples, each of which has a different starting points and different desired end points for each of several variables (Figure 4). Positions on the graph are meant to indicate, in a qualitative manner, results of discussion among the relevant stakeholders designing the game. Using the heuristic, they need to answer three questions: (a) Given the current water governance context, where are we in terms of each of these variables? (benchmarking); (b) given our overall purpose, what should we strive for, for each of these variables? (forecasting); and (c) given our starting situation and purpose, what game-based approach will work best to achieve it? (designing). The purpose of a heuristic is not to be precise or accurate but rather to get people to discuss and dialogue.

4.1.2. Blind Spots or Hotspots

First, reviewing the papers of the SI revealed that few if any game-based approaches currently exist in the "society-driven sustainability" corner (Figure 1, lower right). We found no game-based approaches with either representation of a wide range of worldviews (implicitly involving many citizens) or co-design of the approach. At best, the game-based approach featured different roles and worldviews [67,69], but the number of participants was limited and did not include all affected citizens. One explanation could be that the papers did not focus on the design itself, with one exception which reported that the design of the game-based approach included some local knowledge [70], as a step towards co-design; however, a third party initiated and led the design. The companion modelling approach, which sometimes also includes role-playing games, has a long history of co-design [34].

Second, the game-based approaches reviewed rarely addressed phase 7 of the water governance process (i.e., monitor and evaluate progress). This might indicate windows of opportunity to develop future research and explore the design of effective game-based approaches to support this phase and move towards more "society-driven sustainability" approaches. Two paths are possible to move toward this end, which addresses wicked, ill structured and ill-defined issues, for which no consensus on the sustainable solution exist. One can increase the representation by involving as many participants, and hence worldviews, as possible, and/or follow user-driven development (co-design). Addressing such wicked issues, or moving towards "society-driven sustainability", will require boundary crossing and building new connections. The resulting game-based approaches designed to tackle such issues may differ from the existing, reported approaches.

4.1.3. Windows of Opportunity

Game-based approaches in the "society-driven sustainability" corner (Figure 1, lower right) deal with relatively poorly structured issues that have relatively high dissonances and distances physically, normatively, or both. Few precedents or examples of how a game-based approach might support these cases exist. In "wicked" contexts, issues are often nexus issues: They go well beyond technical water management issues, including phenomena such as boundary-crossing, co-creation of knowledge, and dealing with emergence and associated uncertainty. The question is whether game-based approaches can help within such a "messy" environment, and if so, what kind of game-based approaches and characteristics might work in such contexts. In particular, how do designers transfer the variables of purpose and outcomes ("why" in the variance approach) and of people and processes ("why" in the process approach) (Figure 4) into the game's design and technical aspects ("how") (Figure 5).

Similar to the goal of Dillon et al. [76], the point of positioning the SI papers in a heuristic (Figures 2, 4, and 5) is not to describe them precisely. Instead, it is meant to invite deeper and more thorough discussion about the parameters and dimensions that underlie water governance and help determine the types of game-based approach that may best address or overcome the given water governance challenge. For instance, designers can use a radar chart (Figure 4) to map the starting position of all variables of purpose and outcomes in a given context and then use another radar chart (Figure 5) to translate their agreed-upon configuration into a specific corresponding game design. Game designers may need to find a design (Figure 5) that builds trust between stakeholders who may not trust each other, find a way to arrive at a common language to discuss the issue at hand, or develop game mechanisms that can help participants deconstruct and reconstruct their worldviews and frame the issue–together. Future research could focus on how stakeholders can agree upon how to position or assess each variable represented in the heuristic. The game-based approach will also need to find a way to (a) address emergence and the iterative nature of reflection and co-creation, and (b) treat goals and knowledge as tentative, being subject to revision based on ongoing critical and collaborative dialogue, inquiry, and action.

4.2. Reflecting on Assessment of Game-Based Approaches

As observed in previous studies, assessment studies are diverse. The aim of this paper is to provide a framework to support future research related to game-based approaches. Assessing game-based approaches for water governance would mean assessing (a) whether the specific objectives of the water governance processes are met, (b) whether the game design actually helped reach the specific objectives that were set, and (c) how the game design influenced the process of achieving these objectives. In particular, we recommend developing—simultaneously—game-based approaches and the studies used to assess them, to develop research that enables the former's goals and objectives to be "measured". In addition, we suggest connecting the assessment to existing theories (e.g., participation theory, theories of trust building or behavioral changes) or, if no theories are available, justifying why that is so and helping to develop new ones. Referring to theories is valid not only for the assessment phase but also for the design phase [74].

We hope that this framework for assessing studies of game-based approaches will help develop rigorous and systematic assessment that matches the objectives advertised for a given approach. Alternately, we recommend designing research along the lines described by Landers et al. [18]. Such research is essentially trans- and inter-disciplinary (i.e., involving stakeholders outside academia across sectors, and involving scientists across academic disciplines, respectively [77])—for two reasons. First, gamification researchers are often "scientist-practitioners" [18], since they develop, assess, theorize, and implement their work. Second, game-based approaches for water governance also require anchoring the playfulness in reality and giving it a meaning, to follow the wording of Harteveld about triadic (serious) game design [6,78]. Performing a complete assessment would lead to many smaller assessment studies (e.g., for each focus, for the two approaches). Covering the entire range of questions means covering many disciplines, which seems unrealistic for a single researcher, as highlighted recently in socio-hydrology [79]. Thus, we recommend that future projects gather game designers, psychologists (for the individual focus), sociologists (for the group and society foci), experts from water governance, and experts from the water issue at stake.

Since water governance aims at decision-making for the real world, we call for longer-term assessment of governance processes conducted with game-based approaches in the real world. For instance, a tracer study was able to demonstrate a long-term effect on water resource sharing in rural areas of Tanzania [80]. This point relates to difficulties in transposing outcomes of the game-based approach to the real world, as reported in papers on the use of role-playing games for resource management [81]. Lankford and Watson [21] discuss it using the terminology of cognitive metaphors: The experience in the source domain (the gamified workshop) should be rich enough to trigger actions in the target domain (the complex real world). Others suggest that outcomes of game-based approaches are not reached in the spatial and temporal contexts of the game itself, but in debriefing sessions during and immediately after a game [82]. Thus, assessing game-based approaches for water governance would require assessing whether one-time game-based approaches have (a) a short-term effect or (b) a longer-term effect, and (c) whether and how longer game-based approaches have a long-term effect. In addition, one could assess whether and how spatial context has an effect or study the connection between the safe trial environment created in the game and the real world. As mentioned, how—if at all—do participants transpose their game experience to the real world? How does game design transcribe real-world complexity into a game setting?

Finally, assessing game-based approaches for water governance raises normative considerations. In particular, does the design of game-based approaches strengthen some characteristics of society? If so, which ones? Alternately, does it challenge society, emphasizing transgression? To what end are game-based approaches used?

4.3. Conclusions

Future studies could consist of developing questionnaires to rate characteristics and variables of game-based approaches as objectively as possible. We recommend starting by assessing the purposes, people, and processes of the situation, and then deciding on the game design and technical aspects. Doing so would allow an underlying governance question to drive game design, rather than a given technology. This would help ensure that the most context-relevant game-based approaches are developed. This would also increase the potential effectiveness of game-based approaches, which should support water governance rather than drive it. Clearly, using game-based approaches to address wicked water governance issues—in which knowledge often is unstable and contested, and multiple stakeholders hold multiple value positions, interests, and constructions of reality—poses challenges for game designers. The specific context, understood through talking to stakeholders and immersion, should drive the design in a trans- and inter-disciplinary process.

Again in line with Dillon at al. [76], the point of our framework is not that game-based approaches need to move from instrumental or outcome-based uses (e.g., implementing a preconceived outcome) to more emancipatory or process-oriented uses (e.g., developing certain capacities of players). Instead, the point is that before considering the use of any game-based approach, designers need to reflect on the nature of the water governance problem at stake. Doing so will influence the design and assessment of an appropriate game-based approach. We hope our framework will nourish these thoughts as well as further discussions about the design and assessment of game-based approaches.

Author Contributions: A.H.A. initiated the original idea and message of the paper and was lead author. W.M. and A.E.J.W. thoroughly challenged the original ideas, which evolved to their current state. In particular, W.M. worked on Table 1, and A.E.J.W. suggested the graphical representation of Figure 1. All authors participated in the writing and review process.

Funding: This research is supported by a Swiss National Science Foundation Ambizione grant (project 173973, "Environmental Decision Analysis with Games – Edanaga") to Aubert. This research was also supported through a Social Sciences and Humanities Research Council of Canada (SSHRC) Partnership Development grant (SSHRC 890-2014-0056) to Adamowski and contributed to by Nguyen from the McGill University Brace Centre for Water Resources Management.

Acknowledgments: We thank the two anonymous reviewers of this paper for their constructive feedbacks.

Conflicts of Interest: The authors declare no conflict of interest.

References

1. Jarvis, W.T. Scientific mediation through serious gaming facilitates transboundary groundwater cooperation. *Water Resour. Impact* **2018**, *20*, 21–22.
2. Carruthers, I.; Smith, L. The River Wadu role play—Ten years experience. *Irrig. Drain. Syst.* **1989**, *3*, 281–308. [CrossRef]
3. Meadows, D. A brief and incomplete history of operational gaming in system dynamics. *Syst. Dyn. Rev.* **2007**, *23*, 199–203. [CrossRef]
4. Skinner, C. #EGU18 Geoscience Games Night—List of Games. Available online: https://seriousgeogames.wordpress.com/2018/03/27/egu18-geoscience-games-night-list-of-games/ (accessed on 23 April 2019).
5. Hockaday, S.; Jarvis, W.T.; Taha, F. Serious Gaming in Water. Available online: https://www.mediate.com/articles/HockadayS1.cfm (accessed on 23 April 2019).
6. Aubert, A.H.; Bauer, R.; Lienert, J. A review of water-related serious games to specify use in environmental Multi-Criteria Decision Analysis. *Environ. Model. Softw.* **2018**, *105*, 64–78. [CrossRef]
7. Furber, A.; Medema, W.; Adamowski, J. Assessing the benefits of serious games to support sustainable decision-making for transboundary watershed governance. *Can. Water Resour. J.* **2018**, *43*, 401–415. [CrossRef]
8. Medema, W.; Furber, A.; Adamowski, J.; Zhou, Q.; Mayer, I. Exploring the potential impact of serious games on social learning and stakeholder collaborations for transboundary watershed management of the St. Lawrence river basin. *Water* **2016**, *8*, 175. [CrossRef]
9. Savic, D.; Morley, M.; Khoury, M. Serious gaming for water systems planning and management. *Water* **2016**, *8*, 456. [CrossRef]
10. Larousse. Dictionnaire de Français. 2015. Available online: https://larousse.fr/dictionnaires/francais/jeu/44887?q=jeu#44826 (accessed on 23 April 2019).
11. Deterding, S. Gamification: Designing for motivation. *Interactions* **2012**, *19*, 14–17. [CrossRef]
12. Djaouti, D.; Alvarez, J.; Jessel, J.-P. Classifying Serious Games: The G/P/S Model. In *Handbook of Research on Improving Learning and Motivation through Educational Games: Multidisciplinary Approaches*; Patrick, F., Ed.; IGI Global: Hershey, PA, USA, 2011; pp. 118–136.
13. Abt, C.C. *Serious Games*; Viking Compass Edition: New York, NY, USA, 1970; p. 170.
14. Stanitsas, M.; Kirytopoulos, K.; Vareilles, E. Facilitating sustainability transition through serious games: A systematic literature review. *J. Clean. Prod.* **2019**, *208*, 924–936. [CrossRef]
15. Katsaliaki, K.; Mustafee, N. Edutainment for sustainable development: A survey of games in the field. *Simul. Gaming* **2014**, *46*, 647–672. [CrossRef]
16. Hamari, J.; Koivisto, J.; Sarsa, H. Does Gamification Work? A Literature Review of Empirical Studies on Gamification. In Proceedings of the 47th Hawaii International Conference on System Sciences, Waikoloa, HI, USA, 6–9 January 2014; pp. 3025–3034.
17. Rigby, C.S. Gamification and Motivation. In *The Gameful World*; MIT Press: Cambridge, MA, USA, 2014; pp. 113–138.
18. Landers, R.N.; Auer, E.M.; Collmus, A.B.; Armstrong, M.B. Gamification Science, Its History and Future: Definitions and a Research Agenda. *Simul. Gaming* **2018**, *49*, 315–337. [CrossRef]
19. Seaborn, K.; Fels, D.I. Gamification in theory and action: A survey. *Int. J. Hum.-Comput. Stud.* **2015**, *74*, 14–31. [CrossRef]

20. Morschheuser, B.; Werder, K.; Hamari, J.; Abe, J. How to gamify? Development of a method for gamification. In Proceedings of the 50th Annual Hawaii International Conference on System Sciences (HICSS), Maui, HI, USA, 4–7 January 2017.
21. Lankford, B.; Watson, D. Metaphor in natural resource gaming: Insights from the RIVER BASIN GAME. *Simul. Gaming* **2007**, *38*, 421–442. [CrossRef]
22. Lane, D.C. On a resurgence of management simulations and games. *J. Oper. Res. Soc.* **1995**, *46*, 604–625. [CrossRef]
23. Burke, B. *Gamify: How Gamification Motivates People to Do Extraordinary Things*; Bibliomotion: New York, NY, USA, 2014; p. 188.
24. Hertzog, T.; Poussin, J.-C.; Tangara, B.; Kouriba, I.; Jamin, J.-Y. A role playing game to address future water management issues in a large irrigated system: Experience from Mali. *Agric. Water Manag.* **2014**, *137*, 1–14. [CrossRef]
25. Zhou, Q. The Princess in the Castle: Challenging Serious Game Play for Integrated Policy Analysis and Planning. Ph.D. Thesis, Delft University of Technology (TU Delft), Delft, The Netherlands, 2014.
26. Wesselow, M.; Stoll-Kleemann, S. Role-playing games in natural resource management and research: Lessons learned from theory and practice. *Geogr. J.* **2018**, *184*, 298–309. [CrossRef]
27. Dede, C. Immersive Interfaces for Engagement and Learning. *Science* **2009**, *323*, 66. [CrossRef]
28. Mendler de Suarez, J.; Suarez, P.; Bachofen, C.; Fortugno, N.; Goentzel, J.; Gonçalves, P.; Grist, N.; Macklin, C.; Pfeifer, K.; Schweizer, S.; et al. *Games for a New Climate: Experiencing the Complexity of Future Risks*; The Frederick S. Pardee Center for the Study of the Longer-Range Future, Boston University: Boston, MA, USA, 2012; p. 109. ISBN 978-1-936727-06-3.
29. Plass, J.L.; Homer, B.D.; Kinzer, C.K. Foundations of game-based learning. *Educ. Psychol.* **2015**, *50*, 258–283. [CrossRef]
30. Csikszentmihalyi, M. *Flow: The Psychology of Optimal Experience*; Harper & Row: New York, NY, USA, 1990.
31. Nakamura, J.; Csikszentmihalyi, M. The concept of flow. In *Handbook of Positive Psychology*; Oxford University Press: New York, NY, USA, 2002; pp. 89–105.
32. Sweetser, P.; Wyeth, P. GameFlow: A model for evaluating player enjoyment in games. *Comput. Entertain.* **2005**, *3*, 3. [CrossRef]
33. Simon, C.; Etienne, M. A companion modelling approach applied to forest management planning. *Environ. Model. Softw.* **2010**, *25*, 1371–1384. [CrossRef]
34. Barreteau, O.; Bousquet, F.; Étienne, M.; Souchère, V.; dAquino, P.; Etienne, M. Companion Modelling: A Method of Adaptive and Participatory Research. In *Companion Modelling*; Springer: Dordrecht, The Netherlands, 2014; pp. 13–40.
35. Medema, W.; Adamowski, J.; Orr, C.; Furber, A.; Wals, A.; Milot, N. Building a Foundation for Knowledge Co-Creation in Collaborative Water Governance: Dimensions of Stakeholder Networks Facilitated through Bridging Organizations. *Water* **2017**, *9*, 60. [CrossRef]
36. Ubbels, A.A.; Verhallen, A.J. *Suitability of Decision Support Tools for Collaborative Planning Processes in Water Resources Management (RIZA 99.067)*; Institute for Inland Water Management and Wastewater Treatment (RIZA): Lelystad, The Netherlands, 2000.
37. Lim, C.P.; Nonis, D.; Hedberg, J. Gaming in a 3D multiuser virtual environment: Engaging students in Science lessons. *Br. J. Educ. Technol.* **2006**, *37*, 211–231. [CrossRef]
38. Jean, S.; Medema, W.; Adamowski, J.; Chew, C.; Delaney, P.; Wals, A. Serious games as a catalyst for boundary crossing, collaboration and knowledge co-creation in a watershed governance context. *J. Environ. Manag.* **2018**, *223*, 1010–1022. [CrossRef]
39. den Haan, R.-J.; van der Voort, M. On evaluating social learning outcomes of serious games to collaboratively address sustainability problems: A literature review. *Sustainability* **2018**, *10*, 4529. [CrossRef]
40. Soekarjo, M.; van Oostendorp, H. Measuring effectiveness of persuasive games using an informative control condition. *Int. J. Serious Games* **2015**, *2*, 37–54. [CrossRef]
41. Flood, S.; Cradock-Henry, N.A.; Blackett, P.; Edwards, P. Adaptive and interactive climate futures: Systematic review of 'serious games' for engagement and decision-making. *Environ. Res. Lett.* **2018**, *13*, 063005. [CrossRef]
42. Franco, L.A.; Hämäläinen, R.P. Behavioural operational research: Returning to the roots of the OR profession. *Eur. J. Oper. Res.* **2016**, *249*, 791–795. [CrossRef]

43. Ryan, R. The motivational pull of video game feedback, rules, and social interaction: Another self-determination theory approach. *Comput. Hum. Behav.* **2017**, *73*, 446–450. [CrossRef]
44. Vegt, N.; Visch, V.; Vermeeren, A.; de Ridder, H. Player Experiences and Behaviors in a Multiplayer Game: Designing game rules to change interdependent behavior. *Int. J. Serious Games* **2016**, *3*. [CrossRef]
45. MDPI. Special Issue "Understanding Game-based Approaches for Improving Sustainable Water Governance: The Potential of Serious Games to Solve Water Problems". Available online: https://www.mdpi.com/journal/water/special_issues/Game-based-Water-Governance (accessed on 23 April 2019).
46. Glynn, P.D.; Voinov, A.A.; Shapiro, C.D.; White, P.A. From data to decisions: Processing information, biases, and beliefs for improved management of natural resources and environments. *Earth's Future* **2017**, *5*, 356–378. [CrossRef]
47. Gray, S.; Voinov, A.; Paolisso, M.; Jordan, R.; BenDor, T.; Bommel, P.; Glynn, P.; Hedelin, B.; Hubacek, K.; Introne, J.; et al. Purpose, processes, partnerships, and products: Four Ps to advance participatory socio-environmental modeling. *Ecol. Appl.* **2018**, *28*, 46–61. [CrossRef]
48. Reed, M.S.; Vella, S.; Challies, E.; de Vente, J.; Frewer, L.; Hohenwallner-Ries, D.; Huber, T.; Neumann, R.K.; Oughton, E.A.; del Ceno, J.S.; et al. A theory of participation: What makes stakeholder and public engagement in environmental management work? *Restor. Ecol.* **2018**, *26*, S7–S17. [CrossRef]
49. Arnstein, S.R. A ladder of citizen participation. *J. Am. Inst. Plan.* **1969**, *35*, 216–224. [CrossRef]
50. Jickling, B.; Wals, A.E.J. Globalization and environmental education: Looking beyond sustainable development. *J. Curr. Stud.* **2008**, *40*, 1–21. [CrossRef]
51. Whetten, D.A. What constitutes a theoretical contribution? *Acad. Manag. Rev.* **1989**, *14*, 490–495. [CrossRef]
52. Organization for Economic Co-operation and Development (OECD). Water Governance Programme. Available online: http://www.oecd.org/env/watergovernanceprogramme.htm (accessed on 23 April 2019).
53. Engle, N.L.; Johns, O.R.; Lemos, M.C.; Nelson, D.R. Integrated and adaptive management of water resources: Tensions, legacies, and the next best thing. *Ecol. Soc.* **2011**, *16*. [CrossRef]
54. Medema, W.; McIntosh, B.S.; Jeffrey, P.J. From premise to practice: A critical assessment of integrated water resources management and adaptive management approaches in the water sector. *Ecol. Soc.* **2008**, *13*. [CrossRef]
55. Belton, V.; Elder, M.D. Decision support systems: Learning from visual interactive modelling. *Decis. Support Syst.* **1994**, *12*, 355–364. [CrossRef]
56. Argyris, C. *Organizational Learning: A Theory of Action Perspective*; Addison-Wesley: Reading, MA, USA, 1978; pp. VII, 344S.
57. Reed, M.S.; Evely, A.C.; Cundill, G.; Fazey, I.; Glass, J.; Laing, A.; Newig, J.; Parrish, B.; Prell, C.; Raymond, C.; et al. What is Social Learning? *Ecol. Soc.* **2010**, *15*. [CrossRef]
58. Medema, W.; Adamowski, J.; Orr, C.J.; Wals, A.; Milot, N. Towards sustainable water governance: Examining water governance issues in Québec through the lens of multi-loop social learning. *Can. Water Resour. J.* **2015**, *40*, 373–391. [CrossRef]
59. Ditzler, L.; Klerkx, L.; Chan-Dentoni, J.; Posthumus, H.; Krupnik, T.J.; Ridaura, S.L.; Andersson, J.A.; Baudron, F.; Groot, J.C.J. Affordances of agricultural systems analysis tools: A review and framework to enhance tool design and implementation. *Agric. Syst.* **2018**, *164*, 20–30. [CrossRef]
60. Lienert, J.; Schnetzer, F.; Ingold, K. Stakeholder analysis combined with social network analysis provides fine-grained insights into water infrastructure planning processes. *J. Environ. Manag.* **2013**, *125*, 134–148. [CrossRef]
61. Pelzer, P.; Geertman, S.; Heijden, R.v.d.; Rouwette, E. The added value of Planning Support Systems: A practitioner's perspective. *Comput. Environ. Urban Syst.* **2014**, *48*, 16–27. [CrossRef]
62. Huizinga, J. *Homo Ludens: A Study of the Play-Element in Culture*; Routledge & Kegan Paul Ltd.: London, UK, 1949; p. 220.
63. Poole, M.S.; Van De Ven, A.H. Empirical Methods for Research on Organizational Decision-Making Processes. In *Handbook of Decision Making*; Nutt, P.C., Wilson, D.C., Eds.; Wiley-Blackwell: Oxford, UK, 2010.
64. Ferrero, G.; Bichai, F.; Rusca, M. Experiential learning through role-playing: Enhancing stakeholder collaboration in water safety plans. *Water* **2018**, *10*, 227. [CrossRef]
65. Keijser, X.; Ripken, M.; Mayer, I.; Warmelink, H.; Abspoel, L.; Fairgrieve, R.; Paris, C. Stakeholder engagement in maritime spatial planning: The efficacy of a serious game approach. *Water* **2018**, *10*, 724. [CrossRef]

66. Gomes, S.; Hermans, L.; Islam, K.; Huda, S.; Hossain, A.T.M.; Thissen, W. Capacity building for water management in peri-urban communities, Bangladesh: A simulation-gaming approach. *Water* **2018**, *10*, 1704. [CrossRef]
67. Jean, S.; Gilbert, L.; Medema, W.; Keijser, X.; Mayer, I.; Inam, A.; Adamowski, J. Serious games as planning support systems: Learning from playing maritime spatial planning challenge 2050. *Water* **2018**, *10*, 1786. [CrossRef]
68. Khoury, M.; Gibson, M.J.; Savic, D.; Chen, A.S.; Vamvakeridou-Lyroudia, L.; Langford, H.; Wigley, S. A serious game designed to explore and understand the complexities of flood mitigation options in urban–rural catchments. *Water* **2018**, *10*, 1885. [CrossRef]
69. Magnuszewski, P.; Królikowska, K.; Koch, A.; Pająk, M.; Allen, C.; Chraibi, V.; Giri, A.; Haak, D.; Hart, N.; Hellman, M.; et al. Exploring the role of relational practices in water governance using a game-based approach. *Water* **2018**, *10*, 346. [CrossRef]
70. Rodela, R.; Ligtenberg, A.; Bosma, R. Conceptualizing serious games as a learning-based intervention in the context of natural resources and environmental governance. *Water* **2019**, *11*, 245. [CrossRef]
71. Sušnik, J.; Chew, C.; Domingo, X.; Mereu, S.; Trabucco, A.; Evans, B.; Vamvakeridou-Lyroudia, L.; Savić, D.; Laspidou, C.; Brouwer, F. Multi-stakeholder development of a serious game to explore the water-energy-food-land-climate nexus: The sim4nexus approach. *Water* **2018**, *10*, 139. [CrossRef]
72. Zhou, Q.; Mayer, I. Models, simulations and games for water management: A comparative q-method study in the Netherlands and China. *Water* **2018**, *10*, 10. [CrossRef]
73. Galván-Pérez, L.; Ouariachi, T.; Pozo-Llorente, M.; Gutiérrez-Pérez, J. Outstanding videogames on water: A quality assessment review based on evidence of narrative, gameplay and educational criteria. *Water* **2018**, *10*, 1404. [CrossRef]
74. Marini, D.; Medema, W.; Adamowski, J.; Veissière, S.; Mayer, I.; Wals, A. Socio-psychological perspectives on the potential for serious games to promote transcendental values in IWRM decision-making. *Water* **2018**, *10*, 1097. [CrossRef]
75. Barreteau, O. The joint use of role-playing games and models regarding negotiation processes: Characterization of associations. *J. Artif. Soc. Soc. Simul.* **2003**, *6*. Available online: http://jasss.soc.surrey.ac.uk/6/2/3.html (accessed on 23 April 2019).
76. Dillon, J.; Stevenson, R.B.; Wals, A.E.J. Guest Editors Introduction to the special section: Moving from citizen to civic science to address wicked conservation problems. *Conserv. Biol.* **2016**, *30*, 450–455. [CrossRef] [PubMed]
77. Tress, G.; Tress, B.; Fry, G. Clarifying integrative research concepts in landscape ecology. *Landsc. Ecol.* **2005**, *20*, 479–493. [CrossRef]
78. Harteveld, C. *Triadic Game Design: Balancing Reality, Meaning and Play*; Springer: London, UK, 2011; p. 316.
79. Srinivasan, V.; Sanderson, M.; Garcia, M.; Konar, M.; Blöschl, G.; Sivapalan, M. Moving socio-hydrologic modelling forward: Unpacking hidden assumptions, values and model structure by engaging with stakeholders: Reply to "What is the role of the model in socio-hydrology?". *Hydrol. Sci. J.* **2018**, 1–3. [CrossRef]
80. Rajabu, K.R.M. Use and impacts of the river basin game in implementing integrated water resources management in Mkoji sub-catchment in Tanzania. *Agric. Water Manag.* **2007**, *94*, 63–72. [CrossRef]
81. Dray, A.; Perez, P.; Le Page, C.; D'Aquino, P.; White, I. Who wants to terminate the game? The role of vested interests and metaplayers in the ATOLLGAME experience. *Simul. Gaming* **2007**, *38*, 494–511. [CrossRef]
82. Garcia, C.; Dray, A.; Waeber, P. Learning begins when the game is over: Using games to embrace complexity in natural resources management. *GAIA* **2016**, *25*, 289–291. [CrossRef]

MDPI
St. Alban-Anlage 66
4052 Basel
Switzerland
Tel. +41 61 683 77 34
Fax +41 61 302 89 18
www.mdpi.com

Water Editorial Office
E-mail: water@mdpi.com
www.mdpi.com/journal/water

www.ingramcontent.com/pod-product-compliance
Lightning Source LLC
Chambersburg PA
CBHW061836210926
51850CB00006B/905